Contents

Cambridge Studies in Biological and Evolutionary Anthropology 55

Spider Monkeys
Behavior, Ecology and Evolution of the Genus *Ateles*

Spider monkeys are one of the most widespread New World primate genera, ranging from southern Mexico to Bolivia. Although they are common in zoos, spider monkeys are traditionally very difficult to study in the wild, because they are fast moving, live high in the canopy and are almost always found in small subgroups that vary in size and composition throughout the day. The past decade has seen an expansion in research being carried out on this genus and this book is an assimilation of both published, and new and previously unpublished, research. It is a comprehensive source of information for academic researchers and graduate students interested in primatology, evolutionary anthropology and behavioral ecology and covers topics such as taxonomy, diet, sexuality and reproduction, and conservation.

CHRISTINA J. CAMPBELL is an assistant professor of Anthropology at California State University, Northridge. Her research interests include behavioral ecology and reproductive endocrinology and physiology.

Cambridge Studies in Biological and Evolutionary Anthropology

Series editors

HUMAN ECOLOGY
C. G. Nicholas Mascie-Taylor, University of Cambridge
Michael A. Little, State University of New York, Binghamton
GENETICS
Kenneth M. Weiss, Pennsylvania State University
HUMAN EVOLUTION
Robert A. Foley, University of Cambridge
Nina G. Jablonski, California Academy of Science
PRIMATOLOGY
Karen B. Strier, University of Wisconsin, Madison

Also available in the series

Spider Monkeys

Behavior, Ecology and Evolution of the Genus Ateles

Edited by

Christina J. Campbell

California State University, Northridge

CAMBRIDGE
UNIVERSITY PRESS

CAMBRIDGE UNIVERSITY PRESS
Cambridge, New York, Melbourne, Madrid, Cape Town, Singapore, São Paulo, Delhi

Cambridge University Press
The Edinburgh Building, Cambridge CB2 8RU, UK

Published in the United States of America by Cambridge University Press, New York

www.cambridge.org
Information on this title: www.cambridge.org/9780521867504

First published 2008

Printed in the United Kingdom at the University Press, Cambridge

A catalog record for this publication is available from the British Library

Library of Congress Cataloging in Publication data
Spider monkeys : behavior, ecology and evolution of the genus Ateles / edited by
Christina J. Campbell.
  p. cm.
Includes index.
ISBN 978-0-521-86750-4 (hardback)
1. Spider monkeys. I. Campbell, Christina J. II. Title.
QL737.P915.S65 2008
599.8′58 – dc22   2008015447

ISBN 978-0-521-86750-4 hardback

Contributors

Filippo Aureli
Research Centre in Evolutionary Anthropology and Palaeoecology, School of
Biological and Earth Science, Liverpool John Moores University,
Liverpool L3 3AF, UK

Christina J. Campbell
Department of Anthropology, California State University Northridge,
Northridge, CA 91330, USA

Andrew C. Collins
Department of Conservation Biology, Ada F. and Daniel L. Rice Conservation
Biology and Research Center, Brookfield Zoological Institution, Brookfield,
IL 60513, USA

Siobhán B. Cooke
The Graduate Center, The City University of New York and The New York
Consortium in Evolutionary Primatology (NYCEP), New York, NY 10016,
USA

Loretta A. Cormier
Department of Anthropology, University of Alabama at Birmingham,
Birmingham, AL 35294-3350, USA

J. Lawrence Dew
Department of Biological Sciences, 2000 Lakeshore Drive, University of New
Orleans, New Orleans, LA 70148, USA

Anthony Di Fiore
Center for the Study of Human Origins, Department of Anthropology, New
York University, New York, NY 10003, USA

Annika M. Felton
The Fenner School of Environment and Society (formerly CRES), The
Australian National University, Canberra ACT 0200, Australia

K. Nicole Gibson
Department of Anthropology, Yale University, New Haven, CT 06511, USA

Walter C. Hartwig
Department of Basic Sciences, Touro University College of Osteopathic
Medicine, Vallejo, CA 94592, USA

Lauren Halenar
The Graduate Center, The City University of New York and The New York
Consortium in Evolutionary Primatology (NYCEP), New York, NY 10016,
USA

Kosei Izawa
Department of Animal Sciences, Teikyo University of Science and
Technology, Yamanashi, 409–0193, Japan

Andres Link
Center for the Study of Human Origins, Department of Anthropology,
New York University, New York, NY 10003, USA

Akisato Nishimura
Biological Laboratory, Science and Engineering Research Institute, Doshisha
University, Kyoto, 602-8580, Japan

Gabriel Ramos-Fernández
Centro Interdisciplinario de Investigación para el Desarrollo, Integral
Regional (CIIDIR), Unidad Oaxaca,, Instituto Politécnico Nacional, Santa
Cruz Xoxocotlán, Oaxaca 71230, México

Alfred L. Rosenberger
The New York Consortium in Evolutionary Primatology (NYCEP) and the
Department of Anthropology and Archaeology, Brooklyn College, The City
University of New York, Brooklyn, NY 11210, USA

Colleen M. Schaffner
Department of Psychology, Chester Centre for Stress Research, University of
Chester, Chester CH1 4BJ, UK

Yukiko Shimooka
Laboratory of Human Evolution, Department of Zoology, Kyoto University,
Kyoto 606-8502, Japan

Bernardo Urbani
University of Illinois at Urbana-Champaign, 109 Davenport Hall, 607 South
Mathews Avenue, Urbana, IL 61801, USA, and Centro de Antropología,
Instituto Venezolano de Investigaciones Científicas, Caracas, Venezuela

Laura Greer Vick
Department of Anthropology, Peace College, Raleigh, NC 27604, USA

Robert B. Wallace
Wildlife Conservation Society – Bolivia, San Miguel, La Paz, Bolivia

Dionisios Youlatos
Aristotle University of Thessaloniki, School of Biology, Department of
Zoology, GR-54124 Thessaloniki, Greece

Acknowledgements

The chapters that are included in this volume represent the first ever compilation of such a diversity of papers about spider monkeys and I thank each of the authors for their contribution and their patience during the editorial process. I would also like to thank Colin Groves, Agustin Fuentes, Manuel Lizarralde and Alan Dixson for their useful reviews and comments. Dominic Lewis and Alison Evans at Cambridge have provided assistance in the publication of this volume and I thank them for their help. Lastly, but definitely not least, I would like to thank my family, Tom and Summer Wake for their patience and continued support throughout this process.

1 Introduction

CHRISTINA J. CAMPBELL

For those outside of the primatological community, it may seem surprising that this is the first volume about spider monkeys (genus *Ateles*). However, while they are often seen as the "typical monkey," mischievously hanging from their tails, swinging through the trees and eating fruit, spider monkeys are relatively unstudied in the wild compared with many species of Old World monkey, and other New World genera such as capuchin monkeys (*Cebus* spp.) and howler monkeys (*Alouatta* spp.). Indeed there are more data concerning most aspects of the social lives of the highly endangered muriqui (*Brachyteles* spp.), thanks largely to the long-term studies of Karen Strier, than the more common and much more widely distributed spider monkey.

Spider monkeys are fast moving, wide ranging and high canopy animals whose social system often means that only a few community members can be followed at any one time. Such features make the study of wild spider monkeys notoriously difficult, and are almost certainly the reasons behind the historical dearth of long-term studies of the genus. That said, the number of field studies of spider monkeys has exploded over the last decade (Table 1.1). Studies have been, or are being, carried out by researchers at all levels in academia, and perhaps most importantly there has been an increase in the number of students carrying out their Ph.D. research at various sites throughout the spider monkey range. Although natural history studies are still being carried out, especially on those species and populations we know less about, hypothesis-driven research focusing on many aspects of spider monkey ecology, behavior, physiology, morphology and evolution has increased dramatically.

Given the difficulties in studying spider monkeys in a natural setting, one might postulate that studies carried out on captive animals would be more common as many of the difficulties in studying them in the wild are negated. Unfortunately this does not appear to be the case, as scientific publications on captive spider monkeys are surprisingly rare (Eisenberg and Kuehn, 1966; Klein, 1971; Klein and Klein, 1971; Eisenberg, 1976; Rondinelli and Klein, 1976; Chapman and Chapman, 1990; McDaniel *et al.*, 1993; Watt, 1994; Laska

Spider Monkeys: Behavior, Ecology and Evolution of the Genus Ateles, ed. Christina J. Campbell. Published by Cambridge University Press. © Cambridge University Press 2008.

Table 1.1 *Information on field study sites for Ateles spp.*

Species	Study site	Country	Type of research	Primary researchers (>3 months)	Publications
Ateles geoffroyi	Los Tuxtlas Biological Research Station	Mexico	Population density, conservation	Alejandro Estrada	Estrada and Coates-Estrada (1996)
	Otoch Ma'ax Yetel Koh Reserve (Punta Laguna)	Mexico	Communication, social behavior, development, habitat use	Filippo Aureli, Gabriel Ramos-Fernández, Colleen Schaffner, Kathryn Slater, Alejandra Valero, Laura Vick	Ramos-Fernández (2005); Ramos-Fernández and Ayala-Orozco (2002); Ramos-Fernández et al. (2003); Campbell et al. (2005) Aureli et al. (2006); Valero et al. (2006); Aureli and Schaffner (2007); Slater et al. (2007)
	Parque de la Flora y Fauna Silvestre Tropical	Mexico	Conservation, translocation, ecology, social behavior, sexual behavior, diet, sensory physiology	Laura Hernandez Salazar, Matthias Laska	Laska et al. (2007a)
	Tikal	Guatemala	Locomotion, behavioral ecology	Margaret Baxter, John Cant[a]	Coelho et al. (1976); Cant (1977, 1978, 1990); Fedigan and Baxter (1984)
	Parque Nacional Pico Bonito	Honduras	Ecology, diet	Justin Hines[a]	Hines (2002, 2004a, b, c, 2005)
	El Zota Biological Research Station	Costa Rica	Behavioral ecology, social behavior, conservation, cognition	Stacey Lindshield, Michelle Rodrigues	Luckett et al. (2004); Lindshield (2006); Rodrigues and Lindshield (2007)
	Hacienda Los Inocentes	Costa Rica	Ecology, social behavior	Patricia McDaniel[a]	McDaniel (1994)
	La Selva Biological Station	Costa Rica	Diet, ecology, social behavior, ranging patterns, activity cycles	Aimee Campbell[c]	Campbell and Sussman (1994); Wentz et al. (2003)
	Osa National Wildlife Refuge	Costa Rica	Behavior, conservation	K. Nicole Gibson	None to date
	Punta Rio Claro Wildlife Refuge	Costa Rica	Color vision, diet preference	Pablo Riba-Hernandez	Riba-Hernandez et al. (2004)

	Site	Country	Topics	Researchers	References
	Santa Rosa National Park	Costa Rica	Ecology, seed dispersal, social behavior, color vision	Filippo Aureli, Colin Chapman, Chihiro Hiramatsu, Colleen Schaffner, Patricia Teixidor[a]	Chapman (1987, 1988, 1990a, b); Chapman et al. (1989); Campbell et al. (2005); Hiramatsu et al. (2005)
	Sirena Biological Station; Corcovado National Park	Costa Rica	Ecology, behavior, population survey	Susan Boinksi, Martha Pineros, Jennifer Weghorst[b]	Piñeros (1994); Weghorst (2007)
	Barro Colorado Island	Panama	Ecology, diet, demography, social behavior, reproduction	Christina Campbell, Ronald Dare, Katharine Milton	Eisenberg and Kuehn (1966); Richard (1970); Dare (1974); Milton (1981, 1993); Ahumada (1992); Milton and Hopkins (2005); Campbell (2000, 2003, 2004, 2006a,b); Campbell et al. (2001, 2005); Russo et al. (2005)
Ateles hybridus	El Avila National Park	Venezuela	Population density	Diana Liz Duque Sandoval	None to date
	Serrania de Las Quinchas	Colombia	Behavioral ecology, social behavior	Andres Link	None to date
Ateles belzebuth belzebuth	La Macarena	Colombia	Behavioral ecology	Lewis Klein[a]	Klein (1971, 1972); Klein and Klein (1977)
	La Macarena (Tinigua)	Colombia	Ecology, social behavior, social structure, demography, communication	Jorge Ahumada, Gisel Didier L., Luisa Fernanda, Agumi Inaba, Kosei Izawa, Andres Link, Kohei Matsushita, Akisato Nishimura, Nelson Pinilla, Juan C. Pizarro, Marcela Quiñones, Yukiko Shimooka[a]	Ahumada (1989); Izawa and Mizuno (1990); Izawa (1993, 2000); Izawa et al. (1997); Stevenson et al. (2000); Shimooka (2003, 2005); Russo et al. (2005); Link et al. (2006); Matsuda and Izawa (2007)
	Maracá Ecological Station	Brazil	Diet, ecology	Andrea Nunes	Nunes (1995, 1998); Nunes and Chapman (1997)
	Tawadu Forest	Venezuela	Feeding behavior, ecology	Hernán Castellanos[a]	Castellanos (1995); Castellanos and Chanin (1996)

(cont.)

Table 1.1 (cont.)

Species	Study site	Country	Type of research	Primary researchers (>3 months)	Publications
	Yasuní	Ecuador	Demography, diet, seed dispersal, cognition, communication, genetics, locomotion, social behavior	John Cant, J. Larry Dew, Anthony Di Fiore, Andres Link, Wilmer Pozo-Rivera, Stephanie Spehar, Scott Suarez, Dionisios Youlatos	Dew (2001, 2005); Suarez (2003, 2006); Spehar (2006); Cant et al. (2001, 2003); Pozo Rivera (2004); Russo et al. (2005); Link and Di Fiore (2006)
Ateles belzebuth chamek	Cocha Cashu, Manu National Park	Peru	Seed dispersal, behavioral ecology, demography, male social relationships, habitat use	K. Nicole Gibson, Sabrina Russo, Margaret Symington	Symington (1987a,b, 1988a,b); Russo (2003, 2005); Russo and Augspurger (2004); Russo et al. (2006); Campbell et al. (2005)
	Lago Caiman, Parque Nacional Noel Kempff Mercado	Bolivia	Diet, behavioral ecology, population density	Robert Wallace[a]	Wallace (1998, 2001, 2005, 2006); Wallace et al. (1998); Karesh et al. (1998); Rocha (1999)
	Estacion Biologica del Beni	Bolivia	Population density, ecology	P. Fabiana Mendez, Luis Pacheco[a]	Méndez (1999); Pacheco (1997); Pacheco and Simonetti (1998, 2000)
	Tunquini, Parque Nacional Cotapata	Bolivia	Population density, behavioral ecology	Amira Apaza	Apaza (2002)
	La Chonta, Concesion Forestal	Bolivia	Diet, behavioral ecology, conservation ecology	Annika Felton[b]	None to date
	Oquiriquia, Concesion Forestal	Bolivia	Diet, population density	Lila Sainz	Sainz (1997)
Ateles paniscus	Ralleighvallen National Park	Surinam	Diet, ecology	Marc van Roosmalen[a]	van Roosmalen (1985)
	Station des Nouargues	French Guiana	Locomotion, diet, population density	Bruno Simmen, Dionisios Youlatos	Simmen and Sabatier (1996); Simmen et al. (1998); Youlatos (2002)

[a] Completed Ph.D. research.
[b] Ph.D. thesis in progress.
[c] Uncompleted Ph.D. thesis.

1996, 1998; Laska *et al.*, 1996, 1998, 1999, 2000a, 2000b, 2003, 2006, 2007b; Matisoo-Smith *et al.*, 1997; Hernández-López *et al.*, 1998, 2002; Pastor-Nieto, 2000, 2001; Hernandez-Salazar *et al.*, 2003; Davis *et al.*, 2005; Schaffner and Aureli, 2005; Joshi *et al.*, 2006). While spider monkeys are common in zoological parks they are conspicuously absent from primate research centers, and this may account for the apparent discrepancy. Although the chapters in this book focus on field research, we certainly acknowledge the importance of captive research and urge scientists to look to captive groups for many avenues of research that may be difficult to carry out in the wild.

In bringing together the various chapters in this volume, I had several goals in mind. First and foremost I wanted to provide a single comprehensive source for readers interested in any aspect of spider monkey behavior, ecology and evolution. Second, I wanted to showcase the expansion in research being carried out on this genus in the past decade. The authors who have contributed to this volume are numerous; however, the list does not include many who have contributed greatly to our knowledge of spider monkey behavioral ecology in the past, but who no longer study them today. Additionally absent are many graduate students currently gathering data and whose work will almost certainly broaden our knowledge even further.

The first section of this book deals with the morphology, evolution, phylogeny and taxonomy of spider monkeys. Spider monkeys are often compared to the Hominoidea because of their ability to brachiate, and the morphological features that allow them to do so. Rosenberger and colleagues provide a detailed overview of the cranial, dental and postcranial morphology of spider monkeys. They put this information into an evolutionary context by examining the evolutionary history of *Ateles* and the other members of the tribe, Atelini (*Brachyteles* and *Lagothrix*) using both morphological and genetic data sets. Reference is also made to howler monkeys, *Alouatta* spp., who, along with the atelins, constitute the subfamily Atelinae.

The taxonomy of the various species of *Ateles* has changed multiple times over the years and continues to be debated today. In Chapter 3 Collins covers the history of this taxonomic debate and discusses issues that relate to taxonomic inconsistencies today. The major issue I was confronted with in editing the chapters for this volume is the designation of the Bolivian and Peruvian black spider monkeys (*Ateles belzebuth chamek* versus *Ateles chamek*). I have chosen in this volume to be consistent throughout the chapters and to follow Collins by using *Ateles belzebuth chamek*. Adding to the confusion of this taxon is the fact that the spider monkeys at Cocha Cashu National Park, Peru, have been widely published under the incorrect name of *A. paniscus* (Symington, 1987a, 1987b, 1988a, 1988b). Genetic evidence clearly shows, however, that the morphological similarities they share with *A. paniscus* in Surinam and the

Guianan Shield (i.e. black coat and pink faces) are superficial and they should either be *A. belzebuth chamek*, or *A. chamek* (see Collins, this volume).

The second section of the book deals with the ecological challenges spider monkeys face and the ways they have adapted to these various pressures. Di Fiore and colleagues examine possibly the most well studied aspect of spider monkey life – their diet. After reviewing the different kinds of foods eaten by spider monkeys and the varying proportions of the diet that these food types contribute, they go on to investigate the similarities – or lack thereof – in the fruit genera consumed in 13 different studies. Spider monkeys appear to be quite plastic in the fruits that they consume and adjust well to each environment in which they are found.

In Chapter 5, Wallace investigates the myriad factors that influence spider monkey ranging patterns. While concluding that the availability of ripe fruit is the most important factor, he cautions that most studies investigating ranging patterns have not truly investigated all the possible contributing factors, and calls for a unified method of measuring ranging patterns. He also challenges the idea that female spider monkeys have core home ranges in all populations, and shows that these animals can be quite plastic in their use of varying habitat types.

In Chapter 6, Dew reviews the evidence showing that spider monkeys play an important role as seed dispersers in the forests they inhabit and backs up this review with primary data from his research at Yasuní National Park in Ecuador. Comparing spider monkeys (*Ateles belzebuth belzebuth*) with the closely related and sympatric woolly monkey (*Lagothrix lagothricha poeppigii*), he shows that spider monkeys are highly effective dispersers at this site. They disperse seeds of a wide variety of sizes (including large seeds that woolly monkeys do not disperse), they show low levels of seed predation, they disperse seeds far away from the parent tree and they do not damage the seed by ingesting it. The conservation implications of the importance of spider monkeys for forest renewal and maintenance are clear.

The third section investigates the behavior of spider monkeys. Youlatos provides a detailed explanation of spider monkey locomotion and posture. He begins by outlining the various modes of locomotion and posture, calling for a unified and standard set of definitions so that comparative studies can be more fruitful. With the initial provision that methods are not standardized in the data sets currently available, he provides a comparative investigation of the ways in which spider monkeys run, walk, sit, stand, swing and leap (to name a few), elucidating possible specific-level differences. A major avenue for future research is highlighted in this chapter – the need for fine-scale analyses of environmental features that may help explain why spider monkeys at different sites differ in these locomotor and postural attributes.

In Chapter 8, Ramos-Fernández reviews what is currently known about spider monkey communication, and then focuses on the most intensely studied call, the whinny. He reviews the various hypotheses to explain the function of the call and the evidence supporting and refuting each of these hypotheses. The one aspect of the whinny that does seem to be clear is that it contains information about the individual identity of the caller. He cautions that call "function" may not be easy to tease out as there may be multiple functions, depending on whether the intended recipient is a single individual, all individuals within a subgroup, or anyone that can hear the call.

Aureli and Schaffner examine the social structure and social relationships of spider monkeys. They provide an in-depth comparative discussion of fission–fusion societies and then provide a theoretical framework in which they house the rest of their chapter. They go on to review in great detail what is known about female–female, male–male, and female–male relationships. Perhaps the striking feature of this chapter is how the picture has changed over the last few years with the discovery that male–male relationships in spider monkeys are more complex then we once assumed, and how strikingly similar to common chimpanzees (*Pan troglodytes*) spider monkeys are in their social behavior.

Campbell and Gibson examine the reproductive biology and sexual behavior of spider monkeys, drawing largely on information from their two study sites, Barro Colorado Island, Panama, and Cocha Cashu, Manu National Park, Peru. They examine and discuss possible functions of the unique reproductive morphologies of both female and male spider monkeys – the hypertrophied pendulous clitoris of females and the large, bacculum-free penis of males. Additionally they provide detailed descriptions of sexual behavior in the genus, highlighting the secluded and prolonged nature of the spider monkey copulation.

In her chapter on immature spider monkeys, Vick provides a seminal piece of work examining the challenging world that immature spider monkeys face. It would appear that the life of the immature male spider monkey, in particular, is filled with many perils that may cause their death prior to adulthood. Of particular interest in her chapter is the clear indication that spider monkey juveniles show many of the differences in behavior that adult male and female spider monkeys show.

In Chapter 12 Shimooka and colleagues present the first comprehensive review of spider monkey demographic factors such as interbirth interval, age at dispersal, community size and sex composition. They provide evidence for what appear to be interspecific differences, but caution that even with this data set, the sample size may be too small to truly know if the differences are real. As is echoed in many of the volume's chapters, a call for continued long-term research is called for as the kind of data presented in this chapter can only be garnered from such studies.

In the final section of the volume, two chapters investigate interactions between humans and spider monkeys. Ramos-Fernández and Wallace tackle what is possibly one of the most urgent issues facing many researchers of spider monkeys today – their conservation. They review the current status of the various taxa and then go on to to illuminate and discuss the variety of factors that are contributing to the current decline in populations. Although a few taxa require immediate action to ensure their survival, the picture is not totally bleak for all spider monkeys and many populations appear to be sustainable if efficient management of protected areas is guaranteed.

Finally, Cormier and Urbani discuss the interaction of spider monkeys with that ever-present primate species – *Homo sapiens*. We can never forget that the lives of spider monkeys, in much of their range, is intricately entwined with the lives of our species. The presence of spider monkeys in archaeological data such as faunal assemblages and iconography is reviewed and discussed. The authors also review data concerning the importance of spider monkeys to modern-day peoples – largely in the Amazonian region. Spider monkeys are often considered one of the tastiest primate species and as such hunting can play an important role in the survival of various populations.

It is my hope that the various chapters in this volume will be exciting, informative, and useful to those who read them. It is clear that although spider monkeys are difficult to study in the field, there are many researchers who are willing to face those difficulties head on in order to gain further insight into the lives of these fascinating primates. I hope that the increase in research being carried out on wild populations of spider monkeys will continue in future years.

References

Ahumada, J. A. (1989). Behavior and social structure of the free-ranging spider monkeys (*Ateles belzebuth*) in La Macarena. *Field Studies of New World Monkeys, La Macarena, Colombia*, **2**, 7–31.

Ahumada, J. A. (1992). Grooming behavior of spider monkeys (*Ateles geoffroyi*) on Barro Colorado Island. *Int. J. Primatol.*, **13**(1), 33–49.

Apaza, A. E. (2002). Comportamiento alimentario de *Ateles chamek* (Cebidae) disponibiliad de frutos en época húmeda en los Yungas del PN-ANMI Cotapata. Unpublished Licenciatura thesis, Universidad Mayor de San Andrés, Argentina.

Aureli, F. and Schaffner, C. M. (2007). Aggression and conflict management at fusion in spider monkeys. *Biol. Lett.* **3**, 147–149.

Aureli, F., Schaffner, C. M., Verpooten, J., Slater, K. and Ramos-Fernández, G. (2006). Raiding parties of male spider monkeys: insights into human warfare? *Am. J. Phys. Anthropol.*, **131**, 486–497.

Campbell, A. F. and Sussman, R. W. (1994). The value of radio tracking in the study of neotropical rain forest monkeys. *Am. J. Primatol.*, **32**, 291–301.

Campbell, C. J. (2000). The reproductive biology of black-handed spider monkeys (*Ateles geoffroyi*): integrating behavior and endocrinology. Unpublished Ph.D. thesis, University of California, Berkeley.

Campbell, C. J. (2003). Female directed aggression in free-ranging *Ateles geoffroyi*. *Int. J. Primatol.*, **24**(2), 223–238.

Campbell, C. J. (2004). Patterns of behavior across reproductive states of free-ranging female black-handed spider monkeys (*Ateles geoffroyi*). *Am. J. Phys. Anthropol.*, **124**(2), 166–176.

Campbell, C. J. (2006a). Copulation in free-ranging black-handed spider monkeys (*Ateles geoffroyi*). *Am. J. Primatol.*, **68**, 507–511.

Campbell, C. J. (2006b). Lethal intragroup aggression by adult male spider monkeys (*Ateles geoffroyi*). *Am. J. Primatol.*, 1197–1201.

Campbell, C. J. (2007). Primate sexuality and reproduction. In *Primates in Perspective*, ed. C. J. Campbell, A. F. Fuentes, K. C. MacKinnon, M. Panger and S. Bearder, New York: Oxford University Press, pp. 423–437.

Campbell, C. J., Aureli, F., Chapman, C. A., *et al.* (2005). Terrestrial behavior of spider monkeys (*Ateles* spp.): a comparative study. *Int. J. Primatol.*, **26**, 1039–1051.

Campbell, C. J., Shideler, S. E., Todd, H. E. and Lasley, B. L. (2001). Fecal analysis of ovarian cycles in female black-handed spider monkeys (*Ateles geoffroyi*). *Am. J. Primatol.*, **54**, 79–89.

Cant, J. G. H. (1977). Ecology, locomotion and social organization of spider monkeys (*Ateles geoffroyi*). Unpublished Ph.D. thesis, University of California, Davis.

Cant, J. G. H. (1978). Population survey of the spider monkey *Ateles geoffroyi* at Tikal, Guatemala. *Primates*, **19**(3), 525–535.

Cant, J. G. H. (1990). Feeding ecology of spider monkeys (*Ateles geoffroyi*) at Tikal, Guatemala. *Hum. Evol.*, **5**(3), 269–281.

Cant, J. G. H., Youlatos, D. and Rose, M. D. (2001). Locomotor behavior of *Lagothrix lagothricha* and *Ateles belzebuth* in Yasuní National Park, Ecuador: general patterns and nonsuspensory modes. *J. Hum. Evol.*, **41**, 141–166.

Cant, J. G. H., Youlatos, D. and Rose, M. D. (2003). Suspensory locomotion of *Lagothrix lagothricha* and *Ateles belzebuth* in Yasuní National Park, Ecuador. *J. Hum. Evol.*, **44**, 685–699.

Castellanos, H. G. (1995). Feeding behaviour of *Ateles belzebuth* E. Geoffroy 1806 (Cebidae: Atelinae) in Tawadu Forest southern Venezuela. Unpublished Ph.D. thesis, University of Exeter, UK.

Castellanos, H. G. and Chanin, P. (1996). Seasonal differences in food choice and patch preference of long-haired spider monkeys (*Ateles belzebuth*). In *Adaptive Radiations of Neotropical Primates*, ed. M. A. Norconk, A. L. Rosenberger and P. A. Garber, New York: Plenum Press, pp. 451–466.

Chapman, C. (1987). Flexibility in diets in three species of Costa Rican primates. *Folia Primatol.*, **49**, 90–105.

Chapman, C. (1988). Patterns of foraging and range use by three species of neotropical primates. *Primates*, **29**, 177–194.

Chapman, C. A. (1990a). Association patterns of spider monkeys: the influence of ecology and sex on social organization. *Behav. Ecol. Sociobiol.*, **26**, 409–414.

Chapman, C. A. (1990b). Ecological constraints on group size in three species of neotropical primates. *Folia Primatol.*, **55**, 1–9.

Chapman, C. A. and Chapman, L. J. (1990). Reproductive biology of captive and free-ranging spider monkeys. *Zoo. Biol.*, **9**, 1–9.

Chapman, C. A., Chapman, L. J. and McLaughlin, R. L. (1989). Multiple central place foraging by spider monkeys: travel consequences of using many sleeping sites. *Oecologia*, **79**, 506–511.

Coelho, A. M., Jr., Bramblett, C. A., Quick, L. B. and Bramblet, S. S. (1976). Resource availability and population density in primates: a socio-bioenergetic analysis of the energy budgets of Guatemalan howler and spider monkeys. *Primates*, **17**(1), 63–80.

Dare, R. J. (1974). The social behavior and ecology of spider monkeys, *Ateles geoffroyi*, on Barro Colorado Island. Unpublished Ph.D. thesis, University of Oregon, Eugene.

Davis, N., Schaffner, C. M. and Smith, T. E. (2005). Evidence that zoo visitors influence HPA activity in spider monkeys (*Ateles geoffroyi*). *App. Anim. Behav. Sci.*, **90**(2), 131–141.

Dew, J. L. (2001). Synecology and seed dispersal in woolly monkeys (*Lagothrix lagotricha peoppigii*) and spider monkeys (*Ateles belzebuth belzebuth*) in Parque Nacional Yasuní, Ecuador. Unpublished Ph.D. thesis, University of California, Davis.

Dew, J. L. (2005). Foraging, food choice, and food processing by sympatric ripe-fruit specialists: *Lagothrix lagotricha poeppigii* and *Ateles belzebuth belzebuth*. *Int. J. Primatol.*, **26**(5), 1107–1135.

Eisenberg, J. F. (1976). Communication mechanisms and social integration in the black spider monkey, *Ateles fusciceps robustus* and related species. *Smithson. Contribut. Zool.*, **213**, 1–108.

Eisenberg, J. F. and Kuehn, R. E. (1966). The behavior of *Ateles geoffroyi* and related species. *Smithson. Misc. Coll.*, **151**(8), 1–63.

Estrada, A. and Coates-Estrada, R. (1996). Tropical rain forest fragmentation and wild populations of primates at Los Tuxtlas. *Int. J Primatol.*, **5**, 759–783

Fedigan, L. M. and Baxter, M. J. (1984). Sex differences and social organization in free-ranging spider monkeys (*Ateles geoffroyi*). *Primates*, **25**, 279–294.

Hernández-López, L., Mayagoita, L., Esquivel-Lacroix, C., Rojas-Maya, S. and Mondragón-Ceballos, R. (1998). The menstrual cycle of the spider monkey (*Ateles geoffroyi*). *Am. J. Primatol.*, **44**, 183–195.

Hernández-López, L., Parra, G. C., Cerda-Molina, A. L., *et al.* (2002). Sperm quality differences between the rainy and dry seasons in captive black-handed spider monkeys (*Ateles geoffroyi*). *Am. J. Primatol.*, **57**(1), 35–42.

Hernández Salazar, L. T., Laska, M. and Rodriguez Luna, E. (2003). Olfactory sensitivity for aliphatic esters in spider monkeys, *Ateles geoffroyi*. *Behav. Neurosci.*, **117**, 1142–1149.

Hines, J. J. H. (2002). *Ateles geoffroyi* in northern Honduras: feasibility survey. Unpublished Masters thesis, University of Toronto, Canada.

Hines, J. J. (2004a). Size and composition of foraging groups of black-handed spider monkeys (*Ateles geoffroyi*) in Parque Nacional Pico Bonito, Atlántida, Honduras. *Am. J. Primatol*, **62**(S1), 89.

Hines, J. J. (2004b). Taxonomic status of *Ateles geoffroyi* in northern Honduras. *Am. J. Primatol.*, **62**(S1), 79–80.

Hines, J. J. H. (2004c). Survey of *Ateles geoffroyi* in Parque Nacional Pico Bonito, Honduras. *Folia Primatol.*, **75**(S1), 274.

Hines, J. J. H. (2005). Ecology and taxonomy of *Ateles geoffroyi* in Parque Nacional Pico Bonito, Atlántida, Honduras. Unpublished Ph.D. thesis, Australian National University, Canberra, ACT.

Hiramatsu, C., Tsutsui, T., Matsumoto, Y., *et al.* (2005). Color vision polymorphism in wild capuchins (*Cebus capucinus*) and spider monkeys (*Ateles geoffroyi*) in Costa Rica. *Am. J. Primatol.*, **67**, 447–461.

Izawa, K. (1993). Soil eating by *Alouatta* and *Ateles*. *Int. J. Primatol.*, **14**(2), 229–242.

Izawa, K., ed. (2000). *Adaptive Significance of Fission-Fusion Society in Ateles*. Sendai: Miyagi University of Education (in Japanese).

Izawa, K., Kimura, K. and Nieto, A. S. (1997). Grouping of the wild spider monkey. *Primates*, **20**, 503–512.

Izawa, K. and Mizuno, A. (1990). Chemical properties of special water drunk by wild spider monkeys (*Ateles belzebuth*) in La Macarena, (Colombia). *Field Studies of New World Monkeys, La Macarena, Colombia*, **4**, 38–46.

Joshi, D., Völkl, M., Shepherd, G. M. and Laska, M. (2006). Olfactory sensitivity for enantiomers and their racemic mixtures – a comparative study in CD-1 mice and spider monkeys. *Chem. Senses*, **31**, 655–664.

Karesh, W. B., Wallace, R. B., Painter, R. L. E., *et al.* (1998). Immobilization and health assessment of free-ranging black spider monkeys (*Ateles paniscus chamek*). *Am. J. Primatol.*, **44**(2), 107–123.

Klein, L. L. (1971). Observations on copulation and seasonal reproduction of two species of spider monkeys, *Ateles belzebuth* and *A. geoffroyi*. *Folia Primatol.*, **15**, 233–248.

Klein, L. L. (1972). The ecology and social organization of the spider monkey, *Ateles belzebuth*. Unpublished Ph.D. thesis, University of California, Berkeley.

Klein, L. and Klein, D. (1971). Aspects of social behaviour in a colony of spider monkeys *Ateles geoffroyi* at San Francisco Zoo. *Int. Zoo Yrbk.*, **22**, 175–181.

Klein, L. L. and Klein, D. B. (1977). Feeding behaviour of the Colombian spider monkey. In *Primate Ecology: Studies of Feeding and Ranging Behavior in Lemurs, Monkeys and Apes*, ed. T. H. Clutton-Brock, London: Academic Press, pp. 153–182.

Laska, M. (1996). Manual laterality in spider monkeys *(Ateles geoffroyi)* solving visually and tactually guided food-reaching tasks. *Cortex*, **32**, 717–726.

Laska, M. (1998). Laterality in the use of the prehensile tail in the spider monkey *(Ateles geoffroyi)*. *Cortex*, **34**, 123–130.

Laska, M., Bauer, V. and Hernández-Salazar, L. T. (2007a). Self-anointing behavior in free-ranging spider monkeys (*Ateles geoffroyi*) in México. *Primates*, **48**, 160–163.

Laska, M., Carrera Sanchez, E. and Rodriguez Luna, E. (1998). Relative taste preferences for food-associated sugars in the spider monkey *(Ateles geoffroyi)*. *Primates*, **39**, 91–96.

Laska, M., Carrera Sanchez, E., Rodriguez Rivera, J. A. and Rodriguez Luna, E. (1996). Gustatory thresholds for food-associated sugars in the spider monkey *(Ateles geoffroyi)*. *Am. J. Primatol.*, **39**, 189–193.

Laska, M., Freist, P. and Krause, S. (2007b). Which senses play a role in nonhuman primate food selection? A comparison between squirrel monkeys and spider monkeys. *Am. J. Primatol.*, **69**, 282–294.

Laska, M., Hernández Salazar, L. T. and Rodriguez Luna, E. (2000a). Food preferences and nutrient composition in captive spider monkeys, *Ateles geoffroyi*. *Int. J. Primatol.*, **21**, 671–683.

Laska, M., Hernández Salazar, L. T. and Rodriguez Luna, E. (2003). Successful acquisition of an olfactory discrimination paradigm by spider monkeys *(Ateles geoffroyi)*. *Physiol. Behav.*, **78**, 321–329.

Laska, M., Hernández Salazar, Rodriguez Luna, E. and Hudson, R. (2000b). Gustatory responsiveness to food-associated acids in the spider monkey, *Ateles geoffroyi*. *Primates*, **41**, 213–221.

Laska, M., Rivas Bautista, R. M. and Hernandez Salazar, L. T. (2006). Olfactory sensitivity for aliphatic alcohols and aldehydes in spider monkeys, *Ateles geoffroyi*. *Am. J. Phys. Anthropol.*, **129**, 112–120.

Lindshield, S. M. (2006). The density and distribution of *Ateles geoffroyi* in a mosaic landscape at El Zota Biological Field Station, Costa Rica. Unpublished Masters thesis, Iowa State University, Ames.

Link, A. and Di Fiore, A. (2006). Seed dispersal by spider monkeys and its importance in the maintenance of neotropical rain-forest diversity. *J. Trop. Ecol.*, **22**, 335–346.

Link, A., Palma, A. C., Velez, A. and de Luna, A. G. (2006). Costs of twins in free-ranging white-bellied spider monkeys (*Ateles belzebuth belzebuth*) at Tinigua National Park, Colombia. *Primates*, **47**(2), 131–139.

Luckett, J., Danforth, E., Linsenbardt, K. and Pruetz, J. (2004). Planted trees and corridors for primates at El Zota Biological Field Station, Costa Rica. *Neotrop. Primates*, **12**, 143–146.

Matisoo-Smith, E., Watt, S. L., Allen, J. S. and Lambert, D. M. (1997). Genetic relatedness and alloparental behaviour in a captive group of spider monkeys (*Ateles geoffroyi*). *Folia Primatol.*, **68**(1), 26–30.

Matsuda, I. and Izawa, K. (2007). Predation of wild spider monkeys at La Macarena, Colombia. *Primates*, **49**(1), 65–68.

McDaniel, P. (1994). The social behavior and ecology of the black-handed spider monkey (*Ateles geoffroyi*). Ph.D. dissertation, University of Saint Louis, MO.

McDaniel, P. S., Janzow, F. T., Porton, I. and Asa, C. S. (1993). The reproductive and social dynamics of captive *Ateles geoffroyi* (black-handed spider monkey). *Am. Zool.*, **33**, 173–179.

Méndez, P. F. R. (1999). La influencia de la distribución y abundancia de frutos sobre el tamaño de subgrupo de *Ateles chamek* en la EBB, Bolivia. Unpublished Licenciatura thesis, Universidad Mayor de San Andrés, La Paz, Bolivia.

Milton, K. (1981). Estimates of reproductive parameters for free-ranging spider monkeys, *Ateles geoffroyi*. *Primates*, **22**, 574–579.

Milton, K. (1993). Diet and social organization of a free-ranging spider monkey population: the development of species-typical behavior in the absence of adults. In *Juvenile Primates: Life History, Development, and Behavior*, ed. M. E. Pereira and L. A. Fairbanks, New York: Oxford University Press, pp. 172–181.

Milton, K. and Hopkins, M. E. (2005). Growth of a reintroduced spider monkey population on Barro Colorado Island, Panama. In *New Perspectives in the Study of Mesoamerican Primates: Distribution, Ecology, Behavior and Conservation*, ed. A. Estrada, P. Garber, M. Pavelka and L. Luecke, New York: Springer, pp. 417–436.

Nunes, A. (1995). Foraging and ranging patterns in white-bellied spider monkeys. *Folia Primatol.*, **65**, 85–99.

Nunes, A. (1998). Diet and feeding ecology of *Ateles belzebuth belzebuth* at Maracá Ecological Station, Roraima, Brazil. *Folia Primatol.*, **69**, 61–76.

Nunes, A. and Chapman, C. A. (1997). A re-evaluation of factors influencing the sex ratio of spider monkey populations with new data from Maracá Island, Brazil. *Folia Primatol.*, **68**(1), 31–33.

Pacheco, L. F. (1997). Consecuencias demográficas y genéticas para Inga ingoides (Mimosoideae) de la extinción local de Ateles paniscus (Cebidae), uno de sus dispersores de semillas. Unpublished doctoral thesis, Universidad de Chile, Santiago.

Pacheco, L. F. and Simonetti, J. A. (1998). Consecuencias demográficas para *Inga ingoides* (Mimosoideae) por la pérdida de *Ateles paniscus* (Cebidae), uno de sus dispersores de semillas. *Ecología en Bolivia*, **31**, 67–90.

Pacheco, L. F. and Simonetti, J. A. (2000). Genetic structure of a mimosoid tree deprived of its seed disperser, the spider monkey. *Conserv. Biol.*, **14**(6), 1766–1775.

Pastor-Nieto, R. (2000). Female reproductive advertisement and social factors affecting the sexual behavior of captive spider monkeys. *Lab. Primate Newsl.*, **39**, 5–9.

Pastor-Nieto, R. (2001). Grooming, kinship, and co-feeding in captive spider monkeys. *Zoo Biol.*, **20**, 293–303.

Piñeros, M. C. (1994). Population characteristics of spider monkeys (*Ateles geoffroyi panamensis*) in Parque Nacional Corcovado, Costa Rica. *Am. Phys. Anthropol. Suppl.*, **18**, 160.

Pozo Rivera, W. E. (2004). Agrupación y dieta de *Ateles belzebuth belzebuth* en el Parque Nacional Yasuní, Ecuador. *Anuar Investigación Científica*, **2**, 77–102.

Ramos-Fernández, G. (2005). Vocal communication in a fission-fusion society: do spider monkeys stay in touch with close associates? *Int. J. Primatol.*, **26**, 1077–1092.

Ramos-Fernández, G. and Ayala-Orozco, B. (2002). Population size and habitat use of spider monkeys in Punta Laguna, Mexico. In *Primates in Fragments: Ecology and Conservation*, ed. L. K. Marsh, New York: Kluwer/Plenum Press, pp. 191–209.

Ramos-Fernández, G., Vick, L. G., Aureli, F., Schaffner, C. and Taub, D. M. (2003). Behavioral ecology and conservation status of spider monkeys in the Otoch Ma'ax Yetel Kooh protected area. *Neotrop. Primates*, **11**, 155–158.

Riba-Hernández, P., Stoner, K. E. and Osorio, D. (2004). Effect of polymorphic colour vision for fruit detection in the spider monkey *Ateles geoffoyi*, and its implications for the maintenance of polymorphic colour vision in platyrrhine monkeys. *J. Exp. Biol.*, **207**, 2465–2470.

Rocha, N. F. R. (1999). Censo de primates y evaluación de la metología de líneas de transectas en Lago Caiman, Parque Noel Kempff Mercado. Unpublished licenciatura thesis, Universidad Autónoma Gabriel Rene Moreno, Santa Cruz, Bolivia.

Rodrigues, M. R. and Linshield, S. L. (2007). Scratching the surface: observations of tool use in wild spider monkeys. *Am. J. Phys. Anthropol.*, **S44**, 201.

Russo, S. E. (2003). Responses of dispersal agents to tree and fruit traits in *Virola calophylla* (Myristicaceae): implications for selection. *Oecologia*, **136**, 80–87.

Russo, S. E. (2005). Linking seed fate to dispersal patterns: identifying factors affecting predation and scatter-hoarding of seeds of *Virola calophylla* in Peru. *J. Trop. Ecol.*, **21**, 243–253.

Russo, S. E. and Augspurger, C. K. (2004). Aggregated seed dispersal by spider monkeys limits recruitment to clumped patterns in *Virola calophylla*. *Ecol. Lett.*, **7**, 1058–1067.

Russo, S. E., Campbell, C. J., Dew, J. L., Stevenson, P. R. and Suarez, S. A. (2005). A multiforest comparison of dietary preferences and seed dispersal by *Ateles* spp. *Int. J. Primatol.*, **26**(5), 1017–1037.

Russo, S. E., Portnoy, S. and Augspurger, C. K. (2006). Incorporating animal behavior into seed dispersal models: implications for seed shadows and an example for a primate-dispersed tree. *Ecology*, **87**(12), 3160–3174.

Sainz, L. A. (1997). Censo de primates en un área de explotación forestal del Bajo Paraguá. Unpublished licenciatura, thesis, Universidad Autónoma Gabriel René Moreno, Santa Cruz, Bolivia.

Schaffner, C. M. and Aureli, F. (2005). Embraces and grooming in captive spider monkeys. *Int. J. Primatol.*, **26**(5), 1093–1106.

Shimooka, Y. (2003). Seasonal variation in association patterns of wild spider monkeys (*Ateles belzebuth belzebuth*) at La Macarena, Colombia. *Primates*, **44**, 83–90.

Shimooka, Y. (2005). Sexual differences in ranging of *Ateles belzebuth belzebuth* at La Macarena, Colombia. *Int. J. Primatol.*, **26**, 385–406.

Simmen, B. and Sabatier, D. (1996). Diets of some French Guianan primates: food composition and food choices. *Int. J. Primatol*, **17**, 661–693.

Simmen, B., Julliot, C., Bayart, F. and Pagès-Feuillade, E. (1998). Densités de primates en forêt dense guyanaise: test d'une méthode d'estimation par transect. *CR. Acad. Sci. III Vie*, **321**, 699–704.

Slater K. Y., Schaffner C. M. and Aureli F. (2007). Embraces for infant handling in spider monkeys: evidence for a biological market? *Anim. Behav.*, **74**, 455–461.

Spehar, S. N. (2006). The function of the long call in white-bellied spider monkeys (*Ateles belzebuth*) in Yasuní National Park, Ecuador. Unpublished Ph.D. thesis, New York University.

Stevenson, P. R., Quinones, M. J. and Ahumada, J. A. (2000). Influence of fruit availability on ecological overlap among four neotropical primates at Tinigua National Park, Colombia. *Biotropica*, **32**(3), 533–544.

Suarez, S. A. (2003). Spatio-temporal foraging skills of white-bellied spider monkeys (*Ateles belzebuth belzebuth*) in the Yasuní National Park, Ecuador. Unpublished Ph.D. thesis, State University of New York at Stony Brook.

Suarez, S. A. (2006). Diet and travel costs for spider monkeys in a nonseasonal, hyperdiverse environment. *Int. J. Primatol.*, **27**(2), 411–436.

Symington, M. M. (1987a). Ecological and social correlates of party size in the black spider monkey, *Ateles paniscus chamek*. Unpublished Ph.D. thesis, Princeton University, NJ.

Symington, M. M. (1987b). Sex ratio and maternal rank in wild spider monkeys: when daughters disperse. *Behav. Ecolo. Sociobiol.*, **20**, 421–425.

Symington, M. M. (1988a). Demography, ranging patterns, and activity budgets of black spider monkeys (*Ateles pansicus chamek*) in the Manú National Park, Peru. *Am. J. Primatol.*, **15**, 45–67.

Symington, M. M. (1988b). Food competition and foraging party size in the black spider monkey (*Ateles paniscus chamek*). *Behaviour*, **105**, 117–132.

Valero, A., Schaffner, C. M., Vick, L. G., Aureli, F. and Ramos-Fernandez, G. (2006). Intragroup lethal aggression in wild spider monkeys. *Am. J. Primatol.*, **68**, 732–737.

van Roosmalen, M. G. M. (1985). Habitat preferences, diet, feeding strategy and social organization of the black spider monkey (*Ateles paniscus paniscus* Linnaeus 1758) in Surinam. *Acta Amazonica.* **15**, 1–238.

Wallace, R. B. (1998). The behavioural ecology of black spider monkeys in north-eastern Bolivia. Unpublished Ph.D. thesis, University of Liverpool, UK.

Wallace, R. B. (2001). Diurnal activity budgets of black spider monkeys, *Ateles chamek*, in a southern Amazonian tropical forest. *Neotrop. Primates*, **9**, 101–107.

Wallace, R. B. (2005). Seasonal variations in diet and foraging behavior of *Ateles chamek* in a southern Amazonian tropical forest. *Int. J. Primatol.*, **26**(5), 1053–1076.

Wallace, R. B. (2006). Seasonal variations in black-faced black spider monkey (*Ateles chamek*) habitat use and ranging behavior in a southern Amazonian tropical forest. *Am. J. Primatol.*, **68**(4), 313–332.

Wallace, R. B., Painter, R. L. E and Taber, A. B. (1998). Primate diversity, habitat preferences, and population density estimates in Noel Kempff Mercado National Park, Santa Cruz Department, Bolivia. *Am. J. Primatol.*, **46**(3), 197–211.

Watt, S. (1994) Alloparental behavior in a captive group of spider monkeys (*Ateles geoffroyi*) at the Auckland Zoo. *Int. J. Primatol.*, **15**, 135–151.

Weghorst, J. A. (2007). High population density of black-handed spider monkeys (*Ateles geoffoyi*) in Costa Rican lowland wet forest. *Primates* (Online First), **48**, 108–116.

Wentz, E. A., Campbell, A. F. and Houston, R. (2003). A comparison of two methods to create tracks of moving objects: linear weighted distance and constrained random walk. *Int. J. Geo. Inf. Sci.*, **17**, 623–645.

Youlatos, D. (2002). Positional behavior of black spider monkeys (*Ateles paniscus*) in French Guiana. *Int. J. Primatol*, **23**, 1071–1093.

Part I
Taxonomy, phylogeny and evolution

2 *Morphology and evolution of the spider monkey, genus* Ateles

ALFRED L. ROSENBERGER, LAUREN HALENAR,
SIOBHÁN B. COOKE AND WALTER C. HARTWIG

Introduction

Spider monkeys cast a distinct morphological silhouette – long scrawny arms and a snaky prehensile tail arching from a narrow pot-belly torso, topped by a small round head and blunt face. The commitment of this relatively large-bodied platyrrhine to a large-tree, upper canopy milieu and to ripe fruit foraging is seen throughout its skeletal and craniodental morphology. Spider monkeys are the signature New World suspensory-postured brachiators. Bodily, they are the closest thing to a gibbon that has evolved anywhere else within the Order Primates. Less obvious may be the fact that they are also gibbonesque craniodentally. But in the context of the adaptive array of Latin America's four ateline genera, *Alouatta*, *Lagothrix*, *Brachyteles* and *Ateles*, spider monkeys are not simply the polar end of an adaptive morphocline, standing opposite howlers or even opposite *Lagothrix* if we draw our comparison more narrowly, to encompass only atelins. Spider monkeys are *different by far*. For example, as close as *Brachyteles* is to the visage of a spider monkey with its ungainly limbs and shortness of face, it does not match *Ateles* in the high-energy lifestyle that goes along with eating quickly metabolized fruit and little else. Nor can *Brachyteles* deftly fly and lope through the trees as if gravity and substrate did not matter and hands, feet and tail were octopus tentacles. How ironic that Geoffroy Saint-Hilaire was so impressed with the spider monkey's lone anatomical "deficiency," its missing thumb, that in 1806 he named the genus *Ateles*, meaning imperfect.

General morphology

Spider monkeys are built to roam for ripe fruit in the upper canopy of a stratified tropical rain forest. Their lithe skeleton is designed to suspend and hurl their

Spider Monkeys: Behavior, Ecology and Evolution of the Genus Ateles, ed. Christina J. Campbell. Published by Cambridge University Press. © Cambridge University Press 2008.

Figure 2.1 *Ateles* skeleton (Encyclopedia Britannica, 1893).

body weight, rather than strut it against gravity quadrupedally (Figure 2.1). When dealing with the minimally resistant pulp of a choice fruit, spider monkey anatomy abides as well – in the form of an impressive incisor battery and undistinguished, open-basin molar occlusal surfaces. Additionally, their energy-rich diet allows the spider monkey anatomy to afford modestly enlarged brains.

Historically, spider monkey anatomy, in the context of its membership in the ateline group, has been compared most frequently and favorably to that of hominoids (Erikson, 1963; Rosenberger and Strier, 1989). A more pointed comparison might emphasize resemblances with gibbons, but less empirical work has been done to examine that aspect. Field, museum and genetics studies have progressed from broad inter generic surveys to focused and single question, intra specific projects (e.g. Norconk *et al.*, 1996). These investigations have expanded our sense of the uniqueness of *Ateles* and challenged long-held phylogenetic interpretations (Jones, 2004; Hartwig, 2005), but they have not altered the fundamental ecomorphological depiction of *Ateles* as a ripe-fruit driven, upper canopy suspensory brachiator.

Body size and sexual dimorphism

It has been difficult to develop a clear, consistent assessment of body size and sexual dimorphism in *Ateles*. Some studies of male and female body weight in

Table 2.1 *Body weight and sexual dimorphism in* Ateles

Species	Dimorphism index	Male weight (g)	Female weight (g)
A. belzebuth	1.052	8532	8112
A. fusciceps	1.010	8890	8800
A. geoffroyi	1.101	8210	7456
A. paniscus	0.853	7460	8750

Dimorphism index = male weight/female weight
Sources: Ford (1994); Ford and Davis (1992).

the various species conclude that this genus is most often monomorphic, with one species, *A. paniscus*, showing negative sexual dimorphism in which the females are actually larger than the males (see Table 2.1; Ford and Davis, 1992; Ford, 1994). The most recent survey by Di Fiore and Campbell (2007) obtained similar results using a sample restricted to free-ranging individuals. Their mean values were comparable to those of Ford and Davis (1992) for all species except *A. paniscus*. Di Fiore and Campbell (2007) also summarized published assessments, noting that Smith and Jungers (1997) provided a dimorphism index of 1.08 for *A. paniscus* as compared with the 0.853 value reported by Ford and Davis (1992). The discrepancy represents the difference between categorizing *A. paniscus* as monomorphic or negatively dimorphic. For comparison with the other ateline primates, the dimorphism indices given by Di Fiore and Campbell (2007) for *Alouatta* ranged from 1.2 (*A. seniculus*) to 1.76 (*A. pigra*); *Brachyteles* ranged from 1.13 to 1.2; and *Lagothrix* from 1.24 to 1.57. When these data are taken into consideration, *Ateles* is the closest genus to being monomorphic among the atelines, but it must be noted that comparable data on a variety of other atelin species or populations is not really available.

Cranial morphology

The *Ateles* cranium has a gracile build, characterized by large rounded orbits, a globular braincase, a narrow face ending in a fairly prominent but narrow snout, and a shallow mandible (Figure 2.2). This pattern, especially in the shape of the face, makes spider monkeys easily recognizable and quite distinct from other ateline genera. *Alouatta* skulls have a relatively large uptilted face and relatively small braincase, a lower jaw in which the angle of the mandible is deep and flared posteriorly, and a tall ramus. *Lagothrix* and *Brachyteles* skulls share more general resemblances. They have moderately large, broad faces and braincases that are less rounded in shape than in *Ateles*. The mandible of *Lagothrix* is

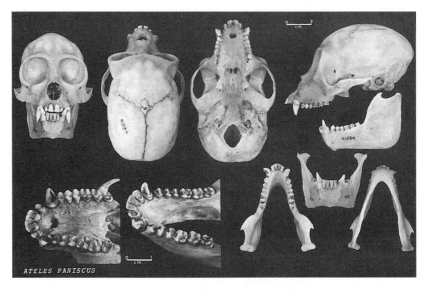

Figure 2.2 *Ateles paniscus* skull. (From Hershkovitz, 1977.)

moderately developed and not very deep and *Brachyteles* has a lower jaw that is inflated postero-inferiorly and carries a ramus that is also quite tall.

In terms of overall size, spider monkey crania show no obvious sexual dimorphism (Masterson and Hartwig, 1998), but there is evidence of differences in growth patterns between males and females. According to Corner and Richtsmeier (1993), in their study of *Ateles geoffroyi*, females in the oldest subadult age groups appeared to be larger because of an earlier onset of maturity. Schultz (1960) also pointed out that even though the male and female crania were of similar overall size in his study sample, the sexes differed somewhat in specific cranial dimensions, including having a shorter postcanine length, longer facial breadth and taller facial height in males than females.

The overall shape of the *Ateles* skull is strongly influenced by regional growth patterns and the packaging requirements of the face and braincase, especially compared with other atelines (Hartwig, 1993). Inside the globular neurocranium of *Ateles* is a brain typically over 100 grams (Armstrong and Shea, 1997). When compared with body size, the brain is slightly above regression lines based on other genera (Hartwig, 1993, 1996). In general, dimensions of the spider monkey facial skeleton scale with other relatively "unspecialized" New World monkey genera, but as the neurocranium becomes more globular and frontally disposed, the facial skeleton develops more orthognathically. *Ateles* crania thus display the relatively frontated orbital alignment and facial recession typical of small-snouted, orthognathic primates, with the emphasis here on *relative within New World monkeys* (Hartwig, 1993).

The external morphology of the *Ateles* brain has been studied, with special attention given to features related to innervating the prehensile tail. Hershkovitz (1970, 1977) provides comparative schematic figures of the cerebral cortex and volumetric measurements. The regions of the brain that have both sensory and motor control of tail function are larger in *Ateles* than in other species, reflecting the remarkably sensitive, flexible prehensile tail (see below) and altering the spatial arrangement of sulci and gyri of the brain's lateral surface. This specialization also leads to more direct and efficient nerve endings in the lower levels of the spinal cord itself that are related to tail function (Armstrong and Shea, 1997).

The cranial anatomy of *Ateles* is also notable for what it does *not* display. The narrow facial skeleton is relatively shallow, gracile and unremarkable, in keeping with the general expectations of a highly frugivorous taxon. The braincase is also relatively simple in design, rounded as might be expected in a modestly encephalized form and lacking marked temporal and nuchal lines, crests, or rugosities on the outer table. This simplistic picture is not meant, however, to imply that the *Ateles* head is also primitive in design. On the contrary – it combines a variety of traits not expected in the ancestral morphotype of atelins or atelines, which we believe more closely resembled a more robust architecture similar to *Lagothrix*.

Dental morphology

The overall morphology of the *Ateles* dentition is consistent with its relatively small face, but blunt snout tip. The cheek teeth are unimpressive in size, but the incisor teeth are well developed (Figure 2.3a). This pattern stands in contrast to *Alouatta*, with large cheek teeth and small incisors; *Brachyteles*, also with larger cheek teeth and small incisors (Figure 2.3b); and *Lagothrix*, with large cheek teeth and proportionately large incisors (Figure 2.3a). All three of these latter taxa have larger faces, although each is built somewhat differently.

The upper dental arcade of the spider monkey is parabolic with the palate broadening posteriorly and the molars set farther apart than the canines. The lower dental arcade is more U-shaped, with the cheek teeth rows set closer together and running more parallel to one another anteroposteriorly. Kinzey (1970) attempted to quantify the length:breadth proportions of primate lower jaws, given the purported importance of this measure for diagnosing hominins. He developed an index, the "basic rectangle of the mandible." The mean value for *Ateles fusciceps* (= *A. geoffroyi fusciceps*) was 140.5 ± 1.5, placing its

(a)

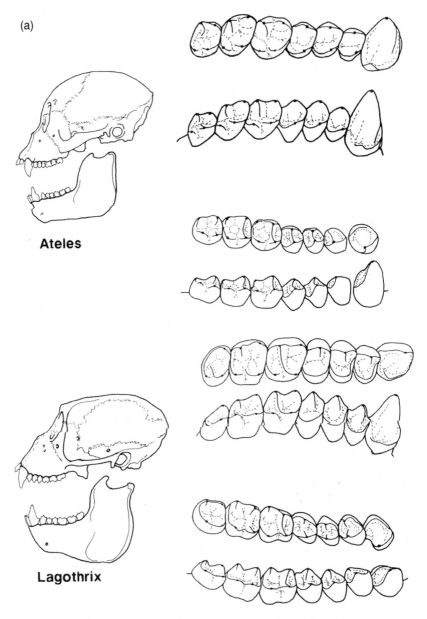

Ateles

Lagothrix

Figure 2.3 Comparative craniodental anatomy of *Ateles* and *Lagothrix* (a) and *Brachyteles* and *Alouatta* (b). Skulls drawn to approximately the same cranial length. Maxillary (occlusal and $^3/_4$ lingual views) and mandibular (occlusal and $^3/_4$ buccal views) dentitions each drawn to approximately the same length. (From Rosenberger and Strier, 1989, with permission of the authors.)

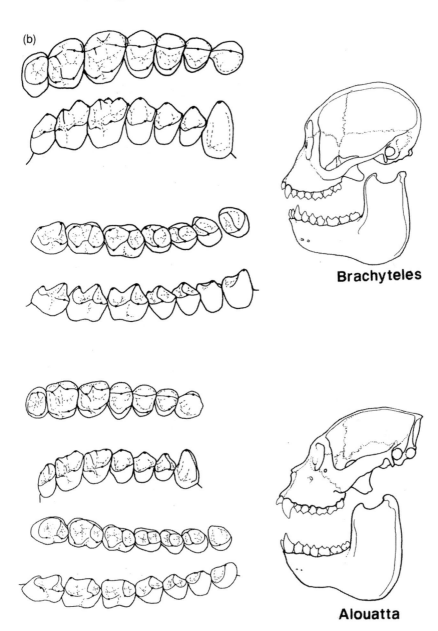

(b)

Brachyteles

Alouatta

Figure 2.3 (*cont.*)

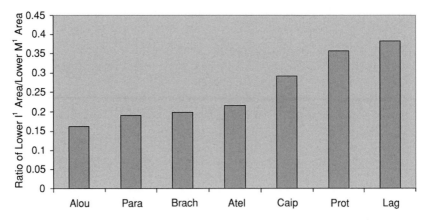

Figure 2.4 Ratio of lower first incisor area relative to lower first molar area in *Ateles* compared with both extant and extinct atelines. Abbreviations to genera, left to right: *Alouatta, Paralouatta, Brachyteles, Ateles, Caipora, Protopithecus, Lagothrix.* (Data augmented from Rosenberger, 1992.)

shape between the extreme parabola of *Homo sapiens* and the extreme U-shape of some other primates.

The incisor teeth are an impressive feature of the *Ateles* dentition. Upper and lower incisors are relatively wide and high-crowned. The incisors do not appear to be disproportionately large relative to molar size (Rosenberger, 1992; Figure 2.4). Upper central incisors are much larger than the lateral incisors and are spatulate in shape, while the upper lateral incisors are more conical. All lower incisors are more or less equal in size and homomorphically spatulate.

The upper canines are long, slender, and recurved in males and slightly shorter, stouter and less projecting in females. The lower canines occlude into a sizable diastema between the upper second incisors and the upper canines. The maxillary canines occlude with an enlarged mesiobuccal surface of the lower second premolars, resembling the canine–P^3 honing complex seen in catarrhines. However, the canines of *Ateles* appear not to be particularly large. Harvey *et al.* (1978) calculated relative canine size for primates, irrespective of sex differences. They determined that canines in *A. geoffroyi* are about 90% of the size expected based on a regression model that sampled 39 species, including nonplatyrrhines.

Sexual dimorphism in canine size has been documented and assessed in different ways. Orlosky (1973) provided useful descriptive statistics for canine length and breadth dimensions in *A. geoffroyi* and *A. belzebuth*, and concluded that metric sexual dimorphism was minimal but varied in its expression in the two. Kay *et al.* (1988), examining *A. geoffroyi, A. fusiceps* (including specimens now known as *A. geoffroyi fusiceps* and *A. g. robustus*) and *A. paniscus* found consistent degrees of dimorphism in each taxa. Using multivariate analysis, they developed a combined measure of tooth diameters, which effectively showed that males were about 10% larger than females. Plavcan and Kay (1988) and Plavcan and van Schaik (1997) found the degrees of canine height dimorphism in *Ateles* consistent with that of other atelines, despite perceived lower levels of male–male competition (see also Masterson and Hartwig, 1998). These interpretations should not be construed with other studies that may emphasize monomorphism, dimorphism or negative dimorphism in *Ateles* species based on other anatomical systems. For example, for *A. geoffroyi*, Corner and Richtsemeier (1993) determined that skulls were monomorphic in overall morphology, but noted that male and females matured at different rates. Schultz (1960) identified negative dimorphism in some cranial dimensions but not in others. Masterson and Hartwig (1998) noted that *A. paniscus* was monomorphic for most cranial metrics. Chapman and Chapman (1990) found negative dimorphism in body weight.

The upper premolars are typical bicuspid teeth, wider buccolingually than they are mesiodistally. The buccal cusp is taller than the lingual cusp and there is a wide deep basin in the center of the tooth. P^2 is slightly smaller than P^3 and P^4, which are equal in size. Generally, crowns of the cheek teeth are of a nonshearing design. The first and second upper molars are relatively square and have relatively low relief, with a prominent ridge-like crista obliqua and marginal crests that are not strongly beveled. The low cusps are spaced far apart on the corners of the crown, creating a wide central basin. The hypocone is well developed and tends to be separated from the central basin by a distinct groove. Upper M^3s are greatly reduced in size and complexity as compared with M^1 or M^2.

The lower premolars are rounded in shape with cusps that are more equal in height. Due to its participation in occlusion and wear of the upper canine, P_2 is somewhat larger than P_3 and P_4, especially in males with larger canines. This creates a sloping wear facet on the buccal surface and a dominant buccal cusp that makes the tooth almost caniniform. All three premolars have a relatively large bulbous buccal surface. The first two lower molars are smaller and narrower buccolingually than the uppers. They also show a lack of shearing crests in favor of deep basins between widely spaced, low-relief cusps. M_3s are reduced but not to the same degree as the other molars, although this could be a variable trait among species and individuals. Both upper and lower molars lack cingula and structures such as accessory cuspules.

Functionally, the *Ateles* dentition is considered well suited for a classically frugivorous diet (Kay, 1975; Hylander, 1979; Rosenberger, 1992) with relatively broad incisors and proportionately small molars (Anthony and Kay, 1993; Anapol and Lee, 1994). The extreme reduction of third molars is connected with the general reduction of the masticatory apparatus (Rosenberger and Strier, 1989), as is common in frugivores. Smith (1978) concluded much the same about *Ateles* in his biomechanical study of the temporomandibular joint in primates. Dental microwear studies of *Ateles* show a scratch-dominated pattern related to the consumption of seeds and pulp of mature fleshy fruits (Kay, 1987). Other pertinent functional studies are Wall (1999), who examined movements of the mandibular condyle in *A. geoffroyi* cineradiographically. Additionally, in a series of classic studies on the dynamics of the chewing cycle of primates, Kay and Hiiemae (1974 *et seq.*) described the dynamics of *Ateles* mastication in relation to molar morphology.

The shoulder girdle

The shoulder girdle of *Ateles* has been of particular interest to many researchers (e.g. Campbell, 1937; Erikson, 1963; Ashton and Oxnard, 1963, 1964; Oxnard, 1963, 1967; Jenkins *et al.*, 1978; Konstant *et al.*, 1982; Turnquist, 1983; Takahashi, 1990; Young, 2003; Jones, 2004), particularly in connection with the comparative anatomy of gibbons and great apes and the functional morphology of suspensory positional behaviors. Spider monkeys do exhibit many resemblances to hominoids, as outlined by Erikson (1963), and *Ateles* and the hylobatids are the most acrobatic arm-swingers among platyrrhines and catarrhines, respectively. So, within the atelins, *Ateles* tends to present the most exaggerated morphologies relating to brachiation-style positional behaviors and *Lagothrix* the least, with *Brachyteles* resembling a bulked-up anatomical personification of *Ateles*. *Alouatta*, which is quite different in its positional behavior, also differs substantially from atelins in many anatomical details.

   In *Ateles*, the scapula is positioned dorsally rather than on the lateral aspect of the thorax, as is common among nonateline platyrrhines, and it is greatly elongated craniocaudally (Figure 2.5). The glenoid fossa points cranially, even when the arms are at rest. The scapular spine of *Ateles* is obliquely oriented relative to the blade's medial border. This has been interpreted as a facilitator of arm-raising and an adaptation to suspensory behaviors aided by the action of the cranial portion of the trapezius, which attaches along the scapular spine and may assist in scapular rotation (Inman *et al.*, 1944; Ashton and Oxnard, 1964). A strongly angled line of attachment would add to the mechanical advantage of the trapezius when the arm is raised. It should be noted, however, that it remains possible that the infraspinatus, which originates on the infraspinous fossa, may be more important for scapular rotation (Larson and Stern, 1986; Larson, 1995). Specialization of the trapezius is also implicated by the morphology of the acromion process, which projects past the glenoid fossa and is generally longer

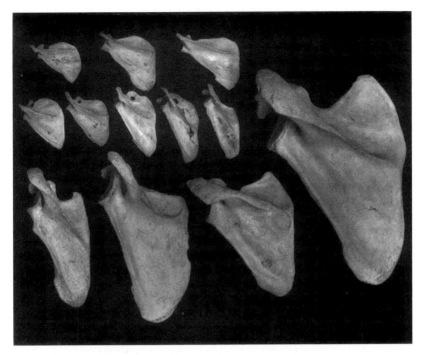

Figure 2.5 Primate scapulae. Top row, left to right: *Cebus, Colobus, Macaca*; middle row, left to right: *Alouatta, Lagothrix, Brachyteles, Ateles, Hylobates*; bottom row, left to right: *Pan, Pongo, Homo, Gorilla*. (From Erikson, 1963, with permission from the publisher.)

than in most other primates (Young, 2003). This feature may provide better mechanical advantage for either the cranial portion of the trapezius during scapular rotation and arm elevation (Jungers and Stern, 1984; Young, 2003), or it may increase the length of the lever arm of the deltoid, which would be advantageous during the elevation of the arm (Larson and Stern, 1986; Young, 2003).

Since the clavicle of *Ateles* spans between the manubrium on the chest and the dorsally positioned scapula, it is obliquely oriented and relatively long (Ashton and Oxnard, 1964; Jenkins *et al.*, 1978). While Ashton and Oxnard (1964) suggest that this morphology enhances the range of motion of the shoulder joint, Erikson (1963) explains it as a correlate of the widening of the thorax and the ventral shift of the vertebral column into the thoracic cavity, a pattern that is typical of atelines and is also hominoid-like. Other details of clavicular morphology that are well developed in *Ateles* involve its torsion and sigmoid shape.

Clavicular torsion occurs when the distal end of the clavicle is directed more cranially than the proximal end. It might be related to the cranial orientation of the glenoid fossa and the dorsal position of the scapula (Ashton and Oxnard, 1964). A sigmoidal shape occurs when the proximal portion of the clavicle curves ventrally while the distal portion curves dorsally. The functional significance of this pattern is unclear. Jenkins *et al.* (1978) suggest that it might allow greater freedom of movement for the humerus in the glenoid fossa, while Voisin (2006) proposes that the proximal curvature may act as a "crank" for the clavicular insertion of the deltoid muscle, to help rotate the glenoid cavity cranially during the elevation of the arm. While such a function has also been emphasized for other mammals, Oxnard (1968) suggests that it is unlikely to serve as a crank in primates. Konstant *et al.* (1982) provide additional information pertinent to the clavicle and shoulder. They give comparative, electromyographic and kinematic information on the subclavius muscle of *Ateles* and other atelines in connection with climbing and arm swinging.

The forelimb

Intermembral indices of the forelimb (humerus + radius/femur + tibia × 100) have established that the upper limb as a whole is relatively longer in *Ateles* than in a variety of other platyrrhine primates (Erikson, 1963). In *Aotus* and *Cebus* the indices are 74 and 80, respectively, falling well outside the range of *Ateles*, which has an index of 105. Among the other atelines, *Brachyteles* also has an index of 105, and the two other genera each have an index of 98. Another measure developed by Erikson compares limb length with trunk length, and this further emphasizes the unique proportions of *Ateles*. He shows that the humerus and radius of *Ateles* is 150% of the length of the trunk. This compares with *Brachyteles* at 140% trunk length, and with *Lagothrix* and *Alouatta* at 91% and 109%, respectively.

More modern efforts to capture limb proportions have examined scaling. With few exceptions, in primates the intermembral index tends to increase with increasing body size, and in most platyrrhines the forelimbs show positive allometry while the hindlimbs show negative allometry (Jungers, 1985). This pattern has been explained as an adaptation to maintaining balance while climbing and traveling on arboreal supports. The increased length of the forelimbs allows an animal to lean away from a support and thus maintain a high level of pedal friction without raising the center of gravity and decreasing overall stability (Cartmill, 1974; Jungers, 1985). Still, *Ateles* has forelimbs that are approximately 36–38% longer than expected for its body size (Jungers, 1985), suggesting a complex functional explanation relating to an acrobatic locomotor

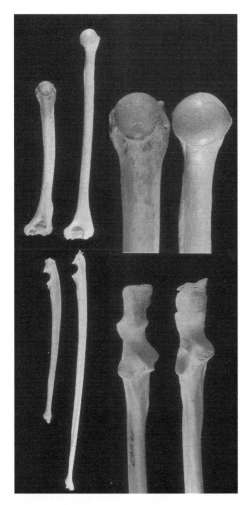

Figure 2.6 Comparing the forelimb. Left, *Alouatta*; right, *Ateles*. Proximal humeri and radii are on the right. (After Gebo, 1996, with permission from the publisher.)

style that includes significant amounts of below-branch suspension, arboreal climbing and quadrupedalism, as well as rapid brachiation (e.g. Mittermeier, 1978; Fontaine, 1990; Defler, 1999; Youlatos, 2002, this volume).

The shaft of the humerus of *Ateles* is fairly long, straight and slender, resembling the upper arm of *Brachyteles* and, to a lesser degree, *Alouatta* and *Lagothrix* (Figure 2.6). It has a large globular head with a small degree of medial torsion at the humeral neck. Overall humeral shape is also a similarity shared by *Ateles* and *Hylobates*. It has been interpreted as an adaptation that minimizes bending and shear stresses under torsion, which would be prevalent during suspensory locomotion (Swartz, 1990). Another osteological correlate to suspension is the relatively distal location of the deltoid tuberosity on the humerus shaft. This more distal insertion of the deltoid muscle provides a greater mechanical advantage for the deltoid when the arm is abducted (Ashton and Oxnard, 1964).

The distal end of the *Ateles* humerus has an enlarged projection of the medial epicondyle, a trait also associated with arm-swinging behavior as it relates to the origin of most of the flexor muscles of the forearm. Takahashi (1990) suggests that strong forearm flexors might also be a key muscle group during climbing, which is extremely important in the behavioral repertoire of *Ateles*. While *Ateles* is known for its dramatic brachiating locomotion, clambering, climbing and quadrupedal walking and running make up roughly 50% of its locomotor repertoire; 23% of locomotion is suspensory (Cant *et al.*, 2001; see also Youlatos, this volume). Overall the flexors are quite well developed in *Ateles*, and all four digits are generally flexed at once to provide a strong grip during suspensory postures (Youlatos, 2000). Unlike the lumbering quadrupedalist *Alouatta*, which has a well-developed olecranon on the proximal ulna, the process is reduced in *Ateles*.

The *Ateles* elbow shows several other features that are convergent on hominoids and contribute to an increased ability to pronate and supinate the forearm, motions which are extensively employed during suspensory activities (Rose, 1988). The radial head is relatively round with a small lateral lip, and the area for articulation with the radial notch of the ulna extends far around the radial head. This results in much less restricted axial rotation of the radius than is seen in other platyrrhine primates that have more flattening along the posterolateral side of the radial head (Rose, 1988).

The wrist shows several unique features also correlated with the *Ateles* locomotor profile. First, the carpal tunnel is quite deep to accommodate the large tendons of the flexors of the forearm (Napier, 1961). Napier (1961) notes that the deep tunnel affects the position of the first metacarpal, which is largely vestigial in *Ateles* and articulates with a trapezium that is steeply angled in toward the palm of the hand. Second, the carpals of *Ateles* allow a large range of motion across the wrist joint. A ball and socket joint is formed between the proximal and distal carpal rows; the capitate and hamate move in the socket formed by the proximal carpal row. This arrangement of the bones of the wrist allows midcarpal supination which permits increased mobility in the midcarpal region. As a result, *Ateles* can rotate the wrist at both the junction between the radius and proximal carpals and between the proximal and distal carpal rows, summing to almost 90 degrees of axial swivel (Jenkins, 1981). This wrist morphology also allows significantly more ulnar deviation of the hand during pronation as well as during suspensory locomotion (Lemelin and Schmitt, 1998). Despite the ball and socket joint, other aspects of the wrist joint remain fairly primitive and *Ateles* retains the primitive synovial septum which separates the ulnar and radial compartments of the joint (Lewis, 1971, 1972).

While the hands of spider monkeys appear superficially to be very long and hook-like, according to Jouffroy *et al*. (1991) spider monkey hands are not disproportionately longer than the hands of other platyrrhine primates in relation to the length of the entire forelimb. Jouffroy *et al*. (1991) show that the hands are approximately 27% of forelimb length; 26–32% is the range for other platyrrhines. Relatively, *Leontopithecus* has the longest hands of all New World monkeys. Regarding the functional axis and grasping pattern of the hand, *Ateles* is paraxonic, with the third and the fourth digits being of equal length. Paraxonic hands are also found in *Lagothrix*, but mesaxonic hands, with the middle digit being best developed, are the norm in the other platyrrhines (Jouffroy *et al*., 1991; Lemelin and Schmitt, 1998).

Of course, the most notable feature of the *Ateles* hand is the absence or great reduction of the pollex (Figure 2.7), a feature shared with *Brachyteles* and *Colobus*. In *Ateles*, the first metacarpal is present but, as in *Colobus*, the proximal phalanx is variably present and it is very variable in size when it does occur (Tague, 1997). We are not aware of a cogent functional argument explaining why the external thumb is lost in these cases. Historically, while convergence is surely behind its joint absence in platyrrhines and catarrhines, we think the loss in *Ateles* and *Brachyteles* is more likely to be homologous.

The hindlimb

Less attention has been given to the *Ateles* hindlimb. Descriptions liken it to hominoids, relating the similarities to comparable suspensory locomotor adaptations and somewhat orthograde body orientations (Stern, 1971; Larson, 1995, 1998; Johnson and Shapiro, 1998). Stern (1971) and Stern and Larson (1993) give detailed descriptions of the muscles of the hip and thigh and their possible relevance to questions related to the evolution of human bipedality. This short list of papers forms the basis of our account.

The femoral head is very round and globular and maintains a distinct articular surface that does not run down onto the femoral neck. The greater trochanter is highly elevated, coming up slightly below the level of the head, with a deep fossa behind it. Posteriorly, the proximal femur exhibits a large knob-like lesser trochanter. Together, this morphology is indicative of a mobile hip joint, which is consistent with the important role played by hindlimb suspension in the positional repertoire of spider monkeys (see Youlatos, this volume).

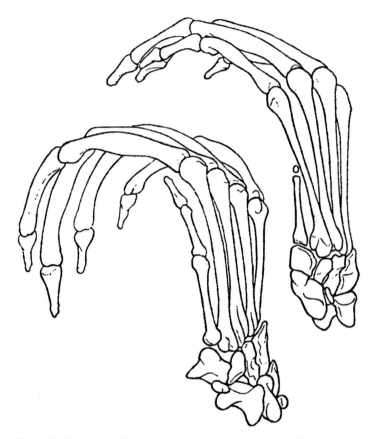

Figure 2.7 The hands of *Hylobates* (bottom) and *Ateles* (top). (Modified from Jouffroy and Lessertisseur, 1977.)

The femoral shaft is slender, long, and mostly straight except for a slight medial bowing towards the distal end. The knee joint is shallow, as expected in a nonleaping taxon, with the medial condyle slightly larger than the lateral condyle and a larger area for articulation with the patella. The articular surface of the proximal tibia matches the femoral condyles; that is, the medial surface is slightly larger than the lateral. The tibial tuberosity is broad and flat and leads onto a tibial shaft that is strongly compressed mediolaterally. Distally, the medial malleolus is robust and knob-like but not very long. The articular surface on the lateral side runs up onto the shaft and the ankle joint is built to be very mobile while being used in quadrumanous climbing and suspension.

Several studies have been done on the interior trabecular bone structure of the *Ateles* postcranial skeleton. The proximal humerus and femur of *A. paniscus* show the expected low degree of structural anisotropy for a suspensory taxon whose bones are under variable, nonrepetitive loading. However, the anisotropy values for *Ateles* and *Hylobates*, another highly suspensory taxon, overlap with those of the more terrestrial *Macaca* and *Papio*, suggesting that the trabeculae vary in a predictable way, but do not reliably discriminate between locomotor groups (Fajardo and Müller, 2001). Similarly, the femoral neck of *A. fusciceps* (= *A. geoffroyi fusciceps*) and *A. paniscus* has a more even distribution of bone with a thicker superior cortex relative to inferior cortex as compared with the rest of a primate-wide sample of 21 species due to the less stereotypical and more generalized loading orientation of quadromanus climbing, another major component of their locomotor repertoire (Rafferty, 1998).

The trunk and spine

The spine of *Ateles* has several unique adaptations to suspensory locomotion, the most interesting of which enables the tail to twist, bend, and curl up on itself, to be used in precision gripping and powerful clasping, and to support and probably propel the full weight of the body during some phases of brachiation. The tail is distinguished from the tails of other atelines in having more advanced features associated with prehensility and acrobatic locomotion. Nevertheless, it is clearly built on an ateline base, which differs from the morphology of the semiprehensile tailed *Cebus*, for example. There are several reasons for thinking that the unique ateline tails evolved prehensile adaptations independently of the semiprehensile tails of *Cebus* (Rosenberger, 1983). For example, ateline tails are relatively longer than expected relative to body weight (Figure 2.8), and the sensorimotor regions of the brain dealing with tail function are morphologically conspicuous.

The trunk, like the trunks of atelines generally, is rather short and stout when compared with the trunks of the other platyrrhine primates. The shortening occurs in the lumbar region, which is reduced to four vertebrae. *Alouatta* has five lumbar vertebrae and other platyrrhines have six to seven vertebrae on average (Schultz, 1961; Erikson, 1963). Shortening and stiffening of the lower back is also effected as the lumbar vertebral bodies are shorter antero-posteriorly and deeper ventrodorsally than in other platyrrhines (Erikson, 1963; Johnson and Shapiro, 1998). Metrically, this pattern is reflected by the ratio of lumbar:thoracic length, with *Ateles* presenting a value of 43%, well outside the range of other platyrrhine primates, at 91–110% (Erikson, 1963).

The functional significance of a shortened lumbar region is a matter of debate. Rose (1975) holds that it is associated with erect postures and compressive forces on the spine. While the atelines do exhibit orthograde postures, these

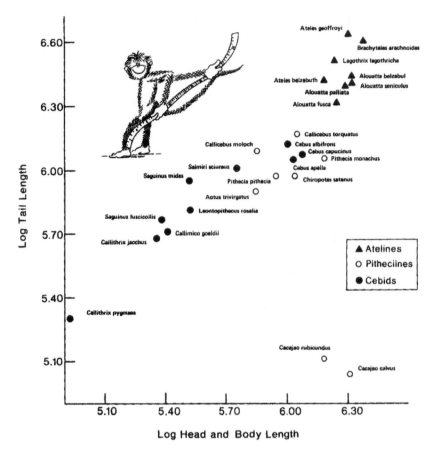

Figure 2.8 Scaling relationships of platyrrhine tails relative to body length. (From Rosenberger, 1983, by permission of the author.)

postures generally occur during suspensory behavior and would not result in high levels of compressive stress on the spine (Johnson and Shapiro, 1998). Johnson and Shapiro (1998) counter that the shortened lumbar region may be an adaptation for reducing bending stress on the lower portion of the spine, which would be pronounced if the animal were supporting most or all of its body weight by the prehensile tail.

Ankel (1972) described a variety of specializations of the sacrum in *Ateles* and other atelines. The sacroiliac joint is larger than in nonprehensile-tailed monkeys, which most likely provides more extensive support during suspensory activity. The sacral canal is of special interest. In primates that lack external tails, the canal ends within the sacrum itself, while in forms with long nonprehensile tails it narrows caudally. In contrast, *Ateles* has a unique canal that widens caudally to allow a relatively large bundle of nerves to pass though. They provide innervation to the sensitive tail tip. In addition to increased innervation in this area, *Ateles* also has an enhanced caudal blood supply. It has two systems of arteries, which are lacking on the nonprehensile-tailed forms. A correlate of this vascular arrangement is the high neural arches of the caudal vertebrae.

The morphology of the sacrocaudal joint of *Ateles* allows for an unusual capacity to "hyperextend" the tail. It is directed distodorsally rather than distally as in most nonprehensile-tailed primates (Turnquist *et al.*, 1999). Enhanced extension allows *Ateles* to grasp branches directly above while hanging vertically (Turnquist *et al.*, 1999) and is particularly suited to tail hanging while the torso is held relatively upright to form a sharp angle with the tail axis. The proximal part of the tail is also very flexible and heavily muscled. This may be a result of an unusual shortening of the proximal vertebrae relative to the patterns found in other primates (Ankel, 1972; Turnquist *et al.*, 1999). However, this interpretation has been contested by German (1982), who shows no statistical differences between the sizes of the proximal vertebrae of nonprehensile- and prehensile-tailed primates. The musculature of the proximal region is also derived in *Ateles*. While most primate species have equally sized dorsal and ventral bundles of muscles, *Ateles* has a much greater number of muscle fibers in the dorsal muscle group than ventral muscle group (Ankel, 1972; German, 1982; Lemelin, 1995).

The proximal portion of the tail ends with a transitional vertebra, which is followed by a longer string of distal tail vertebrae. On average, *Ateles* has 20–27 distal caudal vertebrae, thus bringing the total number to approximately 28–35 (Schultz, 1961). This is a higher count than both *Alouatta* (25–28 total vertebrae) and *Lagothrix* (24–29 total vertebrae) (Schultz, 1961). Distally, the spider monkey tail vertebrae taper by decreasing in both length and width. Also, the most distal vertebrae are relatively flattened dorsoventrally in comparison with those of nonprehensile-tailed monkeys, providing greater areas for muscle attachment and producing a thicker tail overall (Ankel, 1972; Lemelin, 1995). To enhance distal tail flexion, *Ateles* has a greater number of muscle fibers positioned ventrally in the distal portion of the tail, which contrasts with the thicker dorsal bundles in the more proximal portion. The tail tip is also very sensitive. It is primarily used for grasping branches during suspensory postures and locomotion where hand grasps and tail grasps alternate (Turnquist *et al.*, 1999). A naked patch of friction skin on the distal ventral surface assures the grip. It is present in all of the atelines, but not in the semiprehensile tail of *Cebus*, which is fully clothed in fur.

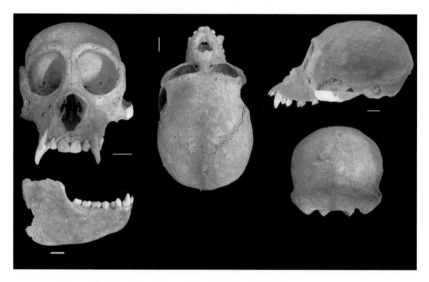

Figure 2.9 The skull and mandible of *Caipora bambuiorum*.

Evolution and phylogeny: the lessons of *Ateles*

Little is known about the morphological evolution of the genus *Ateles* specifically. However, background studies, synthesizing some of evidence for ateline evolution, are provided by Rosenberger and Strier (1989) and Hartwig (2005). There is no fossil record immediately relevant to the spider monkeys as there is for some other modern platyrrhines, which may have evolved as long-lived genera or tribes (e.g. *Alouatta, Saimiri, Aotus, Callimico*; Delson and Rosenberger, 1984; Setoguchi and Rosenberger, 1987; Rosenberger *et al.*, 1990). There are a few extinct forms to provide contextual perspective. For example, a single lower molar tooth representing *Solimoea acrensis* is known from the ~8 million year old (Ma) Acre Formation of western Brazil (Kay and Cozzuol, 2006). It resembles *Ateles* in some ways but is generally more primitive. Kay and Cozzuol place the species as a stem atelin, inferring that it was 5–6 kg in weight and interpreting its diet as *Ateles*-like based on comparable development of molar shearing crests. This is somewhat surprising since the *Ateles* diet is an evolutionary extreme, and other models (Rosenberger and Strier, 1989) infer that a more *Lagothrix*-like molar morphology, suited for a more eclectic frugivorous/folivorous diet, would be basal in atelins. The phylogenetics of this taxon merits further study as more evidence becomes available.

A second fossil form appears to be more closely related to *Ateles, Caipora bambuiorum* from the Pleistocene/Recent of central Brazil (Cartelle and Hartwig, 1996; Figure 2.9). *Ateles* and *Caipora* present numerous shared

Table 2.2 *Body weights of extinct and extant atelines*

Species	Weight (kg)
Alouatta fusca	4.4[a]
Alouatta caraya	5.2[a]
Alouatta belzebul	5.6[a]
Alouatta palliata	6.0[a]
Alouatta seniculus	6.2[a]
Lagothrix lagothricha	6.9[a]
Ateles belzebuth	8.1[a]
Ateles geoffroyi	8.2[a]
Paralouatta varonai	9.6–10.2[b]
Brachyteles arachnoides	13.5[a]
Caipora bambuiorum	20.5[c]
Protopithecus brasiliensis	24.9[d]

Sources: [a]Rosenberger (1992), [b]MacPhee and Meldrum (2006), [c]Cartelle and Hartwig (1996), [d]Hartwig and Cartelle (1996).

derived craniodental similarities but the latter is much larger in body size and may be different in several aspects of the postcranial skeleton, all of which may indicate different microhabitat preferences, and an alternative positional and locomotor profile. Further work is required, but if *Caipora*, at a body weight of 20 kg, is as closely related to *Ateles* as we suspect, it raises interesting possibilities concerning the foraging advantages to spider monkeys of being relatively small in size for an atelin (Table 2.2), a perspective that was lacking in the past (e.g. Rosenberger and Strier, 1989). *Caipora* and its equally large alouattin counterpart from the same fauna, *Protopithecus brasiliensis*, may be autapomorphic giants at 20–25 kg, but there is also the possibility that *Ateles* has become reduced in body size from a larger bodied ancestor as the brachiation complex evolved. Another datum that is pertinent to long-term evolutionary history of *Ateles* is the presence at La Venta, Colombia, at 12–14 Ma, of two *Stirtonia* species, very closely related to *Alouatta*, if not representing the same genus.

This small body of paleontological evidence suggests that the alouattin and atelin clades were probably established in South America by the late Middle Miocene, and that there was dental morphology (e.g. *Solimoea*) of relevance to the evolution of *Ateles* at that time. With the existence of *Caipora*, it is also more evident that evolution produced a "clade of spider monkeys," not simply the one genus that is now split into several species, and that we should not expect all of these close relatives to be adaptively constrained to resemble the modern *Ateles* lifestyle in all its dimensions. Moreover, these extinct spider monkey relatives will offer important clues for piecing together a more detailed picture of *Ateles* evolution, integrating their lessons with the information derived from the living taxa.

Another morphological perspective was presented by Rosenberger and Strier (1989). In the absence of fossils, other than one coming from the alouattin sister taxon of atelines, *Stirtonia*, they took a different approach in employing an adaptational character analysis and cladistic study of the modern genera, combining morphology, ecology and behavior. The branching pattern they endorsed (see Figure 2.10) held that *Ateles* and *Brachyteles* are sister taxa that together link up with *Lagothrix*, with the latter representing the stem atelin lineage. Their conclusions emphasized the soft fruit feeding and hyperactive locomotion of *Ateles* as unique adaptations among the atelins, involving a pattern set that is more derived than the more generalized frugivory/folivory and suspensory locomotor style of the last common ancestor (LCA) shared by *Ateles* and *Brachyteles*.

Molecular evidence is consistent with this very general picture, as Hartwig discussed at length in his re-examination of ateline interrelationships (Hartwig, 2005). There is only one point of discord between the ecomorphological (Rosenberger and Strier, 1989; Rosenberger, 1992; Strier, 1992) and the molecular studies, which is whether *Ateles* is more closely related to *Lagothrix* or to *Brachyteles*. Several laboratories have investigated the interrelationships of platyrrhines using a variety of nuclear and mitochondrial genes. They have come to consistently demonstrate a monophyletic Atelini and a sister-taxon relationship of *Lagothrix* and *Brachyteles* (e.g. Schneider *et al.*, 1993; Harada *et al.*, 1995; Schneider *et al.*, 1996; Horovitz *et al.*, 1998; von Dornum and Ruvolo, 1999; Meireles *et al.*, 1999; Canavez *et al.*, 1999). Collins (2004), on the other hand, who also employed molecular data, suggested that this link is not exceptionally robust since its return via parsimony algorithms is influenced by taxonomic sampling. When different species of *Ateles* were included in his analyses, the relationship among *Ateles*, *Brachyteles*, and *Lagothrix* were shown to be unstable. The same phenomenon has been shown using cranial morphology (Matthews and Rosenberger, in press).

No matter how the relationships of this triad turn out in the long run, the important point about what we have learned regarding *Ateles* phylogenetics is that the genus is part of a small monophyletic group of modern atelins that joins with *Alouatta* to comprise a coherent ecophylogenetic radiation. Prior to the 1980s, it was widely assumed that howlers had little to do with atelins (e.g. Hershkovitz, 1977), and that muriquis were the howler's closest living relatives (Zingeser, 1973). This latter view was promoted for a while even into the 1990s (Kay, 1990; but see Anthony and Kay, 1993). Establishing, with a high degree of confidence, that *Ateles*, *Brachyteles* and *Lagothrix* form a cladistic trio is

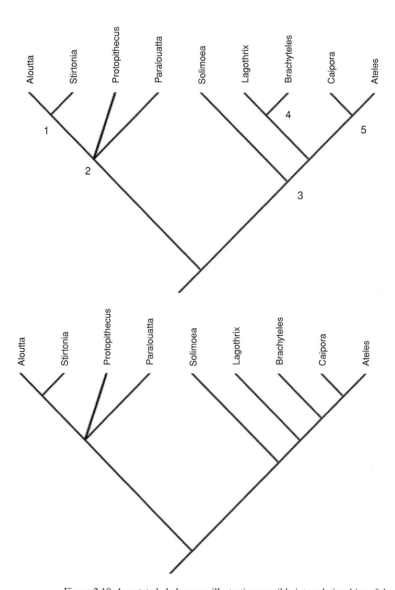

Figure 2.10 Annotated cladograms illustrating possible interrelationships of the
ateline primates, with alternative positions of *Brachyteles* shown to reflect the
predominant views indicated by molecular (top) and morphological (bottom) studies.
1. Evidence for the generic separation of *Alouatta* and *Stirtonia* is meager; *Stirtonia* is
only known from dental remains. 2. While we hold that *Protopithecus* and *Paralouatta*
are alouattins, with *Paralouatta* most likely being more closely related to
Stirtonia–Alouatta clade, their detailed interrelationships are currently under study. 3.
The position of *Solimoea* follows Kay and Cozzuol (2006). Since it is based on a
single molar tooth that has an *Ateles*-like reduction in its shearing crests, which is not
consistent with our interpretation of the atelin morphotype, we caution that additional
information may easily alter this position. 4. The morphological and molecular
interpretations regarding the cladistic position of *Brachyteles* differ: morphology
supports a sister-group relationship with *Ateles*; molecules support the link to
Lagothrix. 5. See Cartelle and Hartwig (1996) for evidence bearing on the relationship
of *Caipora* and *Ateles*.

a major step forward. It is the Big Idea behind all manner of evolutionary hypotheses about the natural history of *Ateles* and the atelins.

Cladistics, however, is only a piece of the puzzle. A broader, more penetrating question asks: How, ecologically, did *Ateles* evolve? The Rosenberger and Strier (1989) scenario, as mentioned, attempted to get at some of the answers by examining how and why anatomical and behavioral features evolved transformationally among atelines. Rosenberger *et al.* (in press) explore a different aspect of the problem by raising questions about community evolution and the biogeographic history of platyrrhines in the New World. Monkeys have been scattered about South America in varying degrees of geographic isolation, occupying habitats of varying quality, even as grasslands, now covering 60% of the landscape, have been the predominant terrestrial biome on the continent for almost 30 million years. Rosenberger *et al.* (in press) point out that over the past 26 million years (at least), platyrrhine evolution has taken place in four distinct provinces in South America: Amazonia, the Atlantic Coastal forest, Patagonia, and the Caribbean, in addition to Central America. They propose that several of the 14 living platyrrhine genera now inhabiting Amazonia may actually have arisen elsewhere. *Ateles* is not one of these, nor is *Lagothrix*. But like the eastern Brazilian endemic *Brachyteles*, which likely evolved in and along with the Atlantic coastal forest, there is a chance that the pan-Neotropical *Alouatta* also emerged originally outside of Amazonia. Among nonatelines, *Cebus*, *Callicebus*, and *Aotus* are other candidates for extra-Amazonian origins. While *Ateles* is ensconced in continental South and Middle America (and has also been recorded as a prehistoric relic in Cuba, though possibly as an import), its crossing of the Panamanian land bridge is a recent phenomenon, related to uplift of the isthmus about 3 million years ago. Thus, effectively, *Ateles*, a devotee of high, mature, wet, large-treed habitat – and decidedly not an ecological marginalist – is an Amazonian endemic. It may have arisen in that biogeographic province as the lowland region evolved its distinctive modern properties during the last 15 million years, according to one hypothesis of Amazonia's evolution (Campbell *et al.*, 2006).

If *Ateles* is a creation of the lowland rain forest, its long Amazonian fixation may be a vivid clue to its history by hinting at a community rationale for the spider monkey's adaptive differentiation. The Amazonian province, by comparison with the Atlantic province and also Middle America, supports the highest degree of primate endemism on the continent, the richest assemblages of sympatric primate diversity, and the most varied cornucopia of pertinent foodstuffs. *Ateles* has succeeded by evolving a unique niche among Amazonia's large biomass of platyrrhine frugivores. This niche revolves around eating an array of ripe soft fruit that perhaps only *Ateles* can afford, given the expensive time and energy budget involved in being so choosy when targeted foods are widely dispersed in space and also attractive to others. The capacity of spider monkeys to extract an inordinate amount of its protein needs from fruit sources, apparently, is also a matter of interest and may be part of its sympatric niche-differentiating strategy. But the connection between this sugary fruit diet and an acrobatic locomotor style is not clear. Easy rapid travel through the complex substrate web of the treetops may provide a distinct advantage if the cost of racing to patchily distributed foods is balanced by the anatomical autapomorphies enabling *Ateles* to do so in an energetically efficient manner. On the other hand, perhaps the acrobatics are basic to mating as well as other social behaviors. Rapid brachiation may be important in managing long-distance intertroop dispersal, maintaining group spatial cohesion as a counterbalance to having widely dispersed, small fission–fusion parties, or enabling exuberant, noisy agonistic displays as an affectation related to mate attraction, or to dissuade food competitors. A rationale for the *Ateles* postcranial skeleton that ignores these "nonmorphological" domains of behavior can only be incomplete, at best. If mating strategies and sociality were not having a profound effect on the evolution of the *Ateles* body, the genus would not be prone to sexual monomorphism.

One might also ask: Why quadrumanous, tail-mediated brachiation in the first place? That is, why center the locomotor pattern on *hanging*? Another form of quadrupedalism would seem to do just as well, as any arboreal cercopithecoid would testify. The explanation for this must go to the very foundations of atelines. Pedal hanging may have been a basal adaptation in atelids (Meldrum, 1993), inherited and abetted by tail-assisted hanging in atelines as body size increased in their LCA (see Rosenberger and Strier, 1989). Originally, this is likely to have been more important as a postural adaptation rather than manifestly and narrowly locomotive. Principally, foot and tail hanging provide a longer reach than simple forelimb extension, and in trees, a longer stable reach. Forelimb elongation may have originally been part of this syndrome, useful for climbing, for reaching foods and grabbing branches sturdy enough to support a relatively heavy primate. Writ on a much smaller size scale, the unusually long-forearmed *Leontopithecus* accomplishes similar objectives: clinging to vertical supports of large diameter and probing huge arboreal bromeliads where prey is cached. Where, precisely, in the canopy environment these biological roles became prominent and of high selective value for *Ateles* is difficult to say. Most likely it was universally important, in the terminal branch milieu as well on larger boughs and branches. The value of tail hanging to augment reaching below the canopy also should not be underestimated. There is a whole range of fruit-bearing angiosperms whose niche is that of a small below-canopy tree, perhaps small enough so that a tail-hanging ateline might prefer to forage them from above, by suspension, rather than climb into them. With many simultaneous selective benefits accrued through semistatic hanging from limbs and tail, the final step in the spider monkey direction, taken partly by *Brachyteles* and fully fledged by *Ateles*, was the addition of the brachiation dynamic as a new locomotor pattern with multiple biological roles.

The varied hallmarks of spider monkeys and gibbons – acrobatic suspensory positional behaviors, rapid long distance travel, preference for ripe pulpy fruit, perhaps a somewhat small body size within their respective clades, size monomorphism and a small foraging-group propensity – whether enforced by a monogamous mating system or facilitated by a fission–fusion social system, appear to be syndromes involving many parallelisms. The complete elegance of this evolutionary package is far from what Geoffroy Saint-Hilaire had in mind when he dubbed spider monkeys genus *Ateles*.

Acknowledgements

We thank the Tow Travel Fellowship of Brooklyn College and the Professional Staff Congress of the City University of New York for providing funding that made this research possible.

References

Anapol, F. and Lee, S. (1994). Morphological adaptation to diet in platyrrhine primates. *Am. J. Phys. Anthropol.*, **94**, 239–261.

Ankel, F. (1972). Vertebral morphology of fossil and extant primates. In *Functional and Evolutionary Biology of the Primates*, ed. R. Tuttle, Chicago: Aldine and Atherton, pp. 223–240.

Anthony, M. R. L. and Kay, R. F. (1993). Tooth form and diet in ateline and alouattine primates: reflections on the comparative method. *Am. J. Sci.*, **293A**, 356–382.

Armstrong, E. and Shea, M. A. (1997). Brains of New World and Old World monkeys. In *New World Primates: Ecology, Evolution, and Behavior*, ed. W. G. Kinzey, New York: Aldine de Gruyter, pp. 25–44.

Ashton, E. H. and Oxnard, C. E. (1963). The musculature of the primate shoulder. *Trans. Zool. Soc. Lond.*, **29**, 553–650.

Ashton, E. H. and Oxnard, C. E. (1964). Functional adaptations in the primate shoulder girdle. *Proc. Zool. Soc. Lond.*, **142**, 49–66.

Campbell, B. (1937). The shoulder musculature of the Platyrrhine monkeys. *J. Mammal.*, **18**(1), 66–71.

Campbell, K. E., Frailey, C. D. and Romero-Pittman, L. (2006). The Pan-Amazonian Ucayali Peneplain, late Neogene sedimentation in Amazonia, and the birth of the modern Amazon River system. *Palaeogeog., Palaeoclimat., Palaeoecol.*, **239**, 166–219.

Canavez, F. C., Moreira, M. A. M., Ladoski, J. L., *et al.* (1999). Molecular phylogenetics of New World primates (Platyrrhine) based on b2-Microglobulin DNA sequences. *Mol. Phylogen. Evol.*, **12**, 74–82.

Cant, J. G. H., Youlatos, D. and Rose, M. D. (2001). Locomotor behavior of *Lagothrix lagothricha* and *Ateles belzebuth* in Yasuní National Park, Ecuador: general patterns and non-suspensory modes. *J. Hum. Evol.*, **41**, 141–166.

Cartmill, M. (1974). Pads and claws in arboreal locomotion. In *Primate Locomotion*, ed. F. A. Jenkins, New York: Academic Press, pp. 45–83.

Cartelle, C. and Hartwig, W. C. (1996). A new extinct primate among the Pleistocene megafauna of Bahia, Brazil. *Proc. Natl. Acad. Sci.*, **93**, 6405–6409.

Chapman, C. A. and Chapman, L. J. (1990). Reproductive biology of free-ranging spider monkeys. *Zoo Biol.*, **9**, 1–9.

Collins, A. C. (2004). Atelinae phylogenetic relationships: the trichotomy revived? *Am. J. Phys. Anthropol.*, **124**, 285–296.

Corner, B. D. and Richtsmeier, J. T. (1993). Cranial growth and growth dimorphism in *Ateles geoffroyi*. *Am. J. Phys. Anthropol.*, **92**, 371–394.

Defler, T. R. (1999). Locomotion and posture in *Lagothrix lagotricha*. *Folia Primatol.*, **70**, 313–327.

Delson, E. and Rosenberger, A. L. (1984). Are there any anthropoid primate "living fossils"? In *Living Fossils*, ed. N. Eldredge and S. Stanley, New York: Fischer, pp. 50–61.

Di Fiore, A. and Campbell, C. J. (2007). The atelines: variation in ecology, behavior, and social organization. In *Primates in Perspective*, ed. C. J. Campbell, A. Fuentes, K. C. MacKinnon, M. Panger and S. K. Beader, New York: Oxford University Press, pp. 155–185.

Erickson, G. E. (1963). Brachiation in New World monkeys and in Anthropoid apes. *Symp. Zool. Soc. Lond.*, **10**, 135–164.

Fajardo, R. J. and Müller, R. (2001). Three-dimensional analysis of nonhuman primate trabecular architecture using micro-computed tomography. *Am. J. Phys. Anthropol.*, **115**, 327–336.

Fontaine, R. (1990). Positional behavior in *Saimiri boliviensis* and *Ateles geoffroyi*. *Am. J. Phys. Anthropol.*, **82**, 485–508.

Ford, S. E. (1994). Evolution of sexual dimorphism in body weight in platyrrhines. *Am. J. Primatol.*, **34**, 221–244.

Ford, S. E. and Davis, L. (1992). Systematics and body size: implications for feeding adaptations in New World monkeys. *Am. J. Phys. Anthropol.*, **88**, 415–468.

Gebo, D. L. (1996). Climbing, brachiation, and terrestrial quadrupedalism: historical precursors of hominid bipedalism. *Am. J. Phys. Anthropol.*, **101**, 55–92.

German, R. (1982). Functional morphology of caudal vertebrae in new world monkeys. *Am. J. Phys. Anthropol.*, **58**, 453–459.

Harada, M. L., Schneider, H., Schneider, M. P. C., *et al.* (1995). DNA evidence on the phylogenetic systematics of New World monkeys: support for the sister-grouping of *Cebus* and *Saimiri* from two unlinked nuclear genes. *Mol. Phylogen. Evol.*, **4**, 331–349.

Hartwig, W. C. (1993). *Comparative Morphology, Ontogeny and Phylogenetic Analysis of the Platyrrhine Cranium*. Ann Arbor: University Microfilms International.

Hartwig, W. C. (1996). Perinatal life history traits in New World monkeys. *Am. J. Primatol.*, **40**, 99–130.

Hartwig, W. C. (2005). Implications of molecular and morphological data for understanding ateline phylogeny. *Int. J. Primatol.*, **26**, 999–1015.

Hartwig, W. C. and Cartelle, C. (1996). A complete skeleton of the giant South American primate *Protopithecus. Nature* **381**, 307–311.

Harvey, P., Kavanagh, M. and Clutton-Brock, T. H. (1978). Sexual dimorphism in primate teeth. *J. Zool. (Lond.)*, **186**, 475–485.

Hershkovitz, P. (1970). Cerebral fissural patterns in platyrrhine monkeys. *Folia Primatol.*, **13**, 213–240.

Hershkovitz, P. (1977). *Living New World Primates (Platyrrhini)*, Vol. 1. Chicago: University of Chicago Press.

Horovitz, I., Zaradoya, R. and Meyer, A. (1998). Platyrrhine systematics: a simultaneous analysis of molecular and morphological data. *Am. J. Phys. Anthropol.*, **106**, 261–281.

Hylander, W. L. (1979). Functional significance of primate mandibular form. *J. Morphol.*, **160**, 223–239.

Inman, V. T., Saunders, M. and Abbott, L. C. (1944). Observations on the function of the shoulder joint. *J. Bone Joint Surg.*, **26**, 1–30.

Jenkins, F. A. (1981). Wrist rotation in primates: a critical adaptation for brachiators. *Symp. Zool. Soc. Lond.*, **48**, 429–451.

Jenkins, F. A., Dombrowski, P. J. and Gordon, E. P. (1978). Analysis of the shoulder in brachiating spider monkeys. *Am. J. Phys. Anthropol.*, **48**, 65–76.

Johnson, S. E. and Shapiro, L. J. (1998). Positional and vertebral morphology in atelines and cebines. *Am. J. Phys. Anthropol.*, **105**, 333–354.

Jones, A. (2004). The evolution of brachiation in atelines: a phylogenetic comparative study. Unpublished Ph.D. thesis. University of California, Davis.

Jouffroy, F. K. and Lessertisseur, J. (1977). Processus de réduction des doigts (main et pied) chez les primates. Modalités, implications génétiques. *Colloques internationaux C.N.R.S. No. 266 Méchanisme de la Rudimentation de Organes chez les Embryons de Vertébrés*, pp. 381–389.

Jouffroy, F. K., Godinot, M. and Nakano, Y. (1991). Biometrical characteristics of primate hands. *Hum. Evol.*, **6**, 269–306.

Jungers, W. L. (1985). Allometry of primate limb proportions. In *Size and Scaling in Primate Biology*, ed. W. L. Jungers, New York: Plenum Press, pp. 345–381.

Jungers, W. L. and Stern, J. T. (1984). Kinesiological aspects of brachiation in lar gibbons. In *The Lesser Apes*, ed. H. Preuschoft, D. J. Chivers, W. Y. Brockelman and N. Creel. Edinburgh: Edinburgh University Press, pp. 119–134.

Kay, R. F. (1975). The functional adaptations of primate molar teeth. *Am. J. Phys. Anthropol.*, **43**, 195–216.

Kay, R. F. (1987). Analysis of primate dental microwear using image processing techniques. *Scanning Microscopy*, **1**, 657–662.

Kay, R. F. (1990). The phyletic relationships of extant and fossil Pitheciinae (Platyrrhini, Anthropoidea). *J. Hum. Evol.*, **19**, 175–208.

Kay, R. F. and Cozzuol, M. A. (2006). New platyrrhine monkeys from the Solimões Formation (late Miocene, Acre State, Brazil). *J. Hum. Evol.*, **50**, 673–686.

Kay, R. F. and Hiiemae, K. M. (1974). Jaw movement and tooth use in recent and fossil primates. *Am. J. Phys. Anthropol.*, **40**, 227–256.

Kay, R. F., Plavcan, J. M., Glander, K. E. and Wright, P. C. (1988). Sexual selection and canine dimorphism in New World monkeys. *Am. J. Phys. Anthropol.*, **77**, 385–397.

Kinzey, W. G. (1970). Basic rectangle of the mandible. *Nature*, **228**, 289–290.

Konstant, W., Stern, J. T., Fleagle, J. G. and Jungers, W. L. (1982). Function of the subclavius muscle in a nonhuman primate, the spider monkey (*Ateles*). *Folia Primatol.*, **38**, 170–182.

Larson, S. G. (1995). New characters for the functional interpretation of primate scapulae and proximal humeri. *Am. J. Phys. Anthropol.*, **98**, 13–35.

Larson, S. G. (1998). Unique aspects of quadrupedal locomotion in nonhuman primates. In *Primate Locomotion: Recent Advances*, ed. E. Strasser, J. Fleagle, A. Rosenberger and H. McHenry, New York: Plenum Press, pp. 157–173.

Larson, S. G. and Stern, J. T. (1986). EMG of scapulohumeral muscles in the chimpanzee during reaching and "arboreal" locomotion. *Am. J. Anat.*, **176**, 171–190.

Lemelin, P. (1995). Comparative and functional myology of the prehensile tail in New World monkeys. *J. Morphol.*, **224**, 351–368.

Lemelin, P. and Schmitt, D. (1998). The relation between hand morphology and quadrupedalism in primates. *Am. J. Phys. Anthropol.*, **105**, 185–197.

Lewis, O. J. (1971). Brachiation and the early evolution of the Hominoidea. *Nature*, **230**, 577–578.

Lewis, O. J. (1972). Osteological features characterizing the wrists of monkeys and apes, with a reconsideration of this region in *Dryopithecus (Proconsul) africanus*. *Am. J. Phys. Anthropol.*, **36**, 45–58.

MacPhee, R. D. E. and Meldrum, D. J. (2006). Postcranial remains of the extinct monkeys of the greater Antilles, with evidence of semiterrestriality in *Paralouatta. Am. Mus. Novitates*, **3516**, 1–65.

Masterson, T. J. and Hartwig, W. C. (1998). Degrees of sexual dimorphism in *Cebus* and other New World monkeys. *Am. J. Phys. Anthropol.*, **107**, 243–256.

Matthews, L. and Rosenberger, A. (in press). An object lesson for primate systematics: parsimony analysis (PAUP*) and the taxonomy of the yellow-tailed woolly monkey, *Lagothrix flavicauda. Am. J. Phys. Anthropol.*

Meireles, C. M., Czelusniak, J., Schneider, M. P. C., *et al.* (1999). Molecular phylogeny of ateline New World monkeys (Platyrrhini, Atelinae) based on g-Globin gene sequences: evidence that *Brachyteles* is the sister group of *Lagothrix. Mol. Phylogen. Evol.*, **12**, 10–30.

Meldrum, D. J. (1993). Postcranial adaptations and positional behavior in fossil platyrrhines. In *Postcranial Adaptation in Nonhuman Primates*, ed. D. L. Gebo, DeKalb: Northern Illinois University Press, pp. 235–251.

Mittermeier, R. A. (1978). Locomotion and posture in *Ateles geoffroy* and *Ateles paniscus. Folia Primatol.*, **30**, 161–193.

Napier, J. R. (1961). Prehensility and opposability in the hands of primates. *Symp. Zool. Soc. Lond.*, **5**, 115–132.

Norconk, M. A., Rosenberger, A. L. and Garber, P. A., eds. (1996). *Adaptive Radiations of Neotropical Primates*. New York: Plenum Press.

Orlosky, F. (1973). *Comparative Dental Morphology of Extant and Extinct Cebidae.* Ann Arbor: University Microfilms.

Oxnard, C. E. (1963). Locomotor adaptations in the primate forelimb. *Proc. Zool. Soc. Lond.*, **10**, 165–182.

Oxnard, C. E. (1967). The functional morphology of the primate shoulder as revealed by comparative anatomical, osteometric and discriminant function techniques. *Am. J. Phys. Anthropol.*, **26**, 219–240.

Oxnard, C. E. (1968). A note on the Olduvia clavicular fragment. *Am. J. Phys. Anthropol.*, **29**, 429–432.

Plavcan, J. M. and Kay, R. F. (1988). Sexual dimorphism and dental variability in platyrrhine primates. *Int. J. Primatol.*, **9**, 169–178.

Plavcan, J. M. and van Schaik, C. (1997). Intrasexual competition and body weight dimorphism in anthropoid primates. *Am. J. Phys. Anthropol.*, **103**, 37–68.

Rafferty, K. L. (1998). Structural design of the femoral neck in primates. *J. Hum. Evol.*, **34**, 361–384.

Rose, M. D. (1975). Functional proportions of primate lumbar vertebral bodies. *J. Hum. Evol.*, **4**, 21–38.

Rose, M. D. (1988). Another look at the anthropoid elbow. *J. Hum. Evol.*, **15**, 333–367.

Rosenberger, A. L. (1983) Tail of tails: parallelism and prehensility. *Am. J. Phys. Anthropol.*, **60**, 103–107.

Rosenberger, A. L. (1992). Evolution of feeding niches in New World monkeys. *Am. J. Phys. Anthropol.*, **88**, 525–562.

Rosenberger, A. L. and Strier, K. (1989). Adaptive radiation of the ateline primates. *J. Hum. Evol.*, **18**, 717–750.

Rosenberger, A. L., Setoguchi, T. and Shigerhara, N. (1990). The fossil record of callitrichine primates. *J. Hum. Evol.*, **19**, 209–236.

Rosenberger, A. L., Tejedor, M., Cooke, S. B., Pekar, S. (in press). Platyrrhine ecophylogenetics in space and time. In *South American Primates: Comparative Perspectives in the Study of Behavior, Ecology and Conservation*, ed. P. Garber, New York: Springer.

Schneider, H., Sampaio, I., Harada, M. L., *et al.* (1996). Molecular phylogeny of the New World monkeys (Platyrrhini, Primates) based on two unlinked nuclear genes: IRBP intron 1 and ε-globin sequences. *Am. J. Phys. Anthropol.*, **100**, 153–179.

Schneider, H., Schneider, M. P. C., Sampaio, M. I. C., *et al.* (1993). Molecular phylogeny of the New World monkeys (Platyrrhini, Primates). *Mol. Phylogen. Evol.*, **2**, 225–242.

Schultz, A. H. (1960). Age changes and variability in the skull and teeth of the Central American monkeys *Alouatta, Cebus* and *Ateles. Proc. Zool. Soc. Lond.*, **133**, 337–390.

Schultz, A. H. (1961). Vertebral column and thorax. *Primatologia*, **4**, 1–66.

Setoguchi, T. and Rosenberger, A. L. (1987). A fossil owl monkey from La Venta, Colombia. *Nature*, **326**, 692–694.

Smith, R. J. (1978). Mandibular biomechanics and temporomandibular joint function in primates. *Am. J. Phys. Anthropol.*, **49**, 341–349.

Smith, R. J. and Jungers, W. L. (1997). Body mass in comparative primatology. *J. Hum. Evol.*, **32**, 523–559.

Stern, J. T. and Larson, S. G. (1993). EMG of the supinator and pronators in the gibbon and chimpanzee. *Am. J. Phys. Anthropol., Suppl.*, **16**, 187–188.

Stern, J. T. (1971). *Functional Myology of the Hip and Thigh of Cebid Monkeys and its Implications for the Evolution of Erect Posture (Bibilotheca primatologica).* Chicago: S. Karger.

Strier, K. B. (1992). Atelinae adaptations: behavioral strategies and ecological constraints. *Am. J. Phys. Anthropol.*, **88**, 515–524.

Swartz, S. M. (1990). Curvature of the forelimb bones of anthropoid primates: overall allometric patterns and specializations in suspensory species. *Am. J. Phys. Anthropol.*, **83**, 477–498.

Tague, R. G. (1997). Variability of a vestigial structure: first metacarpal in *Colobus guereza* and *Ateles geoffroyi*. *Evolution*, **51**, 595–605.

Takahashi, L. K. (1990). Morphological basis of arm-swinging: multivariate analyses of the forelimbs of *Hylobates* and *Ateles*. *Folia Primatol.*, **54**, 70–85.

Turnquist, J. E. (1983). Forelimb musculature and ligaments in *Ateles*, the spider monkey. *Am. J. Phys. Anthropol.*, **62**, 209–226.

Turnquist, J. E., Schmitt, D., Rose, M. D. and Cant, J. G. H. (1999). Pendular motion in the brachiation of captive *Lagothrix* and *Ateles*. *Am. J. Primatol.*, **48**, 263–281.

von Dornum, M. and Ruvolo, M. (1999). Phylogenetic relationships of the New World monkeys (Primates, Platyrrhini) based on nuclear G6PD DNA sequences. *Mol. Phylogen. Evol.*, **11**, 459–476.

Voisin, J. (2006). Clavicle, a neglected bone: morphology and relation to arm movements and shoulder architecture in primates. *Anat. Rec. A*, **288A**, 944–953.

Wall, C. E. (1999). A model of temporomandibular joint function in anthropoid primates based on condylar movements during mastication. *Am. J. Phys. Anthropol.*, **109**, 67–88.

Youlatos, D. (2000). Functional anatomy of forelimb muscles in Guianan atelines (Platyrrhini: Primates). *Annales des Sciences Naturelles-Zoologie et Biologie Animale*, **21**, 137–151.

Youlatos, D. (2002). Positional behavior of black spider monkeys (*Ateles paniscus*) in French Guiana. *Int. J. Primatol.*, **23**, 1071–1094.

Young, N. M. (2003). A reassessment of living hominoid postcranial variability: implications for ape evolution. *J. Hum. Evol.*, **45**, 441–464.

Zingeser, M. R. (1973). Dentition of *Brachyteles arachnoides* with reference to alouattine and atelinine affinities. *Folia Primatol.*, **20**, 351–390.

3 The taxonomic status of spider monkeys in the twenty-first century

ANDREW C. COLLINS

Introduction

Spider monkeys are known by multiple names to local cultures, many of which reflect their acrobatic agility in the trees, or their various differences in coat color. Some indigenous groups also apply names that reference the blue eyes exhibited in some forms of spider monkey (Konstant et al., 1985). Within the scientific literature they occupy various positions in the taxonomic relationships among members of the Atelinae subfamily (see Rosenberger et al., this volume), and have been included in vast numbers of upper-level taxonomic surveys. It is thus surprising that their species-level taxonomic status has been the subject of only a few discriminate character-based investigations over the previous century.

Pelage variation in *Ateles* is substantial and has been the primary characteristic considered in previous categorizations of spider monkeys. Given the wide variation in coat colors among spider monkeys, it is not surprising that the first systematic study of *Ateles* taxonomy (Kellogg and Goldman, 1944) relied primarily on pelage. They initially identified four species and 16 different subspecies of *Ateles* (Kellogg and Goldman, 1944). These subspecies represent racially identifiable groups (Kellogg and Goldman, 1944) that can theoretically be identified based on a combination of different pelage characteristics. In this chapter, I provide a review of the various studies of spider monkey taxonomy and the scientific methods used to discern relationships. It is my hope that this review will help to promote reliance on a revised taxonomy based on discriminate character analysis for spider monkeys discounting the erroneous pelage-based scheme still too often referenced.

The most likely taxonomy of *Ateles*, based on discriminate studies focusing on molecular variation, multivariate analysis of skeletal anatomy, and chromosomal variation would suggest four, or possibly only three species of *Ateles* (Froehlich et al., 1991; Collins and Dubach 2000a, 2001; Nieves et al., 2005). These are *A. geoffroyi, A. belzebuth, A. paniscus*, and the fourth probable

Spider Monkeys: Behavior, Ecology and Evolution of the Genus Ateles, ed. Christina J. Campbell. Published by Cambridge University Press. © Cambridge University Press 2008.

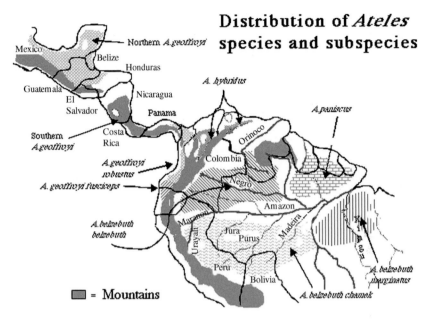

Distribution of *Ateles* species and subspecies

Figure 3.1 Distribution of *Ateles* species and subspecies. Presents the distribution of major *Ateles* species and subspecies including major rivers and mountain ranges. Boundaries are based on Hernández-Camacho and Cooper (1976), Konstant *et al.* (1985), Froehlich *et al.* (1991), and Norconck *et al.* (1996). All South American subspecies are identified following the taxonomy of Collins and Dubach (2000a). In Central America the subspecies are broken into two groups: (1) Northern *A. geoffroyi*, which includes *A. g. yucatanensis*, and *A. g. vellerosus*, and (2) Southern *A. geoffroyi*, including *A. g. frontatus*, *A. g. geoffroyi*, *A. g. panamensis*, *A. g. azuerensis*, *A. g. grisescens*, and *A. g. ornatus*.

species, *A. hybridus*. Figure 3.1 identifies the geographic ranges of these species and most probable subspecies of *Ateles*. This chapter will follow the nomenclature noted here in its discussion of *Ateles* taxonomy, with notes to the specific taxonomic identifications considered by researchers in the various studies reviewed.

Spider monkeys have one of the largest geographical distributions of any primate in the Neotropics (Figure 3.1). They can be found from the Yucatan peninsula and coastal regions of Vera Cruz State in Mexico to northern Bolivia and from the Pacific coast of Ecuador to regions of northeastern South America in Guyana and Suriname (Kellogg and Goldman, 1944; Rowe, 1996). In spite of this vast geographic distribution, spider monkeys have restricted habitat preferences. They prefer the top canopy layers of low, humid rain forest below 800 meters in elevation (Hernandez-Camacho and Cooper, 1976; Klein and Klein, 1977; van Roosmalen, 1980; Konstant *et al.*, 1985). Spider monkeys select primary, evergreen, never-flooded forest (Wolfheim, 1983; Wallace, this volume) and are frugivores specializing in soft-fruits, while supplementing their diet with young leaves and flowers (Hernandez-Camacho and Cooper, 1976; Klein and Klein, 1977; van Roosmalen, 1980; Milton, 1981; Konstant *et al.*, 1985; Symington, 1987, 1988; Fedigan *et al.*, 1988; Russo *et al.*, 2005; Di Fiore and Campbell, 2007; Di Fiore *et al.*, this volume).

Spider monkeys have a fission–fusion social system similar to chimpanzees (van Roosmalen, 1980; Symington, 1987, 1988, 1990; Aureli and Schaffner, this volume) with dispersal of females upon maturation (Symington, 1987, 1990; Shimooka *et al.*, this volume; Vick, this volume). These habitat preferences, dietary specializations, and social system result in large home ranges for spider monkey groups (Milton, 1981; Fedigan *et al.*, 1988; Di Fiore and Campbell, 2007; Wallace, this volume).

Biogeography and distribution

Biogeographical history

Most Neotropical primates are postulated to have origins in the southern Amazon Basin (Kinzey, 1997). Fossil evidence (Hartwig, 1995) of *Ateles*-like ancestors would support this scenario as well, suggesting spider monkeys evolved approximately 15 million years ago (mya). However, molecular evidence (Schneider *et al.*, 1993; Porter *et al.*, 1997) suggests *Ateles* origins were more recent, at 5 mya. This date would suggest that *Ateles* had to disperse across much of the suitable Neotropical habitat in the last five million years. At the beginning of the Pliocene, biogeographic reconstructions suggest that the river draining the Amazon Basin was not as massive as it is today (Brown, 1987; Colinvaux, 1996). Additionally the Andes Mountains were still in the process of uplifting and had not reached substantial heights in many areas, especially the northern reaches in present-day Colombia (van der Hammen, 1982; Haffer, 1987). The mountains of the Guianan Shield are older and were already present at this time. These mountains would have likely formed a barrier to dispersal in this region through a combination of altitude and unsuitable habitat. The Central American isthmus was still a chain of islands at this time (White, 1986).

Ateles would have migrated across this landscape possibly crossing the Amazon via direct substrate contact or by being effectively transported when an oxbow in the river was cut off leaving the monkeys isolated on the other side. In the rest of the southern and western Amazon Basin it seems likely that *Ateles* would have maintained gene flow during much of the Pliocene until the beginning of the Pleistocene and the associated changes proposed in the habitat due to climatic fluctuations caused by glacial episodes (Haffer, 1982).

It would seem likely that *Ateles* could have crossed the northern Andes cordillera before these mountains became the probable barrier to gene flow they presently represent. The modern isthmus in Central America formed around 3.5–3.1 mya (Coates and Obando, 1996) and *Ateles'* colonization of this region would not likely have begun in earnest until that time. Since then, they would likely have been forced to contend with climatic fluctuations as this region has been continually modified since its formation.

Biogeographical mechanisms responsible for the majority of the speciation events among spider monkey populations occurred during the middle to late Pliocene and early Pleistocene (Collins and Dubach, 2000b). They were the result, primarily, of major vicariant events caused by geological factors such as the continued uplift of the northern ranges of the Andes and the formation of the Amazon River as the basin's primary drainage. These findings are similar to those reported by molecular phylogenetic studies of other Neotropical organisms (Smith and Patton, 1993; Riddle, 1996; Engel *et al.*, 1998) and provide little evidence in support of Pleistocene refugia formation or riverine barriers (with the exception of the Amazon) as primary mechanisms in spider monkey speciation.

The specific habitat requirements and large territories required by *Ateles* may have prohibited them from surviving in smaller, less stable refugia formed during Pleistocene fluctuations in habitat. This would mean that only larger populations, with a higher level of genetic variation, survived in theorized refugia. Considering the long interbirth intervals and associated life-spans of *Ateles*, it appears probable that, in many instances, when populations from large refuges were reconnected, their gene pools had not diverged enough to prohibit interbreeding, and thus speciation did not occur. For a more complete review of the role of biogeography in *Ateles* speciation, please refer to Collins and Dubach (2000b). Table 3.1 summarizes numerous reports on *Ateles'* distribution and habitat utilization to provide a more precise view of the status of this genus at specific locations throughout its range.

Current distribution

Spider monkeys' geographic distribution and the environments they utilize are key elements to understanding their taxonomy. These factors affect *Ateles'* gene flow and ultimately their likely taxonomic relationships. The following section details the different geographic regions occupied by various species, subspecies and racial variants of *Ateles* with attention to their

Table 3.1 *Reported* Ateles *occurrences throughout the Neotropics*

Location	Taxonomic identity	Forest type	Altitude	I.D.*
1. Santa Martha Mnts, Vera Cruz, Mexico	*Ateles geoffroyi vellerosus*	Not given	Not given	A, B
2. Los Tuxtlas Preserve, Vera Cruz, Mexico	*Ateles geoffroyi vellerosus*	Not given	Not given	A
3. Sierra Madres Mnts, Chiapas, Mexico	*Ateles geoffroyi vellerosus*	Not given	Up to 1200 m	C
4. Sian Ka'an Reserve, Quintana Roo, Mexico	*Ateles geoffroyi yucatanensis*	Not given	Below 300 m	A
5. Mayan Mnts, Chiquebul forest, Belize	*Ateles geoffroyi yucatanensis*	Not given	Up to 650 m	D
6. Mayan Mnts, Monkey River, Belize	*Ateles geoffroyi yucatanensis*	Not given	Below 250 m	D
7. Macal River, Belize	*Ateles geoffroyi yucatanensis*	Humid semi-deciduous	Not given	E
8. Tikal National Park, Guatemala	*Ateles geoffroyi yucatanensis*	Primary low, humid, tropical forest	Not given	F, G, H
9. Santa Marta, Guatemala	*Ateles geoffroyi vellerosus*	Not given	Not given	I
10. Bartola–Indio–Maizarea Reserve, Nicaragua	*Ateles geoffroyi geoffroyi*	Primary lowland humid forest	Not given	J
11. Palo Verde Park, Guanacaste, Costa Rica	*Ateles geoffroyi ornatus*	Deciduous tropical dry forest with riparian forest and marsh	Under 200 m	K
12. Santa Rosa Park, Costa Rica	*Ateles geoffroyi panamensis, Ateles geoffroyi ornatus*	Not given	Not given	L
13. Barro Colorado Island, Panama	*Ateles geoffroyi panamensis*	Tropical dry	Not given	Z
14. Punta Leona Private Wildlife Refuge, Costa Rica	*Ateles geoffroyi geoffroyi*	Tropical moist forest	Not given	AA
15. Choco Region, Colombia	*Ateles geoffroyi robustus*	Primary – dry forest	Sea level to 2500 m	M
16. Cotacachi–Cayapas Reserve, Ecuador	*Ateles geoffroyi fusciceps*	Low hygrotrophytic to Andean cloud forest	300–1200 m	N
17. Tinigua National Park west of La Macarena Mnts, along Rio Duda, Colombia	*Ateles belzebuth belzebuth*	Pre-montane wet forest	350–400 m	Y, AB
18. Along Rio Uraricoera, Brazil	*Ateles belzebuth belzebuth*	Tropical primary rain forest	Not given	O
19. Rio Alto Yavari, Peru	*Ateles belzebuth chamek*	Transitional zone between Amazon evergreen and Guianan highland savanna	Not given	P
20. Pacaya–Saimiria National Reserve, Peru	*Ateles belzebuth chamek*	Not given	Not given	P

No.	Location	Species	Habitat	Elevation	I.D.[*]
21.	Manu National Park, Peru	*Ateles belzebuth chamek*	Lowland, humid, tropical moist forest with rainy seasons	Not given	P, Q
22.	Parque Nacional Yasuni, Napo Province, Ecuador	*Ateles belzebuth chamek*	Terra-firme primary rain forest	Not given	AC
23.	Noel Kempff Mercado National Park, Santa Cruz Dept., Bolivia	*Ateles belzebuth chamek*	Lowland subhumid forest	Not given	R
24.	Flor de Oro Region of Guapore/Itenez River, Bolivia	*Ateles belzebuth chamek*	Not given	Not given	S
25.	Mountainous area of Huanacha Santa Cruz Dept., Bolivia	*Ateles belzebuth chamek*	Not given	Below 1000 m	T
26.	Pimenta Bueno Municipal Park, Rio Jiparana, Rondonia State, Brazil	*Ateles belzebuth chamek*	Not given	Not given	U, AD
27.	Guajara–Mirim State Park, Rio Madeira, Rondonia State, Brazil	*Ateles belzebuth chamek*	Not given	Below 200 m	V, AD
28.	Confluence of Rios Madeira and Jiparana, Rondonia, Matto Grosso and Amazonas States, Brazil	*Ateles belzebuth chamek*	Primary and possibly secondary lowland, humid, terra-firme forests	Below 200 m	W
29.	Twenty-five locations in Rondonia State, Brazil	*Ateles belzebuth chamek*	See source for details	Not given	AD
30.	Petit–Saut dam, Courcibo River, French Guiana	*Ateles paniscus*	Lowland primary rainforest	Not given	X
31.	Nouragues Reserve, French Guiana	*Ateles paniscus*	Terra-firme primary rain forest	Not given	AC

Provides a description of various research articles detailing the presence of *Ateles* and the associated habitat and conservation status at each location. Nomenclature follows Collins and Dubach (2000a).

*I.D. references correspond to letters as follows: A, Estrada and Coates-Estrada, 1988; B, Garcia–Orduna *et al.*, 1993; C, Yanez, 1991; D, Dahl, 1986; E, Personal observation, 1996; F, Coelho *et al.*, 1976; G, Bramblett *et al.*, 1980; H, Cant, 1990; I, Silva-Lopez *et al.*, 1996; J, Crockett *et al.*, 1997; K, Massey, 1987; L, Freese, 1976; M, Hernandez-Camacho and Cooper, 1976; N, Madden and Albuja, 1987; O, Nunes, 1998; P, Aquino and Encarnacion, 1994; Q, Symington, 1987; R, Karesh *et al.*, 1998; S, Wallace *et al.*, 1996; T, Braza and Garcia, 1988; U, Ferrari *et al.*, 1996; V, Ferrari and Lopes, 1992; X, Granjon *et al.*, 1996; Y, Stevenson *et al.*, 1998; Z, Campbell, 2003; AA, Timock and Vaughan, 2002; AB, Shimooka, 2003; AC, Youlatos, 2003; AD, Iwanaga and Ferrari, 2002.

habitat and distribution. This discussion follows the taxonomic identifications noted above. Across the vast geographic range of spider monkeys, ecological pressures and behavioral aspects vary to different degrees. The primary factors to consider, however, are the potential for gene flow between populations and barriers to gene flow through geographic, behavioral and life history variation.

Ateles geoffroyi

Ateles geoffroyi is distributed throughout Central America and along the western coast of South America west of the Andes. Recent discriminate character taxonomic studies have unfortunately not extensively sampled the many pelage variants found in this region. Thus, the subspecies status of the many different racial populations suggested by Kellogg and Goldman (1944) remains unclear. Populations of spider monkeys identified as *A. g. vellerosus* and *A. g. yucatanensis* (Kellogg and Goldman, 1944) occur throughout Mexico, Guatemala, El Salvador and Honduras. *Ateles* occurs in Mexico in primary forest of two varying types: medium-high mountainous perennial forest and lowland perennial forest (Estrada and Coates-Estrada, 1988). Kellogg and Goldman (1944) reported specimens at elevations up to 1200 meters in Mexico.

Two racial variants of spider monkeys are found in Nicaragua, *A. g. frontatus* and *A. g. geoffroyi* (Kellogg and Goldman, 1944). *A. g. frontatus* is distributed narrowly from northwestern Costa Rica to extreme western and northern Nicaragua along the Pacific coast (Kellogg and Goldman, 1944). *A. g. geoffroyi* is reported to occur in coastal regions of eastern and southeastern Nicaragua through the lowlands to the Pacific coast of central Costa Rica (Kellogg and Goldman, 1944). However, a 1997 survey of Nicaraguan primates by Crockett *et al.* (1997) could only identify spider monkeys from the Bartola–Indio–Maizarea reserve in extreme southern Nicaragua.

Both Costa Rica and Panama support a large number of racial variants of *A. geoffroyi* (Kellogg and Goldman, 1944). *A. g. ornatus* is reported to occur in northwestern Costa Rica (Kellogg and Goldman, 1944; Konstant *et al.*, 1985), while *A. g. panamensis* occurs throughout Panama west of the Cordillera San Blas and into central Costa Rica (Kellogg and Goldman, 1944). *A. g. azuerensis* is thought to occur in the Azuero Peninsula of Panama west to the Burica Peninsula (Kellogg and Goldman, 1944). *A. g. grisescens* is reported from extreme southwestern Panama in the valley of the Rio Tuyra (Kellogg and Goldman, 1944) to northwestern Colombia in the vicinity of Jurado (Konstant *et al.*, 1985) and is restricted by the Baudo Mountains to a narrow coastal strip, which may extend as far south as Cabo Corrientes (Hernandez-Camacho and Cooper, 1976).

The Darien Peninsula region of Panama is within the range of *A. g. robustus*. In Colombia *A. g. robustus* is distributed along the lowlands between the Pacific Ocean and the Andes Mountains and in northern Bolivar Department eastward to the west bank of the lower Cauca River (Hernandez-Camacho and Cooper, 1976). The subspecies *A. g. fusciceps* is historically reported from extreme southern Colombia (Kellogg and Goldman, 1944) into Ecuador, but may be restricted to the Pacific coastal regions of Ecuador today (Hernandez-Camacho and Cooper, 1976). Kellogg and Goldman (1944) report most specimens of this subspecies occur at altitudes over 500 meters up to 1500 meters.

Ateles belzebuth

The subspecies *A. b. belzebuth* is primarily located in the Amazon Basin of northern Peru, Ecuador, Colombia, Venezuela and Brazil. This form is distributed north of a line formed by the upper Amazon and Rio Maranon in the Peruvian Departments of Loreto, San Martin and Amazonas along the eastern cordillera of the Andes from Peru into Colombia (Kellogg and Goldman, 1944; Hernandez-Camacho and Cooper, 1976; Konstant *et al.*, 1985; Aquino and Encarnacion, 1994; Norconck *et al.*, 1996). In Colombia this subspecies occurs from the Amazon Basin northward to the Guiviare River and along the Macarena Mountains of the eastern Andes cordillera to the Upia River in southern Boyaca Department (Hernandez-Camacho and Cooper, 1976). Along this side of the mountains *Ateles'* distribution is limited to elevations below 1300 meters (Hernandez-Camacho and Cooper, 1976). *A. b. belzebuth* ranges northeastward in Venezuela to the Caura River south of the Orinoco (Kellogg and Goldman, 1944; Castellanos and Chanin, 1996; Norconck *et al.*, 1996). The distribution is limited to a general line along the western slopes of the Guianan shield down into Brazil, but its eastern distribution in Brazil is poorly defined. Spider monkeys are not found in the Guianan Highlands and associated black-water drainage areas of Venezuela (Froehlich *et al.*, 1991; Norconck *et al.*, 1996).

A. b. chamek occurs in the Amazon Basin of Peru, Brazil, and Bolivia south of the Solimoes/Japura river system (Iwanaga and Ferrari, 2002) along the eastern slopes of the Andes and eastward to the western bank of the middle portion of the Rio Tapajos (Kellogg and Goldman, 1944; Froehlich *et al.*, 1991; Aquino and Encarnacion, 1994). It is reported at elevations of approximately 1000 meters along the eastern slopes of the Andes (Kellogg and Goldman, 1944). *A. b.*

chamek's southern distribution in Bolivia includes most of Santa Cruz State (Braza and Garcia, 1988; Wallace *et al.*, 1996; Karesh *et al.*, 1998).

The distribution of the population Kellogg and Goldman (1944) categorized as *A. b. marginatus* is unclear. It occurs east of the Rio Tapajos possibly to the Rio Tocantins (Kellogg and Goldman, 1944; Konstant *et al.*, 1985; Nunes, 1995; Ferrari and Lopes, 1996) south of the Amazon River in Para State, Brazil. Efforts to locate this subspecies between the Rio Xingu and the Rio Tocantins have not been successful (Nunes, 1995; Ferrari and Lopes, 1996). However, the type specimen for *A. b. marginatus* is from the western bank of the Rio Tocantins (Kellogg and Goldman, 1944) indicating this area was within their recent historical range.

Ateles paniscus

Ateles paniscus is a monotypic species found north of the Amazon River in Para State, Brazil, French Guiana, Suriname and Guyana (Kellogg and Goldman, 1944; Konstant *et al.*, 1985; Baal *et al.*, 1988; Sussman and Phillips-Conroy, 1995; Norconck *et al.*, 1996). *Ateles paniscus* is not found in coastal regions in this area, which consist of mud flats and mangrove swamps followed by savannas (Baal *et al.*, 1988; Norconck *et al.*, 1996). It is limited to primary interior forests (Baal *et al.*, 1988; Norconck *et al.*, 1996). *Ateles paniscus* is common at many locations in Suriname, but found only in undisturbed high forest in the interior (Baal *et al.*, 1988).

The existence of *A. paniscus* west of the Esquibo River in Guyana has been reported (Sussman and Phillips-Conroy, 1995; Norconck *et al.*, 1996), but *Ateles*' distribution ends just west of the river due to a complex terrain composed of the Koraima Mountains on the Guyana–Venezuela border, the Parakaima Mountains, and widespread savanna regions of the Kanuka Mountains in southwestern Guyana (Norconck *et al.*, 1996). This is a region composed of a mosaic of savanna, montane forest, and dry forest (Sussman and Phillips-Conroy, 1995). In Para State, Brazil, the distribution seems to be broadly limited to the region east of the Rio Branco (Konstant *et al.*, 1985; Froehlich *et al.*, 1991), but several reported sightings in this area suggest possible parapatry with *A. b. belzebuth* (Froehlich *et al.*, 1991; Nunes, 1998).

Ateles hybridus

Ateles hybridus is a monotypic species (Collins and Dubach, 2000a, 2000b; Nieves *et al.* 2005) that ranges from the eastern bank of the lower Cauca River and throughout the lower regions of the Magdalena River valley, but does not extend to the Caribbean coast (Hernandez-Camacho and Cooper, 1976). It can be found up the Magdalena River valley to the northern regions of the Departments of Caldas and Cundinamarca just north of Bogotá (Hernandez-Camacho and Cooper, 1976). Along the middle Magdalena River it has been reported at elevations as high as 2500 meters (Hernandez-Camacho and Cooper, 1976). Presently isolated populations also occur in the Catatumbo River Basin of northern Santander Department and the northeastern piedmont in Comisaria and Arauca Departments in northeastern Colombia (Hernandez-Camacho and Cooper, 1976). *Ateles hybridus* can also be found in the Maracaibo Basin and mountains of northern Miranda State in northwestern Venezuela (Norconck *et al.*, 1996).

Summary of *Ateles* taxonomic schemes

Pelage variation

The original *Ateles* taxonomy of Kellogg and Goldman (1944) is based primarily on variations in pelage augmented by cranial and body size measurements. The specific traits used by Kellogg and Goldman (1944) to group populations into species and subspecies are discussed for each form in the following section. To see these racial forms in color please refer to Konstant *et al.* (1985) directly.

As initially described by Kellogg and Goldman (1944), the possession of dark black heads, hands, and wrists unites the subspecies of *A. geoffroyi. A. g. yucatanensis* has a light brown coat on the back and hips, contrasting with a silvery underside (Konstant *et al.*, 1985). *A. g. vellerosus* ranges from individuals similar to this, to animals with a very dark back and lighter undersides (Konstant *et al.*, 1985). *A. g. geoffroyi* is silvery to brownish – gray on the back and chest, while the abdomen may be somewhat golden (Konstant *et al.*, 1985). *A. g. frontatus* is very similar to *A. g. geoffroyi*, but slightly darker (Konstant *et al.*, 1985). *A. g. ornatus* has a golden back and underside with dark black head, face, forearms, and outer legs (Konstant *et al.*, 1985). *A. g. panamensis* is brown with a black face, a red abdomen, and with red interspersed in the back also (Konstant *et al.*, 1985). *A. g. azuerensis* is poorly described with a grayish-brown back and somewhat darker underside similar to *A. g. panamensis* (Kellogg and Goldman, 1944; Konstant *et al.*, 1985). *A. g. grisescens* is variable in coloration with specimens described as "rufescent," "brownish," "rusty-colored," "grizzled – gray", "dusky or sooty" (Kellogg and Goldman, 1944). *A. geoffroyi robustus* (identified as *A. fusciceps robustus* by Kellogg and Goldman, 1944), includes individuals with black furred bodies and dark skin on the exposed portion of the face (Konstant *et al.*, 1985). *A. g. robustus* has a relatively short tail compared with other spider monkeys (Konstant *et al.*, 1985). *A. g. fusciceps* is very similar, but has a brown head contrasting with its black body (Kellogg and Goldman, 1944; Konstant *et al.*, 1985).

A. *belzebuth* is represented by three subspecies with noncontinuous geographic distributions in the Kellogg and Goldman (1944) scheme. These populations possess a distinctive triangular or strip-shaped patch of hair on their foreheads that is light-colored, often white, and may also be accompanied by light-colored sideburns (Kellogg and Goldman, 1944; Konstant *et al.*, 1985). *Ateles hybridus* (*A. b. hybridus* in the Kellogg and Goldman scheme) ranges in coloration from light brown to a rich mahogany on the upper surfaces, limbs and head with a contrastingly lighter abdomen (Kellogg and Goldman, 1944; Hernandez-Camacho and Cooper, 1976; Konstant *et al.*, 1985). Its eyes are either light brown or blue in color (Konstant *et al.*, 1985). It is distinguished from certain subspecies of *A. geoffroyi* by the presence of the forehead patch and side burns, along with lighter colored hands and wrists (Konstant *et al.*, 1985). *A. b. belzebuth* has a black or dark brown dorsal surface with a golden or yellowish-white undersurface (Konstant *et al.*, 1985). *A. b. marginatus* is all black except for light-colored skin around the muzzle and eyes and the white forehead patch and side burns (Kellogg and Goldman, 1944).

A. *paniscus* has an all-black coat without a forehead patch or sideburns. They differ from *A. g. robustus* due to bare pink skin around the eyes, nose and muzzle (Kellogg and Goldman, 1944; Konstant *et al.*, 1985). *Ateles belzebuth chamek* (labeled as *A. p. chamek* by Kellogg and Goldman, 1944) is the taxonomic label applied to populations in the southwestern Amazon Basin, which exhibited this all-black pelage and the associated bare skin around their eyes. The coat of *A. paniscus* exhibits longer hair compared with other spider monkeys with a very long and thick tail (Konstant *et al.*, 1985).

Thus, Kellogg and Goldman (1944) proposed four species of spider monkeys with 16 recognized subspecies. Their taxonomy paid little attention to the distribution of their supposed species in the Amazon Basin and ignored likely geological barriers to gene flow that would preclude this taxonomic arrangement. This observation is best exemplified by the case of *A. belzebuth* where one subspecies was separated from the other two by the Andes Mountains, while the remaining two Amazonian subspecies were separated from one another by the central periodically embayed portion of the Amazon Basin – habitat that is not tolerated by *Ateles*. Thus, the only available pathway for gene flow between *A. b. hybridus* and *A. b. marginatus* is over the eastern cordillera of the Andes and through populations of *A. p. chamek*. This is an extremely unlikely mechanism that would require that two sympatric but unidentifiable species exist in the southwestern Amazon Basin or else long-distance migration through this region between the *A. belzebuth* subspecies.

Konstant *et al.* (1985) note the difficulty in differentiating between several subspecies as identified by Kellogg and Goldman (1944). They also note that due to *A. b. marginatus'* geographic location and general similarity to *A. p. chamek*, it might be more appropriately placed as a subspecies of *A. paniscus* (Konstant *et al.*, 1985). Finally, they question the validity of several of the subspecies of *A. geoffroyi* in Central America based on pelage similarities or the lack of representatives alive today (Konstant *et al.*, 1985). Konstant *et al.* (1985) note similarities between *A. g. vellerosus* and *A. g. yucatanensis* and question the continued existence of *A. g. azuerensis* (Konstant *et al.*, 1985).

Noting the many similarities between different populations of spider monkeys in coloration and the disjunct distribution of species supposedly united by color patterns, Hershkovitz (1968, 1969) applied his concept of heterochromatism to suggest that *Ateles* were likely one polytypic species.

In a recent examination of pelage variation among subspecies of Central American spider monkeys, Silva-Lopez *et al.* (1996) present evidence for pelage variants that further question the use of pelage as a primary mechanism for taxonomic identification in *Ateles*. Silva-Lopez *et al.* (1996) report populations of *A. g. vellerosus* where the white forehead patch, utilized to categorize *A. belzebuth*, is quite common. Silva-Lopez *et al.* (1996) further question the ability to use pelage to differentiate between *A. g. vellerosus* and *A. g. yucatanensis*, demonstrating several individuals from both populations that might be misclassified based on their pelage.

The use of metachromism as a phylogenetic characteristic was questioned by Shedd and Macedonia (1991) who suggest that at the species level and above the cessation of gene flow allows the evolution of pelage colors along different lineages. It seems evident that the use of pelage variation is not a valid mechanism for determining taxonomic status among *Ateles*. Pelage forms that are supposedly diagnostically useful in identifying species and subspecies have been shown to occur in other species/subspecies. This is especially true with the forehead patches and sideburns of *A. belzebuth*. With the exception of the longer fur and tail in *A. paniscus* all the various pelage traits reported by Kellogg and Goldman (1944) can be demonstrated in other racial variants, suggesting these are simply polymorphic variants present in all spider monkeys, which occur in greater frequencies in some populations than others.

In 1989, Groves proposed a new taxonomy for *Ateles*, still based primarily on pelage, but accounting for the impossible species arrangement in Amazonia proposed by Kellogg and Goldman (1944). Groves (1989) suggested each of the four Amazonian subspecies represented a separate species. Thus, six species of *Ateles* were supported: *A. geoffroyi*, in Central America; *A. fusciceps*, along the Pacific coast; *A. belzebuth* with two subspecies, one along the Magdalena River valley and one in northwestern Amazonia; *A. chamek*, which is the former subspecies *A. p. chamek*, in southwestern Amazonia; *A. marginatus*, which is the former subspecies *A. b. marginatus*, in southeastern Amazonia; and finally *A. paniscus* in northeastern Amazonia (Groves, 1989).

Chromosomal variation

The first systematic attempts to investigate *Ateles* taxonomy based on analysis of chromosomes involved comparisons among several forms by Garcia *et al.* (1975) and Kunkel *et al.* (1980). By examining the karyotypes of spider monkeys based on both C and G banding in *A. geoffroyi* subspp., *A. f. robustus, A. b. hybridus, A. b. belzebuth* and *A. p. chamek* (following Kellogg and Goldman's nomenclature) Kunkel *et al.* (1980) discovered three chromosomes of the 34 diploid chromosomes present in *Ateles* – pairs five, six, and seven – that could be used to detect differences between karyotypes (Kunkel *et al.*, 1980). Pair five was similar in all samples described as form "a," except among *A. f. robustus*, where it was form "b" (Kunkel *et al.*, 1980). Pair six varies in karyotype pattern between most subspecies examined (Kunkel *et al.*, 1980). Form "a" is found in *A. hybridus*, while form "b" occurs in *A. b. chamek*; form "c" occurs in *A. geoffroyi* and *A. b. belzebuth*; form "d" occurs in *A. g. robustus* and *A. b. belzebuth* (Kunkel *et al.*, 1980). For pair seven the trans-Andean forms *A. hybridus, A. g. robustus* and *A. geoffroyi* share metacentric forms described as form "b," while *A. b. belzebuth* and *A. b. chamek* share submetacentric forms described as form "a" (Kunkel *et al.*, 1980).

Kunkel *et al.* (1980) suggest the chromosomal differences between taxa seem to support the taxonomy of Kellogg and Goldman (1944). However, they also suggest that the intraindividual heteromorphisms discovered could be interpreted to support the single species hypothesis of Hershkovitz (1968, 1969). The fact that *A. b. belzebuth* shares two of these three chromosomal forms with *A. p. chamek* and only one with *A. b. hybridus* seems to suggest the former two might be more closely related than the latter two and contradicts Kellogg and Goldman's (1944) taxonomy.

Pieczarka *et al.* (1989) discovered that *A. paniscus* has a different diploid chromosome number than all other spider monkeys. DeBoer and deBruijn (1990) determined that *A. b. chamek* and *A. paniscus* karyotypes represented definitively separate species based on this evidence.

Recent research (Medeiros *et al.*, 1997; Nieves *et al.*, 2005) extends the previous cytogenetic analyses to include karyotypes from all of the possible species suggested by Groves (1989). Medeiros *et al.* (1997) discovered a new form, "c,"

Table 3.2 Ateles *chromosome forms in various subspecies*

Ateles subspecies	Pair 5	Pair 6	Pair 7	Pair 13	Pair 14
Ateles paniscus	Form c	Forms a and e	Form b	Form b	Form b
Ateles belzebuth marginatus	Form a	Form d	Form a	Form a	Form a
Ateles belzebuth chamek	Form a	Forms b and e	Form a	Form a	Form a
Ateles belzebuth belzebuth	Form a	Forms c and d	Form a	Form a	Form a
Ateles hybridus	Form a	Forms a and c	Form b	Form a	Form b
Ateles geoffroyi robustus	Form b	Form d	Form b	Form a	Form a
Ateles geoffroyi subspp.	Forms a and b	Form c	Form b	Form a	Form a

This table recreates the various forms of chromosomal variants discovered by Kunkel *et al.* (1980), Medeiros *et al.* (1997) and Nieves *et al.* (2005) in investigations of spider monkey karyology. Taxonomy follows Collins and Dubach (2000a). The table is modified from the one presented by Medeiros *et al.* (1997) to use the format for identifying various chromosomal forms presented by Kunkel *et al.* (1980) and follows the one found in Nieves *et al.* (2005).

for *A. paniscus'* pair five chromosomes. Nieves *et al.* (2005) corroborated this. For pair five *A. b. marginatus* has form "a". Karyotypes of *A. paniscus* have two forms "a," and "e," for pair six, while *A. b. marginatus* has form "d," similar to *A. b. belzebuth* for pair six (Medeiros *et al.*, 1997). Pair seven was found to exhibit form "b" in *A. paniscus* and form "a" in *A. b. marginatus* (Medeiros *et al.*, 1997). Nieves *et al.* (2005) found form "b" in a sample of *A. b. chamek*. Pair 14 is reported as a new form for discrimination among some forms, where *A. paniscus* and *A. hybridus* share form "b," while all others share form "a" (Medeiros *et al.*, 1997). See Table 3.2 for chromosome patterns.

Medeiros *et al.* (1997) employed a phenetic analysis to conclude that only two definitive species can be ascribed in *Ateles*: *A. paniscus* becomes a monotypic species; and all other populations constitute another species. They do suggest *A. b. belzebuth, A. b. chamek* and *A. b. marginatus* might be grouped as a species; *A. geoffroyi* subspp. and *A. hybridus* might be linked together; while *A. fusciceps* is proposed as another possible species based on chromosomal similarities and differences (Medeiros *et al.*, 1997).

Nieves *et al.* (2005) used a parsimony analysis to investigate their chromosomal variants as compared with the phenetic model used by Medeiros *et al.* (1997) and came up with quite different results. This parsimony analysis produced considerable consistency between the chromosomal analyses and those based on molecular variation (discussed later in this chapter). Nieves *et al.* (2005) consider *A. paniscus* as the basal form in *Ateles* based on morphological studies (Froehlich *et al.*, 1991) and molecular studies (Collins and Dubach, 2000a, 2001) and use them as the outgroup for their parsimony analysis. Their cladograms do not clearly resolve the relationships among *A. b. belzebuth*,

A. b. chamek, and *A. b. marginatus.* However, they note that *A. b. belzebuth* and *A. b. marginatus* have the same states in all the chromosomic characters analyzed and *A. b. chamek,* which lies geographically between these two, differs only in chromosome six, where it possesses form "b." Their bootstrap analysis 50% majority rule tree fails to group any of these forms more closely to any other presenting a trichotomy, which tends to suggest they are in fact a single species.

Additionally, Nieves *et al.* (2005) propose *A. hybridus* as a separate clade from *A. geoffroyi* and the group formed by *A. b. belzebuth, A. b. chamek,* and *A. b. marginatus.* This tends to support *A. hybridus* as a separate species in accordance with the molecular findings of Collins and Dubach (2000a, 2001), but contradicts that of Medeiros *et al.* (1997).

Protein variation

Sampaio *et al.* (1993) examined the relationship between the supposed subspecies *A. p. chamek* and *A. p. paniscus* (as identified by Kellogg and Goldman, 1944) using isozyme analysis. They examined 20 genetic loci between these two groups discovering four polymorphic loci. The resultant genetic distance of 0.149 between these two groups was much higher than that seen between other New World primate subspecies (Sampaio *et al.,* 1993), suggesting that these are, in fact, separate species.

Morphological variation

Froehlich *et al.* (1991) conducted a systematic investigation of *Ateles* relationships based on morphological variation. They collected 76 morphometric measurements of spider monkey skulls and teeth for 26 pooled populations representing 5 to 20 specimens each (Froehlich *et al.,* 1991). A stepwise multivariate analysis of variance reduced the data set to 25 significant measurements, which were then transformed by factor analysis to account for confounding effects of various body sizes (Froehlich *et al.,* 1991). A linkage cluster analysis was performed from the total discriminant analysis to estimate a hierarchical systematic relationship among the groups (Froehlich *et al.,* 1991).

Froehlich *et al.* (1991) use the resultant phenogram to suggest three species of *Ateles. Ateles paniscus* is identified as a monotypic species, which is the most basal clade in the phenogram (Froehlich *et al.,* 1991). *A. b. marginatus, A.b.chamek* and *A. b. belzebuth* form a ring species around the Amazon Basin according to Froehlich *et al.* (1991). Finally, *A. hybridus, A.g.robustus* and the single sample of *A. geoffroyi* from Central America they included in their analyses are reported as one species (Froehlich *et al.*, 1991). Their combined discriminant analysis appears to separate *A. hybridus* from *A. g. robustus* and *A. g. fusciceps* along the second canonical, but the phenogram links these two groups as sister clades (Froehlich *et al.*, 1991). Thus, they support *A. paniscus, A. belzebuth*, and *A. geoffroyi* (which includes *A. hybridus* and *A. fusciceps*) as valid species of spider monkeys.

Molecular phylogenetics

Molecular-based phylogenetic studies examine the genetic code responsible for production of observed morphological and behavioral variation in separate individuals of a population (Table 3.3). As Avise (1989) noted, one area in which molecular evolutionary genetics can be of assistance is in the identification of "taxonomic mistakes" with their following two possible ramifications: (1) the recognition of groups which actually exhibit very little evolutionary differentiation and (2) the lack of recognition of forms which are phylogenetically distinct. Both of these errors can lead to conservation efforts that are misdirected with regard to the protection of biological diversity. Thus, phylogenetic relationships based on large, multilocus data sets gathered from various different geographic regions, representing different populations and subspecies, are the most reliable assessment of the phylogeny of the species, genus or family (Melnick and Hoelzer, 1993; Kimbel and Martin, 1993).

Collins and Dubach (2000a, 2001) studied three genetic regions in an analysis of spider monkey phylogenetics. They sequenced DNA for two regions in the mitochondrial genome: Cytochrome *c* Oxidase Subunit II (COII), a partially conserved region, and the hypervariable I portion of the mitochondrial control region, a quickly evolving region, were utilized to recreate *Ateles* relationships.

It should be noted that evolutionary relationships inferred from mitochondrial DNA trace only the maternal lineage. So, if the organism exhibits male dispersal upon maturation, coupled with limited group division, then the mitochondrial phylogeny may not reflect the actual phylogeny (Melnick and Hoelzer, 1993). This is unlikely to be the case in *Ateles*, however. Female dispersal and their fission–fusion social system promote mitochondrial DNA gene flow between *Ateles* populations (Symington 1987, 1990; Moore, 1993). Thus, the mitochondrial phylogeny is very likely to be highly reflective of the actual phylogeny for this genus. To insure this hypothesis, Collins and Dubach (2001) also sequenced a single-copy nuclear gene, Aldolase A Intron V, to examine relationships among the deepest branches of the spider monkey tree and

Table 3.3 *Comparisons among* Ateles *taxonomies*

Kellogg and Goldman (1944)
A. belzebuth belzebuth/A.belzebuth marginatus
A. belzebuth hybridus
A. fusciceps robustus/A.fusciceps fusciceps
A. paniscus chamek/A.paniscus paniscus
A. geoffroyi azuerensis/A.geoffroyi yucatanensis
A. geoffroyi frontatus/A.geoffroyi vellerosus
A. geoffroyi ornatus/A.geoffroyi panamensis
A. geoffroyi geoffroyi/A.geoffroyi grisescens
A. geoffroyi pan

Hershkovitz (1969)
A. paniscus

Groves (1989)
A. belzebuth belzebuth/A.belzebuth hybridus
A. fusciceps fusciceps/A.fusciceps robustus
A. chamek
A. marginatus
A. paniscus
A. geoffroyi (subspecies of Kellogg and Goldman, 1944)

Froehlich *et al.* (1991)
A. belzebuth belzebuth
A. belzebuth chamek
A. belzebuth marginatus
A. paniscus
A. geoffroyi fusciceps
A. geoffroyi robustus
A. geoffroyi hybridus
A. geoffroyi (subspecies of Kellogg and Goldman, 1944)

Medeiros *et al.* (1997)
A. paniscus
A. belzebuth (subgroup variation was noted)

Collins and Dubach (2000a, 2001)
A. belzebuth belzebuth
A. belzebuth chamek
A. belzebuth marginatus
A. hybridus
A. paniscus
A. geoffroyi fusciceps
A. geoffroyi robustus
A. geoffroyi (Northern? – including *A. g. yucatanensis* and *A. g. vellerosus*)
A. geoffroyi (Southern? – including *A. g. azuerensis, A.g.frontatus, A. g. geoffroyi, A. g.
    grisescens, A.g.ornatus, A.g.panamensis*)

Nieves *et al.* (2005)
A. paniscus
A. belzebuth (subgroup variation was noted)
A. hybridus (chromosomal evidence only)
A. geoffroyi (included *A. fusciceps*, but specific subgroup variation not included)

Compares the various pelage-based taxonomies of Kellogg and Goldman (1944), Hershkovitz (1969) and Groves (1989), with the chromosomal taxonomies of Medeiros *et al.* (1997) and Nieves *et al.* (2005), and craniometric taxonomy of Froehlich *et al.* (1991) to the phylogenetic taxonomy of Collins and Dubach (2000a).

to assure that the mitochondrial phylogeny provided a true picture of *Ateles* relationships.

DNA sequence variation among *Ateles* haplotypes was analyzed with two primary methods to discern phylogenetic relationships; these were parsimony analysis and neighbor-joining analysis, a form of distance-based analysis. The combined mitochondrial DNA consensus tree presented in Figure 3.2 reports bootstrap values (a measure of statistical support for a given grouping) as the average for each branch based on separate parsimony and neighbor-joining phylograms. The dates for the various nodes indicated are based on the use of a local molecular clock and suggest likely times of speciation for the various spider monkey clades (Collins and Dubach, 2006).

The COII gene studies of Collins and Dubach (2000a) were expanded by Nieves *et al.* (2005) to include nine more specimens and produced results in good agreement with the previous study. In addition Collins and Dubach (unpublished data) have added more specimens to their initial study for the D-loop sequences. The updated cladogram is presented in this volume as Figure 3.3. Cladograms from Collins and Dubach (2000a, 2001, unpublished data) and Nieves *et al.* (2005) provide the elements used for the following discussion of spider monkey taxonomy.

Molecular phylogenetics (Collins and Dubach, 2000a, 2001) supports four species of spider monkeys. Most uniquely, *A. hybridus* is supported as a valid, monotypic species. This support was echoed by Rylands *et al.* (2000) and in the chromosomal analysis of Nieves *et al.* (2005), but unfortunately, Nieves *et al.* (2005) did not include *A. hybridus* in their molecular analysis.

Collins and Dubach (2000a, 2001, unpublished data) recognize *Ateles belzebuth* as a species with a high degree of genetic variation among geographically widespread haplotypes including the former species or subspecies *A. b. belzebuth*, *A. b. chamek*, and *A. b. marginatus*, which are found throughout the southern and western Amazon Basin (Kellogg and Goldman, 1944; Groves, 1989; Rylands *et al.*, 2000). There are no significant barriers to gene flow between these populations and morphological analysis supports this relationship (Froehlich *et al.*, 1991). Nieves *et al.* (2005) only included *A. b. chamek* in their sequence analysis, but as noted previously did provide tentative support for this arrangement in their chromosomal study.

Within *A. belzebuth*, in the southern Amazon Basin, distinct genetic forms exist at close geographic proximity, while forms that are more distant are genetically quite similar, suggesting a microscale level of diversity in the southern Amazon Basin. It was not possible to define a geographic boundary between variable haplotypes, even though genetic variation in the combined group is

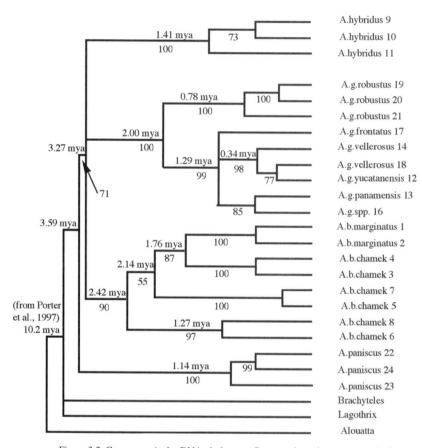

Figure 3.2 Consensus *Ateles* DNA phylogram. Presents the strict consensus *Ateles* phylogeny based on parsimony and neighbor-joining analysis of the two combined mitochondrial regions, Cytochrome *c* Oxidase Subunit II and the control region (Collins and Dubach, 2000a). A phylogeny based on the Aldolase A Intron V region is primarily congruent with this phylogeny (Collins and Dubach, 2001), but incorporated fewer samples, and thus was not incorporated in this consensus tree. All dates in this phylogram found above branches represent the average of the date for that node from mitochondrial and nuclear sequences (Collins and Dubach, 2000b). Bootstrap values, based on the combined mitochondrial phylograms, averaged for all forms of analysis, are presented below each corresponding branch for assessment of confidence in the groupings. For complete information, see the primary publications (Collins and Dubach, 2000a, 2000b, 2001). Primary data used to construct the table has been deposited in GenBank under the following accession numbers: Control region = AF213940 – AF213966; COII = AF216225 – AF216253.

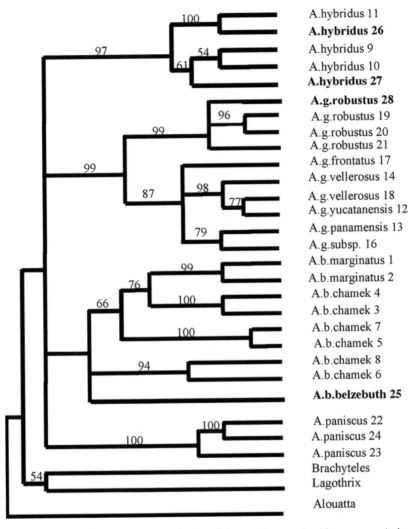

Figure 3.3 *Ateles* DNA control region phylogram. Presents the strict consensus *Ateles* phylogeny based on parsimony and neighbor-joining analysis of the mitochondrial control region including four key new spider monkey samples, which are noted in bold type (Collins and Dubach, unpublished data). The four new samples represent a specimen of *A. belzebuth belzebuth* (#25) from Ecuador, two *A. hybridus* samples (#26 and #27), one from the Magdalena River valley of Colombia and the other from the Lake Maracaibo region of Venezuela and a sample of *A. g. robustus* (#28) from the Darien Peninsula region of Colombia. Bootstrap values, based on the combined mitochondrial phylograms, averaged for both forms of analysis, are presented above each corresponding branch for assessment of confidence in the groupings.

more than among any other suggested *Ateles* species for the mitochondrial control region (Collins and Dubach, 2000a). It is thus probable that subspecies exist, but their genetic and geographic boundaries remain unclear based on molecular studies.

A. paniscus (Groves, 1989; Froehlich *et al.*, 1991; Collins and Dubach, 2000a; Rylands *et al.*, 2000, Nieves *et al.*, 2005) represents the former subspecies *A. p. paniscus* (Kellogg and Goldman, 1944; Konstant *et al.*, 1985).

A. geoffroyi consists of two or more subspecies ranging throughout most of Central America and the Choco Region along the Pacific coast of South America to northern Ecuador (Froehlich *et al.*, 1991; Collins and Dubach, 2000a; Rylands *et al.*, 2000) based on mitochondrial DNA studies. This species includes the former species *A. fusciceps* and 11 previously identified subspecies (Kellogg and Goldman, 1944; Konstant *et al.*, 1985; Groves, 1989).

The molecular studies carried out by Collins and Dubach (2000a) included spider monkeys from eight different locations in Central America. Samples from Honduras, Guatemala, Belize and Mexico represent racial variants of *A. g. yucatanensis* and *A. g. vellerosus*. The sample from Nicaragua is most likely *A. g. frontatus*. *A. g. panamensis*, plus another different pelage type from Panama, which cannot be racially identified, were also included in the molecular analyses (Collins and Dubach, 2000a, 2001). Using these data, no conclusive support for any valid subspecies in *A. geoffroyi* was found, except for the two previous species populations. The various phylograms tended to group spider monkeys from Central America into two groups, one including haplotypes from the northern portions of Central America and the other containing the rest of the racial variants from southern Central America (Figure 3.1).

Molecular evidence separated *A. hybridus* from *A. g. robustus* (Collins and Dubach, 2000a) along a likely boundary formed by the Rio Cauca and the western cordillera of the northern Andes in Colombia (Kellogg and Goldman, 1944; Garcia *et al.*, 1975; Hernandez-Camacho and Cooper, 1976). It is probable that *A. fusciceps fusciceps* is genetically similar to *A. g. robustus* and analysis of samples from this population, which have now been secured, will hopefully confirm such a link.

Collins and Dubach (2000b) used their sequence data and a local molecular clock to conclude that the majority of the speciation events among spider monkey populations occurred during the middle to late Pliocene and very early Pleistocene and were primarily the result of major disturbance–vicariance events such as the final uplift of the Andes Mountains and development of the llanos savannas. These findings provide little evidence to support Pleistocene refugia formation or riverine barriers as primary mechanisms for speciation, as previously speculated (Bush, 1994; Haffer, 1997).

Based on biogeography (Collins and Dubach, 2000b), *Ateles paniscus* is the most basal clade among all spider monkey species. It was isolated by the separation of the Guianan region from Amazonia and the emergence of the Amazon River as the major drainage of the basin more than 3 mya. *Ateles belzebuth*, in the southern and northwestern Amazon Basin, has a complex pattern of genetically related and geographically distinct haplotypes that share a common origin 2.4 mya. The relationships among these clades and subclades are generally consistent with the continent-wide studies of Cracraft (1988) and daSilva and Oren (1996). The trans-Andean species *Ateles geoffroyi* and *A. hybridus* were separated between 3.1 and 2.0 mya. However, *Ateles hybridus* haplotypes share a last common ancestor 1.4 mya during the Pleistocene, while *A. geoffroyi* was separated into at least two recognizable subspecies approximately 2.0 mya, during the very late Pliocene and early Pleistocene.

The biogeographical mechanisms, such as geological vicariance between trans- and cis-Andean populations, and between Guianan and Amazon Basin populations, proposed by Froehlich *et al.* (1991) to be important in the formation of spider monkey species are supported in large part by Collins and Dubach (2000b). It appears that refuge formation, coupled with the geographical barrier of the western cordillera of the Andes, resulted in speciation of *A. hybridus*. It seems probable that riverine barriers to dispersal do not strongly impact *Ateles* populations, because spider monkeys avoid the central periodically embayed portions of the Amazon Basin, thus limiting populations to regions where rivers are smaller and can readily be crossed by spanning substrates.

The specific habitat requirements of *Ateles* are postulated to have prohibited them from surviving in smaller, less stable Pleistocene refugia. This would mean that only larger populations, with a higher level of genetic variation, survived in theorized refugia. These biogeographic factors are not unlike the habitat fragmentation and parks (refugia) being produced by the encroachment of modern humans on the habitats of spider monkeys throughout the Neotropics, creating new risks for this primate. Determination of accurate taxonomic relationships and the biogeographic mechanisms that produced those populations has resulted in a considerably altered view of the evolutionary process affecting spider monkeys.

Conclusions

This review clearly indicates that the still frequently used taxonomy of Kellogg and Goldman (1944) is not valid for spider monkeys. Instead, a consensus based on the taxonomic relationships determined by the various systematic studies of *Ateles'* molecular, morphological, and chromosomal variation would better serve today's researcher. Examination of the various taxonomic studies supports the following arrangement of spider monkeys.

A. paniscus is supported as a valid monotypic species by Groves (1989), Froehlich *et al.* (1991), Sampaio *et al.* (1993), Medeiros *et al.* (1997), Collins and Dubach (2000a, 2001), and Nieves *et al.* (2005). *A. belzebuth* is supported as a valid species with the three subspecies *A. belzebuth marginatus, A. belzebuth chamek* and *A. belzebuth belzebuth* by Froehlich *et al.* (1991). Collins and Dubach (2000a, 2001) discovered high levels of genetic variation in this species, but could not distinguish geographic regions inhabited by discrete subspecies. Medeiros *et al.* (1997) and Nieves *et al.* (2005) also provide tentative support for the three *A. belzebuth* subspecies arrangement. Additionally, Sampaio *et al.* (1993) by eliminating *A. paniscus chamek* from *A. paniscus* would seem to place it in *A. belzebuth* since it occurs geographically between *A. belzebuth belzebuth* and *A. belzebuth marginatus*.

A. hybridus is supported as a valid species primarily by Collins and Dubach (2000a, 2001). However, the chromosomal phylogeny of Nieves *et al.* (2005), based on parsimony analysis, provides additional support for this arrangement. Rylands *et al.* (2000) also endorse this viewpoint. It is viewed as a subspecies of either *A. belzebuth* by Groves (1989) based on its pelage similarities to other *A. belzebuth*, or a subspecies of *A. geoffroyi* by Froehlich *et al.* (1991).

Froehlich *et al.* (1991) and Collins and Dubach (2000a, 2001) support *A. geoffroyi* as a valid species including the former species *A. fusciceps*. It is noted by Medeiros *et al.* (1997) that hybrids between *A. geoffroyi* and *A. fusciceps* are known to exist. Rossan and Baerg (1977) also documented this. It has also been widely observed in the zoological community (Collins and Dubach, unpublished data). Rylands *et al.* (2000) also support the union of these two former species.

This review demonstrates that primary studies based on analysis of discriminate information in spider monkeys – the morphological analyses of Froehlich *et al.* (1991), molecular studies of Collins and Dubach (2000a, 2001, unpublished data), and Nieves *et al.*'s (2005) chromosomal analysis – all concur on the three species, *A. belzebuth, A.paniscus*, and *A. geoffroyi*. The latter two studies also support the species status of *A. hybridus*.

Thus, it appears that while the spider monkey's taxonomy would benefit from further research, especially concerning subspecies status and variation in Central America, the four-species taxonomy supported by Collins and Dubach (2000a) and Nieves *et al.* (2005) seems to be the one which will serve those researching spider monkeys best in the foreseeable future. The American Zoological Society has adopted this taxonomy for use in its management of captive spider monkey populations. Use of this taxonomy as a central reference by researchers in various other behavioral and life-history studies when they refer to the populations under analysis will provide a consensus for use in comparative studies and conservation efforts aimed at protecting spider monkeys. Use of this phylogenetic arrangement in studies of spider monkey behavior and development of conservation plans will help increase the survivability of all forms of this jungle acrobat into the twenty-first century.

References

Aquino, R. and Encarnacion, F. (1994). Primates of Peru. *Primate Rep.*, **40**, 1–127.

Avise, J. C. (1989). A role for molecular genetics in the recognition and conservation of endangered species. *Trends Ecol. Evol.*, **4**, 279–281.

Baal, F. L. J., Mittermeier, R. A. and van Roosmalen, M. G. M. (1988). Primates and protected areas in Suriname, *Oryx*, **22**, 7–14.

Bramblett, C. A., Bramblett, S. S., Coelho, A. M. and Quick, L. B. (1980). Party composition in spider monkeys of Tikal, Guatemala: a comparison of stationary vs. moving observers. *Primates*, **21**, 123–127.

Braza, F. and Garcia, J. E. (1988). Rapport preliminaire sur les singes de la region montangneuse de Hunachaca, Bolivie. *Folia Primatol.*, **49**, 182–186.

Brown, K. S. (1987). Conclusions, synthesis, and alternative hypotheses. In *Biogeography and Quaternary History in Tropical America*, ed. T. C. Whitmore and G. T. Prance, Oxford: Clarendon Press, pp. 175–210.

Bush, M. B. (1994). Amazonian speciation: a necessarily complex model. *J. Biogeogr.*, **21**, 5–17.

Campbell, C. J. (2003). Patterns of behavior across reproductive states of free-ranging female black-handed spider monkeys (*Ateles geoffroyi*). *Am. J. Phys. Anthropol.*, **124**, 166–176.

Cant, J. G. H. (1990). Feeding ecology of spider monkeys (*Ateles geoffroyi*) at Tikal, Guatemala. *Hum. Evol.*, **5**, 269–281.

Castellanos, H. G. and Chanin, P. (1996). Seasonal differences in food choice and patch preference of long-haired spider monkeys (*Ateles belzebuth*). In *Adaptive Radiations of Neotropical Primates*, ed. M. A. Norconk, A. L. Rosenberger and P. A. Garber, New York: Plenum Press, pp. 451–466.

Coates, A. G. and Obando, J. A. (1996). The geologic evolution of the Central American isthmus. In *Evolution and Environment in Tropical America*, ed. J. B. C. Jackson, A. F. Budd and A. G. Coates, Chicago: University of Chicago Press, pp. 21–56.

Coelho, A. M., Coelho, L. S., Bramblett, C. A., Bramblett, S. S. and Quick, L. B. (1976). Ecology, population characteristics, and sympatric association in primates: a socio-bioenergetic analysis of howler and spider monkeys in Tikal, Guatemala. *Yrbk. Phys. Anthropol.*, **20**, 96–135.

Colinvaux, P. A. (1996). Quaternary environmental history and forest diversity in the Neotropics. In *Evolution and Environment in Tropical America*, ed. J. B. C.

Jackson, A. F. Budd and A. G. Coates, Chicago: University of Chicago Press, pp. 359–406.

Collins, A. C. and Dubach, J. (2000a). Phylogenetic relationships of spider monkeys (*Ateles*) based on mitochondrial DNA variation. *Int. J. Primatol.*, **21**, 381–420.

Collins, A. C. and Dubach, J. (2000b). Biogeographic and ecological forces responsible for speciation in *Ateles*. *Int. J. Primatol.*, **21**, 421–444.

Collins, A. C. and Dubach, J. M. (2001). Nuclear DNA variation among spider monkeys (*Ateles*). *Mol. Phylo. Evol.*, **19**, 67–75.

Cracraft, J. (1988). Deep-history biogeography: retrieving the historical pattern of evolving continental biotas. *Syst. Zool.*, **37**, 221–236.

Crockett, C. M., Brooks, R. D., Crockett-Meacham, R., Crockett-Meacham, S. and Mills, M. (1997). Recent observations of Nicaraguan primates and a preliminary conservation assessment. *Neotrop. Primates*, **5**, 71–74.

Dahl, J. F. (1986). The status of howler, spider and capuchin monkey populations in Belize, Central America. *Primate Rep.*, **14**, 161.

daSilva, J. M. C. and Oren, D. C. (1996). Application of parsimony analysis of endemicity in Amazonian biogeography: an example with primates. *Biol. J. Lin. Soc.*, **59**, 427–437.

deBoer, L. E. M. and deBruijn, M. (1990). Chromosomal distinction between the red-faced and black-faced black spider monkeys (*Ateles paniscus paniscus* and *A. p. chamek*). *Zoo Biol.*, **9**, 307–316.

Di Fiore, A. and Campbell, C. J. (2007). The atelines: variation in ecology, behavior, and social organization. In *Primates in Perspective*, ed. C. J. Campbell, A. Fuentes, K. C. MacKinnon, M. Panger and S. K. Bearder, Oxford: Oxford University Press, pp. 155–185.

Engel, S. R., Hogan, K. M., Taylor, J. F. and Davis, S. K. (1998). Molecular systematics and paleobiogeography of the South American sigmodontine rodents. *Mol. Biol. Evol.*, **15**, 35–49.

Estrada, A. and Coates-Estrada, R. (1988). Tropical rain forest conversion and perspectives in the conservation of wild primates (*Alouatta* and *Ateles*) in Mexico. *Am. J. Primatol.*, **14**, 315–327.

Fedigan, L. M., Fedigan, L., Chapman, C. and Glander, K. E. (1988). Spider monkey home ranges: a comparison of radio telemetry and direct observation. *Am. J. Primatol.*, **16**, 19–29.

Ferrari, S. F. and Lopes, M. A. (1992). New data on the distribution of primates in the region of the confluence of the Jiparana and Madeira rivers in Amazonas and Rondonia, Brazil. *Goeldiana Zoologia*, **11**, 1–13.

Ferrari, S. F. and Lopes, M. A. (1996). Primate populations in eastern Amazonia. In *Adaptive Radiations of Neotropical Primates*, ed. M. A. Norconck, A. L. Rosenberger and P. A. Garber, New York: Plenum Press, pp. 53–55.

Ferrari, S. F., Iwanga, S. and Lourenco da Silva, J. (1996). Platyrrhines in pimenta bueno, Rondonia, Brazil. *Neotrop. Primates*, **4**, 151–153.

Ferrari, S. F., Lopes, M. A., Cruz-Neto, E. H., *et al.* (1995). Primates and conservation in the Guajara-Mirim state park, Rondonia, Brazil. *Neotrop. Primates*, **3**, 81–82.

Freese, C. (1976). Censusing *Alouatta palliata*, *Ateles geoffroyi*, and *Cebus capucinus* in the Costa Rican dry forest. In *Neotropical Primates: Field Studies and*

Conservation, ed. R. W. Thorington and P. G. Heltne, Washington DC: National Academy of Sciences, pp. 4–9.

Froehlich, J. W., Supriantna, J. and Froehlich, P. H. (1991). Morphometric analyses of *Ateles*: systematic and biogeographic implications. *Am. J. Primatol.*, **25**, 1–22.

Garcia, M., Caballin, M. R., Aragones, J., Goday, C. and Egozcue, J (1975). Banding patterns of the chromosomes of *Ateles geoffroyi* with description of two cases of pericentric inversion. *J. Med. Primatol.*, **4**, 108–113.

García-Orduña, F., Rodríguez-Luna, E. and Canales-Espinosa, D. (1993). Effects of habitat fragmentation on howler monkey (*Alouatta palliata mexicana*) and spider monkey (*Ateles geoffroyi vellerosus*) populations in Veracruz state, Mexico. *Am. J. Primatol.*, **30**, 312–313.

Granjon, L., Cosson, J. F., Judas, J. and Ringuet, S. (1996). Influence of tropical rainforest fragmentation on mammal communities in French Guiana: short-term effects. *Acta Oecol.*, **17**, 673–684.

Groves, C. P. (1989). *A Theory of Human and Primate Evolution*. Oxford: Clarendon Press, pp. 127–131.

Haffer, J. (1982). General aspects of the refuge theory. In *Biological Diversification in the Tropics*, ed. G. T. Prance, New York: Columbia University Press, pp. 5–22.

Haffer, J. (1987). Quaternary history of tropical America. In *Biogeography and Quaternary History in Tropical America*, ed. T. C. Whitmore and G. T. Prance, Oxford: Clarendon Press, pp. 1–18.

Haffer, J. (1997). Alternative models of vertebrate speciation in Amazonia: an overview. *Biodiv. Conserv.*, **6**, 451–476.

Hartwig, W. C. (1995). A giant new world monkey from the Pleistocene of Brazil. *J. Hum. Evol.*, **28**, 189–195.

Hernandez-Camacho, J. and Cooper, R. W. (1976). The nonhuman primates of Colombia. In *Neotropical Primates: Field Studies and Conservation*, ed. R. W. Thorington and P. G. Heltne, Washington, DC: National Academy of Sciences, pp. 35–69.

Hershkovitz, P. (1968). Metachromism or the principle of evolutionary change in mammalian tegumentary colors. *Evolution*, **22**, 556–575.

Hershkovitz, P. (1969). The evolution of mammals in southern continents. VI. The recent mammals of the neotropical region: a zoogeographic and ecological review. *Quart. Rev. Biol.*, **44**, 1–70.

Hillis, D. M., Moritz, C. and Mable, B. K. (1996). *Molecular Systematics*, 2nd edn. Sunderland, MA: Sinauer Associates, pp. 243–245.

Iwanaga, S. and Ferrari, S. F. (2002). Geographic distribution and abundance of woolly (*Lagothrix cana*) and spider (*Ateles chamek*) monkeys in southwestern Brazilian Amazonia. *Am. J. Primatol.*, **56**, 57–64.

Karesh, W. B., Wallace, R. B., Painter, R. L. E., *et al.* (1998). Immobilization and health assessment of free-ranging black spider monkeys (*Ateles paniscus chamek*). *Am. J. Primatol.*, **44**, 107–123.

Kellogg, R. and Goldman, E. A. (1944). Review on the spider monkeys. *Proc. US Mus. Nat. Hist.*, **96**, 1–45.

Kimbel, W. H. and Martin, L. B. (1993). Species and speciation: conceptual issues and their relevance for primate evolutionary biology. In *Species, Species Concepts, and Primate Evolution*, ed. W. H. Kimbel and L. B. Martin, New York: Plenum Press.

Kinzey, W. G. (1997). *New World Primates: Ecology, Evolution and Behavior.* New York: Aldine de Gruyter, pp. 192–199.

Klein, L. L. and Klein, D. B. (1977). Feeding behavior of the Colombian spider monkey. In *Primate Ecology*, ed. T. H. Clutton-Brock, London: Academic Press, pp. 153–185.

Konstant, W., Mittermeier, R. A. and Nash, S. D. (1985). Spider monkeys in captivity and in the wild. *Primate Conserv.*, **5**, 82–109.

Kunkel, L. M., Heltne, P. G. and Borgaonkar, D. S. (1980). Chromosomal variation and zoogeography in *Ateles*. *Int. J. Primatol.*, **1**, 223–232.

Madden, R. H. and Albuja, L. (1987). Conservation status of *Ateles fusciceps fusciceps* in northwestern Ecuador. *Int. J. Primatol.*, **8**, 513.

Massey, A. (1987). A population survey of *Alouatta palliata*, *Cebus capucinus* and *Ateles geoffroyi* at Palo Verde, Costa Rica. *Revista de Biologia Tropical*, **35**, 345–347.

Medeiros, M. A., Barroso, R. M. S., Pieczarka, *et al.* (1997). Radiation and speciation of spider monkeys, genus *Ateles* from the cytogenetic viewpoint. *Am. J. Primatol.*, **42**, 167–178.

Melnick, D. J. and Hoelzer, G. A. (1993). What is mtDNA good for in the study of primate evolution? *Evol. Anthropol.*, **2**, 2–10.

Milton, K. (1981). Estimates of reproductive parameters for free-ranging *Ateles geoffroyi*. *Primates*, **22**, 574–579.

Moore, J. (1993). Inbreeding and outbreeding in primates: what's wrong with "the dispersing sex"? In *The Natural History of Inbreeding and Outbreeding*, ed. J. Moore, Chicago: University of Chicago Press.

Nieves, M., Ascunce, M. S., Rahn, M. I. and Mudry, M. D. (2005). Phylogenetic relationships among some *Ateles* species: the use of chromosomic and molecular characters. *Primates*, **46**, 155–164.

Norconck, M. A., Sussman, R. W. and Phillips-Conroy, J. P. (1996). Primates of Guyana shield forests. In *Adaptive Radiations of Neotropical Primates*, ed. M. A. Norconck, A. L. Rosenberger and P. A. Garber, New York: Plenum Press, pp. 69–83.

Nunes, A. (1995). Status, distribution and viability of wild populations of *Ateles belzebuth marginatus*. *Neotrop. Primates*, **3**, 17–18.

Nunes, A. (1998). Diet and feeding ecology of *Ateles belzebuth belzebuth* at Maraca ecological station, Roraima, Brazil. *Folia Primatol.*, **69**, 61–76.

Pieczarka, J., Nagamachi, C. Y. and Barroso, R. M. S. (1989). The karyotype of *Ateles paniscus paniscus* (Cebidae, Primates): 2n = 32. *Revista Brasilia Genetica*, **12**, 543–551.

Porter, C. A., Page, S. L., Czelusniak, J., *et al.* (1997). Phylogeny and evolution of selected primates as determined by sequences of the e-globin locus and 5′ flanking regions. *Int. J. Primatol.*, **18**, 261–295.

Riddle, B. R. (1996). The molecular phylogeographic bridge between deep and shallow history in continental biotas. *Trends Ecol. Evol.*, **11**, 207–212.

Rossan, P. N. and Baerg, D. C. (1977). Laboratory and feral hybridization of *Ateles geoffroyi panamensis* Kellogg and Goldman, 1944, in Panama. *Primates*, **18**, 235–237.

Rowe, N. (1996). *The Pictorial Guide to the Living Primates*. East Hampton, NY: Pogonias Press, pp. 107–117.

Russo, S. E., Campbell, C. J., Dew, J. L., Stevenson, P. R. and Suarez, S. A. (2005). A multiforest comparison of dietary preferences and seed dispersal by *Ateles* spp. *Int. J. Primatol.*, **26**(5), 1017–1037.

Rylands, A. B., Schneider, H., Mittermeier, R. A., Groves, C. and Rodriguez- Luna, E. (2000). An assessment of the diversity of new world primates. *Neotrop. Primates*, **8**, 61–93.

Sampaio, M. I., Schneider, M. P. C. and Schneider, H. (1993). Contribution of genetic distances studies to the taxonomy of *Ateles*, particularly *Ateles paniscus paniscus* and *Ateles paniscus chamek*. *Int. J. Primatol.*, **14**, 895–903.

Schneider, H., Schneider, M. P. C., Sampaio, I., *et al.* (1993). Molecular phylogeny of the new world monkeys. *Mol. Phylogen. Evol.*, **2**, 225–242.

Shedd, D. H. and Macedonia, J. M. (1991). Metachromism and its phylogenetic implications for the genus *Eulemur* (Prosimii: Lemuridae). *Folia Primatol.*, **57**, 221–231.

Shimooka, Y. (2003). Seasonal variation in association patterns of wild spider monkeys (*Ateles belzebuth belzebuth*) at La Macarena, Colombia. *Primates*, **44**, 83–90.

Silva-Lopez, G., Motta-Gill, J. and Hernandez, A. I. (1996). Taxonomic notes on *Ateles geoffroyi*. *Neotrop. Primates*, **4**, 41–44.

Smith, M. F. and Patton, J. L. (1993). The diversification of South American murid rodents: evidence from mitochondrial DNA sequence data for the Akodontine tribe. *Biol. J. Linn. Soc.*, **50**, 149–177.

Stevenson, P. R., Quiñones, M. J. and Ahumada, J. A. (1998). Effects of fruit patch availability on feeding subgroup size and spacing patterns in four primate species at Tinigua National Park, Colombia. *Int. J. Primatol.*, **19**, 313–324.

Sussman, R. W. and Phillips-Conroy, J. E. (1995). A survey of the distribution and density of the primates of Guyana. *Int. J. Primatol.*, **16**, 761–790.

Symington, M. M. (1987). Sex ratio and maternal rank in wild spider monkeys: when daughters disperse. *Behav. Ecol. Sociobiol.*, **20**, 421–425.

Symington, M. M. (1988). Demography, ranging patterns, and activity budgets of black spider monkeys (*Ateles paniscus chamek*) in the Manu national park, Peru. *Am. J. Primatol.*, **15**, 45–67.

Symington, M. M. (1990). Fission-fusion social organization in *Ateles* and *Pan*. *Int. J. Primatol.*, **11**, 47–61.

Timock, J. and Vaughan, C. (2002). A census of mammal populations in Punta Leona private wildlife refuge, Costa Rica. *Revista de Biologia Tropical*, **50**, 134–141.

Van der Hammen, T. (1982). Paleoecology of tropical South America. In *Biological Diversification in the Tropics*, ed. G. T. Prance, New York: Columbia University Press, pp. 60–66.

van Roosmalen, M. G. M. (1980). Habitat preferences, diet, feeding strategy, and social organization of the black spider monkey (*Ateles paniscus paniscus*) in Surinam. Unpublished Ph.D. thesis, University of Wageningen, Netherlands.

Wallace, R. B., Painter, R. L. E., Taber, A. B. and Ayres, J. M. (1996). Notes on a distributional river boundary and southern range extension for two species of Amazonian primates. *Neotrop. Primates*, **4**, 149–151.

White, B. N. (1986). The isthmian link, antitropicality and American biogeography: distributional history of the Atherinopsinae (*Pisces: Atherinidae*). *Syst. Zool.*, **35**, 176–194.

Wolfheim, J. H. (1983). *Primates of the World: Distribution, Abundance, and Conservation*. Seattle, WA: University of Washington Press, pp. 246–256.

Yanez, A. H. (1991). Conservation of Mexican spider monkey (*Ateles geoffroyi vellerosus*) on the west end of the Sierra Madre of Chiapas, Mexico. *Am. J. Primatol.*, **24**, 108.

Youlatos, D. (2003). Multivariate analysis of organismal and habitat parameters in two Neotropical primate communities. *Am. J. Phys. Anthropol.*, **123**, 181–194.

Part II
Ecology

4   *Diets of wild spider monkeys*

ANTHONY DI FIORE, ANDRES LINK AND
J. LAWRENCE DEW

Introduction

The first comprehensive field study of wild spider monkeys was undertaken in
Panama in the early 1930s by C. R. Carpenter. In discussing the diet of *Ateles
geoffroyi*, Carpenter (1935) wrote, "Red spider monkeys have been classed correctly as frugivorous. It is estimated that about 90 percent of their food consists
of fruit or nuts" (p. 174). Since that time, field studies of wild *Ateles* have consistently confirmed Carpenter's early assessment of the highly frugivorous diet
of spider monkeys, to the point where they are now treated as a classic example
of a frugivorous primate and are often considered to be "ripe fruit specialists"
(Cant, 1977; Klein and Klein, 1977; van Roosmalen, 1985; van Roosmalen and
Klein, 1988; Cant, 1990; Dew, 2005; Wallace, 2005; Di Fiore and Campbell,
2007). In this chapter, we first review what is known of the diet of wild *Ateles*,
paying particular attention to data from long-term ecological studies. In doing
so, we address the physiological and morphological adaptations for frugivory
that spider monkeys have evolved, as well as the connections among diet, food
resource distribution and foraging behavior that are relevant to understanding
the characteristic "fission–fusion" social organization of *Ateles*.

From there, we move on to discussing the interesting variation seen in the
diets of spider monkeys across tropical forest sites, and we address in more detail
the diets of two populations of white-bellied spider monkeys (*Ateles belzebuth
belzebuth*) which we and our colleagues and collaborators have studied over
multiple years in Colombia and Ecuador. These longitudinal data provide an
interesting perspective on intra- and interannual variation in the foraging strategies of individuals in these two populations. Finally, we consider the ecological
significance of *Ateles* diets and dietary diversity, and we compare and contrast
the diets and dietary strategies of spider monkeys with those of other frugivorous
primates.

Spider Monkeys: Behavior, Ecology and Evolution of the Genus Ateles, ed. Christina J. Campbell.
Published by Cambridge University Press. © Cambridge University Press 2008.

Overall diet and dietary diversity

All studies of *Ateles* covering approximately 12 months (roughly one phenological cycle in most Neotropical forests) have reported that fruits, particularly
ripe fruits, constitute the bulk of the diet, ranging from 55% to more than 90% of
annual feeding time (mean across studies = 77%: Table 4.1). This is true for populations sampled from across the geographic range of the genus, from eastern
Bolivia, to central Brazil, to Surinam, to Guatemala (Figure 4.1), and likewise
characterizes *Ateles* from a number of other sites that have been sampled for
shorter periods of time (Simmen and Sabatier, 1996; Iwanaga and Ferrari, 2001).
Spider monkeys complement this heavily frugivorous diet with other plant
parts such as leaves (mainly young leaves), flowers, seeds, aerial roots, palm
hearts, and the liquid endosperm or "milk" from inside immature palm fruits
(Table 4.1). In some sites, spider monkeys are also reported to consume other
items, including a few species of invertebrates (mainly caterpillars, meliponid
bees, and termites), fungi, decaying wood, and soil from mineral licks
("salados" or "saladeros") and arboreal termite nests (Klein and Klein, 1977;
Symington, 1987; van Roosmalen and Klein, 1988; Izawa, 1993; Castellanos
and Chanin, 1996; Simmen and Sabatier, 1996; Link, 2003; Dew, 2005; Suarez,
2006). Although spider monkeys obtain most of their water requirements
directly from the fruits, leaves and flowers they consume, they will occasionally
drink water directly from tree holes, arboreal bromeliads, and small streams
around mineral licks; some populations of *Ateles geoffroyi* have also been
reported to directly drink from terrestrial water sources (Campbell *et al.*, 2005).

Spider monkeys, especially in equatorial latitudes, feed from a wide variety
of plant taxa (Tables 4.2 and 4.3). For example, in a hyperdiverse western
Amazonian forest in Yasuní National Park, Ecuador, just south of the Equator,
Ateles belzebuth belzebuth eat fruits from more than 250 plant species over
the course of several field seasons (Dew, 2005; Suarez, 2006; A. Di Fiore, A.
Link, and S. Spehar, unpublished data). In less diverse, subtropical sites, such as
Tikal National Park, Guatemala, and Santa Rosa National Park, Costa Rica, *A.
geoffroyi* eat a much smaller number of different fruit species, ~30 to 40 (Cant,
1977; Chapman, 1987). Across sites, a range of variables describing dietary
diversity are related to both proximity to the Equator and to mean annual rainfall,
both of which are known to correlate with local floristic diversity (Table 4.4).

Fruits

The vast majority of fruits eaten by spider monkeys are consumed when they
are "ripe" or "mature," with a fleshy and easily penetrated pulp or aril that

Table 4.1 *Diets of spider monkeys based on studies lasting nine months or more*

Species	Study location	Country	Study dates	% Fruit [2] Mean	Range	% Ripe fruit [2] Mean	Range	% Leaves [3] Mean	Range	% Prey [4] Mean	Range	% Flowers [5] Mean	Range	% Other [6] Mean	Range	Study length (months)	Source(s)
A. geoffroyi	Tikal National Park	Guatemala	1975–1976	54.9	31–84	41	–	15.1	1–34	1.9	0–17	6.3	0–22	21.8	1–66	9	Cant (1977) [7]
A. geoffroyi	Santa Rosa National Park	Costa Rica	1983–1989	71.4	14–100	–	–	12.5	0–86	2.1	0–30	14.0	–	0	0	38	Chapman et al. (1995) [8]
A. geoffroyi	Barro Colorado Island	Panama	1997–1998	82.2	69–91	–	–	17.2	6–32	0.6	0–2	1.0	0–9	0	0	14	Campbell (2000)
A. paniscus	Voltzberg Nature Reserve	Surinam	1976–1978	79.8	54–92	76.6	–	7.9	1–23	<1	–	6.4	1–28	5.6	1–17	26	van Roosmalen (1985)
A. belzebuth belzebuth	Rio Tawadu, Reserva Forestal El Caura	Venezuela	1991–1992	–	–	–	69–90	–	–	–	–	–	–	–	–	16	Castellanos & Chanin (1996); Castellanos (1997)
A. belzebuth belzebuth	Ilha de Maracá, Roraima	Brazil	1987–1989	–	–	88.5	~74–~98	8.3	~0–26	0	0	–	–	3.2	–	12	Nunes (1998) [9]
A. belzebuth belzebuth	Tinigua National Park	Colombia	1990–1991	74	–	–	–	12	–	–	–	5	–	9	–	12	Russo et al. (2005)
A. belzebuth belzebuth	Tinigua National Park	Colombia	2001–2002	73.0	41–96	–	–	13.0	3–25	1.4	0–9	12.0	1–36	1.2	0–4	13	Link, unpublished data
A. belzebuth belzebuth	La Macarena National Park	Colombia	1967–1968	83	78–100	–	–	7	0–22	0	–	<1	–	10	0–18	10	Klein & Klein (1977)
A. belzebuth belzebuth	Yasuní National Park	Ecuador	1995–1996	87	64–100	–	–	9	0–23	<1	–	1	~0–~6	~3	~0–~17	12	Dew (2005)
A. belzebuth belzebuth	Yasuní National Park	Ecuador	1998–1999	78.8	52–92	–	–	7.7	0–22	0	0	3.2	0–24	10.3	0–24	17	Suarez (2006) [10]
A. belzebuth belzebuth	Yasuní National Park	Ecuador	2002–2003	79.4	60–100	79.0	60–100	12.4	0–23	0	–	2.7	0–10	5.5	0–20	11	Link & Di Fiore, unpublished data

(cont.)

Table 4.1 (Cont.)

Species	Study location	Country	Study dates	% Fruit [2] Mean	Range	% Ripe fruit [2] Mean	Range	% Leaves [3] Mean	Range	% Prey [4] Mean	Range	% Flowers [5] Mean	Range	% Other [6] Mean	Range	Study length (months)	Source(s)
A. belzebuth chamek	Cocha Cashu, Manu National Park	Peru	1984–1985	74.7	54–99	—	—	15.5	<1–38	—	—	4.5	~0–~22	4.5	~0–~21	12	Symington (1987, 1988a) [11]
A. belzebuth chamek	Lago Caiman, Noel Kempff National Park	Bolivia	1996–1997	85.7	63–99	81.1	36–98	10.7	1–37	—	—	2.9	0–16	0.6	0–3	11	Wallace (2005) [12]
Average				77.0		73.2		11.4		0.8		4.6		5.5			

[1] Means and ranges for each population are taken or calculated from the data in original source(s). Approximate means and ranges, indicated with '~', are based on extrapolation from figures presented in the original source(s) when raw data are not given. '–' signifies that values were either not reported or could not be calculated or estimated from the original source(s).
[2] Where possible, the proportion of the diet comprising "all fruit" and "ripe fruit" only are presented separately.
[3] Includes leaves and other plant vegetative parts (shoots, petioles, meristems of epiphytic plants, leaf buds).
[4] Includes both adult and larval invertebrates.
[5] Includes nectar, where distinguished in the original source.
[6] Includes seeds, bark, fungi; termitaria soil, ground soil, and other undetermined food items.
[7] The listed range of values for each category is basd on averages for the "wet" and "dry" seasons as noted in the original source. The actual range across the year should be somewhat greater.
[8] Some seeds are included in the "fruit" category, but these cannot be separated out given how the data are presented in the original source.
[9] Unripe fruits and flowers are included in the category "other," but these cannot be separated out given how the data are presented in the original source.
[10] Diets are based on 10 follows of up to two weeks spread among three focal individuals.
[11] Means reported represent the average of means for two study groups, while ranges cover the minimum to maximum percentages noted across both groups. Some animal prey (caterpillars) were consumed by members of one study group in 2 months of the year. These are included in the category "other" as the original source does not allow these to be separated out into "prey."
[12] Some prey were consumed and are included in the category "other," but these cannot be separated out given how the data are presented in the original source.

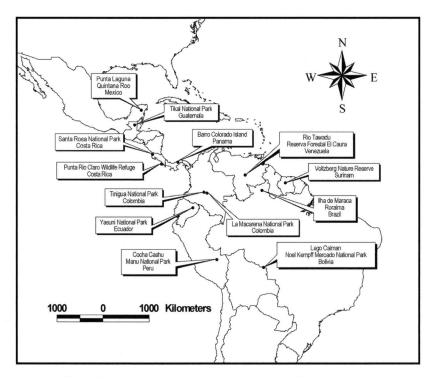

Figure 4.1 Thirteen sites where long-term studies of the diets of various species of wild spider monkeys have been conducted: *Ateles geoffroyi* (Punta Laguna, Tikal, Santa Rosa, Punta Rio Claro, Barro Colorado Island); *Ateles belzebuth belzebuth* (Tinigua, La Macarena, Yasuní, Ilha de Maracá, Rio Tawadu), *Ateles paniscus* (Voltzberg), *Ateles belzebuth chamek* (Cocha Cashu, Lago Caiman).

usually contains a substantial water and soluble sugar component (Tables 4.1 and 4.3). These fruits, and the seeds they contain, tend to be swallowed whole (van Roosmalen, 1985; Dew, 2005; Link and Di Fiore, 2006); seed predation (i.e. mastication and destruction of immature or mature seeds) and seed spitting (i.e. oral processing of fruits to remove the edible portion followed by expulsion of the unswallowed seed under or close to parental trees) are infrequent and only occur regularly for a small number of plant species from which spider monkeys feed (Dew, 2005; Russo *et al.*, 2005; Wallace, 2005). For example, white-bellied spider monkeys, *A. b. belzebuth* in Yasuní National Park, Ecuador, prey upon the seeds of less than 3% of the fruit species they consume ($n = 4$ of 152 species), and masticated seeds comprise less than 1% of the yearly diet (Link and Di Fiore, 2006). At this site, most predated seeds come from immature palm fruits that provide soft "nuts" (e.g. *Iriartea deltoidea*) or significant amounts of palm milk (e.g. *Astrocharyum chambira*, which contained an average of 26.1 ± 3.3 ml of liquid

Table 4.2 *Dietary diversity in spider monkeys* [1]

Species	Study location	Study dates	Latitude (decimal degrees)	Rainfall (mm) [2]	Min # fruit morphospecies in diet	Min # total plant morphospecies in diet	Min # of fruit morphospecies contributing to ≥1% of total diet	Min # of fruit genera contributing to ≥1% of total diet	% of total diet accounted for by top 5 morphospecies genera	% of total diet accounted for by top 5 genera	# genera together accounting for ≥50% of total diet	Source(s)
A. geoffroyi	Tikal National Park	1975–1976	17.22	1350	30	40	–	10	77.3	80.9	1	Cant (1977)
A. geoffroyi	Santa Rosa National Park	1983–1989	10.83	1524	–	–	13	9	61.1	–	3	Chapman et al. (1995)
A. geoffroyi	Punto Rio Claro Wildlife Refuge	1999–2000	8.65	3500	–	–	27	27	54.7	54.7	3	Riba-Hernández et al. (2003) [3]
A. geoffroyi	Barro Colorado Island	1997–1998	9.18	2640	107	137	27	25	36.8	39.8	7	Campbell (2000)
A. paniscus	Voltzberg Nature Reserve	1976–1978	4.68	2200	171	207	21	22	24.3	31.7	12	van Roosmalen (1985)
A. belzebuth belzebuth	Rio Tawadu, Reserva Forestal El Caura	1991–1992	6.36	3300	83	–	–	–	–	–	–	Castellanos & Chanin (1996); Castellanos (1997)
A. belzebuth belzebuth	Ilha de Maraca, Roraima	1987–1989	3.03	2300	62	65	18	16	57.6	63.0	4	Nunes (1998)
A. belzebuth belzebuth	Tinigua National Park	1990–1991	2.67	2782	72	–	30	20	34.6	47.9	6	Stevenson et al. (2002); Russo et al. (2005) [4]
A. belzebuth belzebuth	Tinigua National Park	2001–2002			117	160	21	17	26.1	33.2	10	Link, unpublished data
A. belzebuth belzebuth	La Macarena National Park	1967–1968			–	–	31	19	–	53.0	5	Klein & Klein (1977) [3]

A. belzebuth belzebuth	Yasuni National Park	1995–1996	−0.63	3200	73	–	–	15	–	40.4	8	Dew (2005)
A. belzebuth belzebuth	Yasuni National Park	1998–1999			238	243	28	26	23.1	31.2	11	Suarez (2006)
A. belzebuth belzebuth	Yasuni National Park	2002–2003			113	118	22	22	28.9	35.5	10	Link & Di Fiore, unpublished data
A. belzebuth chamek	Cocha Cashu, Manu National Park	1984–1985	−11.87	2100	125	141	–	–	–	–	–	Symington (1987, 1988a)
A. belzebuth chamek	Lago Caiman, Noel Kempff National Park	1996–1997	−13.60	1637	75	86	17	14	51.5	63.0	4	Wallace (2005)

[1] Dietary diversity is indexed here simply as taxonomic richness – i.e. the number of different taxa (genera, morphospecies) consumed – as other standard diversity measures that take into account relative consumption cannot be calculated for all sites. '–' signifies that values were not reported or could not be calculated or estimated from the original source.

[2] Rainfall data comes from sources listed plus additional publications from the same study site or meteorological databases available online.

[3] The diet list for these sites includes only taxa contributing to the fruit portion of the diet, so it is possible that the values in the last three columns of data would be slightly different if all taxa were included. However, given the low proportion of leaf consumption overall, this is unlikely.

[4] Russo et al. (2005) note that 106 fruit species were consumed by spider monkeys during the same time period as that referenced by Stevenson et al. (2002). We have reported the value from Stevenson et al. (2002) as this source presents primary data, but we note that the 106 value is more in line with that recorded by Link (unpublished data) a decade later.

Table 4.3 *Genera of plants consumed by spider monkeys and parts eaten for 13 long-term study sites*

Family	Genus [2]	Punta Laguna [3]	Tikal [4]	Punta Santa Rosa [5]	Rio Claro [6]	BCI [6]	Rio Tawadu [6]	Voltzberg	Ilha de Maracá	Tinigua	La Macarena [6]	Yasuní [7]	Cocha Cashu [7]	Lago Caiman [8]	# Sites [9]	Fruit size, max diam (mm) [9]	Seed size, max diam (mm) [9]	Fruit color [10]	Fruit protection [10]	% Lipids [11]	% CHO [11]
Acanthaceae	Mendoncia	–	–	–	–	–	FR	FR	–	FR	–	FR	–	–	**4**		8	Blue/Purple	No		
Anacardiaceae	Anacardium	–	–	–	–	–	FR	FR	FR	–	–	–	–	–	**3**						
Anacardiaceae	Antrocaryon	–	–	–	–	–	–	–	–	FR	–	–	–	–	**1**						
Anacardiaceae	Astronium	–	–	–	–	–	–	–	FR	–	–	–	–	–	**1**						
Anacardiaceae	Mangifera	–	–	–	–	FR	–	–	–	–	–	–	–	–	**1**						
Anacardiaceae	Metopium	FR	–	–	–	–	–	–	–	–	–	–	–	–	**1**						
Anacardiaceae	Spondias	FR	FR	–	FR	FR	FR	FR	FR	FR	FR	FR	FR	FR	**12**	35	25	Yellow	No	9.0	45.7
Anacardiaceae	Tapirira	–	FR	–	–	–	FR	–	–	–	–	–	–	–	**2**	21	15	Yellow	No		
Annonaceae	Annona	–	–	–	–	FR	–	–	–	–	–	–	–	–	**1**			Yellow	No		
Annonaceae	Bocageopsis	–	–	–	–	–	FR	–	–	–	–	–	–	–	**1**						
Annonaceae	Duguetia	–	–	–	–	–	–	FR	FR	FR	–	LE, FR	FR	–	**5**	80	18	Red	Yes		
Annonaceae	Ephedranthus	–	–	–	–	–	–	FR	–	–	–	–	–	–	**1**						
Annonaceae	Guatteria	–	–	–	–	FR	–	FR	–	FR	–	FR	FR	–	**5**	19	17	Blue/Purple	No	5.0	
Annonaceae	INDET	–	–	–	–	–	–	–	–	–	–	FL	–	–	**1**						
Annonaceae	Klarobelia	–	–	–	–	–	–	–	–	–	–	FR	–	–	**1**						
Annonaceae	Malmea	–	–	–	–	–	–	FR	–	FR	–	FR	–	–	**3**	50	20	Blue/Purple	Yes		
Annonaceae	Oxandra	–	–	–	–	–	–	–	–	FR	–	FR	FR	–	**3**	14	11	Blue/Purple	No	2.5	28.6
Annonaceae	Porcelia	–	–	–	–	–	–	–	–	–	–	FR	–	–	**1**						

Family	Genus										n			Color			
Annonaceae	Pseudomalmea	–	–	–	–	–	–	–	–	–	1						
Annonaceae	Rollinia	–	–	–	–	–	–	FR	–	FR	3	29	24	Green	Yes		
Annonaceae	Ruizodendron	–	–	–	–	–	–	FR	FR	FR, FL	2		7	Black	No		
Annonaceae	Unonopsis	–	–	FR	–	–	FR	FR	–	LE, FR	3	29	28	Green	No		
Annonaceae	Xylopia	–	–	FR	–	–	–	FR	FR	LE, FR	3			Red	No		
Apocynaceae	Aspidosperma	FL, SE, LE, FR	–	–	–	–	–	–	–	–	1						
Apocynaceae	Couma	–	–	FR	–	–	–	FR	–	–	1						
Apocynaceae	Geissospermum	–	–	FR	FR	–	–	FR	–	–	1						
Apocynaceae	Lacmellea	–	–	FR, LE	FR	–	–	–	–	–	2					9.4	54.3
Apocynaceae	Pacouria	–	–	–	FR	–	FR	FR	–	–	2			Yellow	Yes		
Apocynaceae	Parahancornia	–	–	FR	–	–	FR	–	–	–	1						
Apocynaceae	Stemmadenia	FR, FL	–	–	–	–	–	–	–	–	1			Mixed	No		
Apocynaceae	Tabernaemontana	–	–	–	FR	–	–	–	–	–	1			Mixed	No		
Araceae	Anthurium	–	LE, FR	–	LE, FR	–	–	–	–	LE, FR	3			Variable	No		
Araceae	Heteropsis	–	–	–	FR	–	FR, LE	–	–	–	2						
Araceae	INDET	FR	–	–	–	–	–	–	–	–	1						
Araceae	INDET (3 morphotypes)	LE	–	–	–	–	–	–	–	–	1						
Araceae	Monstera	–	LE, FR	–	FR	–	FR	LE	LE	–	3			Variable	No		
Araceae	Philodendron	LE	LE, FR	FR	LE, FR	–	LE, FR	FR	LE	–	5			Variable	No	13.5	38.8

(cont.)

Table 4.3 (*cont.*)

Family	Genus [2]	Punta Laguna [3]	Tikal [4]	Punta Santa Rosa [5]	Rio Claro [6]	BCI [6]	Rio Tawadu [6]	Voltzberg [6]	Ilha de Maracá	Tinigua	La Macarena [6]	Yasuní [7]	Cocha Cashu [7]	Lago Caiman [8]	# Sites [9]	Fruit size, max diam (mm) [9]	Seed size, max diam (mm) [9]	Fruit color [10]	Fruit protection [10]	% Lipids [11]	% CHO [11]
Araceae	Syngonium	–	–	–	–	LE	–	–	–	LE	–	–	–	–	2			Yellow	Yes		
Araliaceae	Dendropanax	–	FR, FL	FR	–	LE, FR	–	–	–	FR	–	–	–	–	4	13	6	Black	No		
Araliaceae	Oreopanax	–	–	–	–	LE	–	–	–	–	–	–	–	–	1					**21.0**	
Araliaceae	Schefflera (Didymopanax)	–	–	–	–	FL, LE	–	–	FR	–	–	FR	FR	–	4	10	3	Mixed	No		
Arecaceae	Astrocaryum	–	–	–	–	FR	–	–	FR	FR, FL	–	FR, INFL	–	–	4	80		Yellow	No		
Arecaceae	Attalea (Maximiliana)	–	–	–	FR	–	–	LE	FR	–	–	FR	–	FR	5			Variable	No	14.9	50.2
Arecaceae	Bactris	–	–	–	–	–	–	–	–	FR	–	–	–	–	1						
Arecaceae	Desmoncus	–	FL, FR	–	–	–	–	–	–	–	–	FR	–	–	2						
Arecaceae	Euterpe	–	–	–	–	FR	–	FR	–	FR	FR	–	FR	FR	6			Black	No	11.8	36.9
Arecaceae	INDET	–	LE, FR	–	–	–	–	–	–	LE	–	–	–	–	1						
Arecaceae	INDET	–	–	–	–	–	–	–	–	–	–	–	–	–	1						
Arecaceae	Iriartea	–	–	–	–	–	–	–	–	FR	FR	FR, INFL	FR, SE	–	4	40	25	Mixed	No	**25.0**	
Arecaceae	Mauritia	–	–	–	–	–	–	FR	FR	–	–	–	–	–	1			Mixed	No		
Arecaceae	Oenocarpus	–	–	–	FR	FR	–	–	FR	FR	FR	FR	–	–	7	50	40	Mixed	No		
Arecaceae	Scheelia	–	–	–	–	–	–	–	–	–	–	FR	FR	–	1						
Arecaceae	Socratea (Iriartea)	–	–	–	FR	FR	FR	FR	–	FR, SE	–	FR	FR	–	6	35	25	Mixed	No	0.8	75.0
Arecaceae	Syagrus	–	–	–	–	–	–	–	–	FR, FL	–	FR	–	–	2	35	32	Orange	No		

(cont.)

Family	Genus	1	2	3	4	5	6	7	8	9			N
Bignoniaceae	Adenocalymna	–	–	–	–	FL	–	–	–	–			**1**
Bignoniaceae	Anomocteriun	–	–	–	–	LE, FL, SE	–	–	–	–			**1**
Bignoniaceae	Arrabidaea	–	–	–	FL	FL	–	FL	FR, FL	–			**3**
Bignoniaceae	Distictella	–	–	–	–	FL	–	–	–	–			**1**
Bignoniaceae	INDET	–	–	–	–	–	–	LE	FL	–			**1**
Bignoniaceae	INDET	–	–	–	–	–	–	FL	–	–			**1**
Bignoniaceae	INDET	–	–	–	–	–	SE	–	–	–			**1**
Bignoniaceae	Jacaranda	–	–	–	–	–	–	–	–	–			**1**
Bignoniaceae	Memora cf.	–	–	–	–	–	–	–	FR	–			**1**
Bignoniaceae	Mussatia	LE	–	–	–	–	–	–	–	FR			**1**
Bignoniaceae	Paragonia	–	–	–	–	–	–	FL, LE	–	–			**1**
Bignoniaceae	Phryganocydia	–	–	–	FL	–	–	–	–	–			**1**
Bignoniaceae	Stizophyllum	–	–	–	FL, LE	–	–	–	–	–			**2**
Bignoniaceae	Tabebuia	–	FL	–	SE	–	–	–	–	–			**1**
Bignoniaceae	Tanaecium	–	–	–	FL	–	–	LE	–	–			**2**
Bignoniaceae	Xylophragma	FR	–	–	–	–	–	–	–	–			**2**
Bombacaceae	Bernoullia	LE, FL	–	–	LE, FL	–	–	LE, FL	–	–			**2**
Bombacaceae	Bombacopsis	FR	–	–	–	–	–	LE	–	–			**1**
Bombacaceae	Catostemma	–	FL, LE	FL, LE	LE	–	–	LE	–		1.7	70.4	**4**
Bombacaceae	Ceiba	–	–	–	–	–	–	FL, SE	–	–			**1**
Bombacaceae	Chorisia	–	–	–	–	–	–	FL, LE, SE	–	–			
Bombacaceae	Eriotheca	–	–	LE	–	–	LE	–	–	–			**2**
Bombacaceae	Huberoendron	–	–	–	–	–	–	FR, FL	–	–			**1**
Bombacaceae	INDET	–	–	–	–	–	LE	–	–	–			**1**

91

Table 4.3 (cont.)

Family	Genus [2]	Punta Laguna [3]	Tikal [4]	Punta Santa Rosa [5]	Rio Claro [6]	BCI [6]	Rio Tawadu [6]	Voltzberg	Ilha de Maracá	Timigua	La Macarena [6]	Yasuní [7]	Cocha Cashu [7]	Lago Caiman [8]	# Sites	Fruit size, max diam (mm) [9]	Seed size, max diam (mm) [9]	Fruit color [10]	Fruit protection [10]	% Lipids [11]	% CHO [11]
Bombacaceae	Matisia	–	–	–	–	–	–	–	–	–	–	FR	FR, LE	–	2	25	13	Variable	No	5.0	48.1
Bombacaceae	Ochroma	–	–	–	FL	–	–	–	–	–	–	–	–	–	1						
Bombacaceae	Pachira	–	–	–	–	–	FL, LE, SE	–	–	–	–	LE	–	–	2			Green	No		
Bombacaceae	Phragmotheca	–	–	–	–	–	–	–	–	–	–	FL	–	–	1						
Bombacaceae	Pseudobombax	–	–	–	–	FL	–	–	–	–	–	–	–	–	1						
Bombacaceae	Quararibea	–	–	–	–	FR	–	–	–	FR, LE	–	FR, FL	LE, FR, FL	–	4	61	29	Variable	No	14.0	
Boraginaceae	Cordia	–	–	–	–	FR	FR	FR	FR	–	–	FR	–	–	5	13	9	Variable	No		
Bromeliaceae	Achmea	–	–	–	–	LE	–	–	–	–	–	–	–	–	1						
Bromeliaceae	Billbergia	–	–	–	–	LE	–	–	–	–	–	–	–	–	1						
Bromeliaceae	Guzmania	–	–	–	–	FL	–	–	–	–	–	–	–	–	1						
Bromeliaceae	INDET	–	LE	–	–	–	–	–	–	–	–	–	–	–	1						
Burseraceae	Bursera	–	LE	FR	–	–	–	–	–	FR	FR	–	–	–	4			Blue/Purple	No	1.8	15.9
Burseraceae	Crepidospermum	–	–	–	–	–	–	–	FR	FR	–	–	–	–	2	25	15	Variable	No	0.4	8.2
Burseraceae	Dacryodes	–	–	–	–	FR	–	–	–	–	–	–	–	–	1			Black	No	1.8	34.2
Burseraceae	INDET	–	–	–	–	–	–	FR	–	–	–	–	–	–	1						
Burseraceae	Protium	FR	–	–	FR	FR	FR	FR	FR	FR	FR	FR	–	–	9	35	25	Mixed	No	3.4	41.4
Burseraceae	Tetragastris	–	–	–	–	FR	–	–	FR, SE	–	–	FR	–	–	4					3.6	56.5
Burseraceae	Trattinickia	–	–	–	–	–	–	FR, SE	FR	–	FR	–	–		3						

92

(cont.)

Family	Genus												Count	N1	N2	Color	Dom.	V1	V2
Cactaceae	Epiphyllum	–	–	–	FL	–	–	–	–	–	–	–	1			Red	No		
Cactaceae	Hylocereus	–	–	–	–	–	FR, FL	–	–	–	–	–	1			Red	No		
Cactaceae	INDET	LE	–	–	–	–	–	–	–	–	–	–	1						
Cactaceae	Pereskia	–	–	FR, FL	FL	–	FR, FL	–	–	–	–	–	2			Yellow	No		
Capparaceae	Capparis	–	–	–	–	FR	–	–	–	–	–	FR	1			Variable	No		
Capparaceae	Crateva	–	–	–	–	FR	–	–	–	–	–	–	1			Yellow	Yes		
Caricaceae	Jacaratia	FR	FR	–	FR	FR, LE	–	FR, FL, LE	–	FR	–	FL, LE, FR	6	110+	6	Orange	No	10.4	24.6
Cecropiaceae	Cecropia	LE, FR	FR, LE	FR	FR, LE	FR, LE	–	FR, LE, FL	–	FR, LE, FL	–	FR, LE, FR	9	20	2	Variable	No		
Cecropiaceae	Coussapoa	–	–	–	FR	FR	–	FR	–	FR	–	FR	5	27	2	Variable	No	8.1	33.7
Cecropiaceae	Pourouma	FR	–	FR	FR	FR	–	FR	–	FR, FL	–	FR	6	25	20	Blue/Purple	Variable	8.0	6.6
Celastraceae	Amplizoma	–	–	–	–	FR	–	–	–	FR	–	–	1						
Celastraceae	Cheiloclinium	–	FR	FR, SE	FR, SE	SE	–	FR	–	FR	–	FR	4	76	37	Mixed	No	8.0	70.9
Celastraceae	Maytenus	–	–	FR	FR	FR, SE	SE, FR	FR	–	–	–	FR	3						
Chrysobalanaceae	Couepia	–	–	–	FR	SE, FR	–	FR	–	–	–	–	2						
Chrysobalanaceae	Hirtella	–	FR, LE	–	–	–	–	–	–	–	–	–	1		10	Black	No		
Chrysobalanaceae	Licania	FR	–	FR	FR, LE, WD	FR, LE	FR	FR	–	FR	–	FR	6		9	Variable	No		
Chrysobalanaceae	Parinari	–	FR	FR	FR	–	–	–	–	–	–	–	2						
Clusiaceae	Calophyllum	–	–	FR	–	–	FR	–	FR	–	–	–	2						
Clusiaceae	Clusia	FR	–	FR, FL	FR, FL	–	FR	–	FR	–	–	FR	4	17	1	Mixed	No	5.0	
Clusiaceae	Garcinia	FR	FR	FR	–	FR	–	FR	–	FR	–	FR	4	80	30	Yellow	Yes	15.0	
Clusiaceae	Platonia	–	–	–	FR	–	–	–	–	–	–	–	1						
Clusiaceae	Rheedia	–	FR	FR	FR	–	FR	–	–	–	–	–	3						
Clusiaceae	Symphonia	–	–	–	FL	–	–	FL	–	–	–	–	1						

Table 4.3 (*cont.*)

Family	Genus [2]	Punta Laguna [3]	Tikal [4]	Punta Santa Rosa [5]	Rio Claro [6]	BCI [6]	Rio Tawadu [6]	Voltzberg	Ilha de Maracá	Tinigua	La Macarena [6]	Yasuní [7]	Cocha Cashu [7]	Lago Caiman [8]	# Sites [9]	Fruit size, max diam (mm) [9]	Seed size, max diam (mm) [9]	Fruit color [10]	Fruit protection [10]	% Lipids [11]	% CHO [11]
Combretaceae	Buchenavia	–	–	–	–	–	–	–	–	–	–	FR	–	–	1			Yellow	No		
Combretaceae	Combretum	–	–	–	–	–	–	SE	–	–	–	–	FL	–	2						
Combretaceae	Terminalia	–	–	–	–	LE	–	–	–	–	–	–	–	–	1						
Connaraceae	Cnestidium	–	–	–	–	LE	–	–	–	–	–	–	–	–	1			Mixed	No		
Convolvulaceae	Dicranostyles	–	–	–	–	–	FR	–	–	–	–	–	FR	FR	–	3	21		Green	No		
Convolvulaceae	INDET (2 morphotypes)	–	–	–	FR	–	–	–	–	–	–	–	–	–	1						
Convolvulaceae	Ipamoea	–	–	–	–	–	–	–	–	–	–	–	FR	–	1						
Convolvulaceae	Maripa	–	–	–	–	FR	–	SE	–	–	–	–	FR	–	3			Yellow	Yes		
Convolvulaceae	Operculina	–	–	–	–	–	–	SE	–	–	–	–	–	–	1						
Cucurbitaceae	Cayaponia	–	–	–	–	–	–	FR	–	FR	–	FR	–	–	3	22	11	Variable	Variable		
Cucurbitaceae	INDET (3 morphotypes)	–	–	–	–	–	–	–	–	–	–	FR	–	–	1						
Cucurbitaceae	Psiguria	–	–	–	–	FR	–	–	–	–	–	–	–	–	1			Green	Yes		
Cylcanthaceae	Asplundia	–	–	–	–	–	–	–	–	LE	–	–	–	–	1			Green	Yes		
Cylcanthaceae	Carludovica	–	–	–	–	LE	–	–	–	–	–	–	–	–	1			Red	No		
Dichapetalaceae	Tapura	–	–	–	–	–	–	–	–	–	–	FR	–	–	1			Green	No		
Dilleniaceae	Doliocarpus	–	–	–	–	FR	FR	–	–	–	–	FR	–	–	3			Blue/ Purple	No	3.9	68.3
Dioscoreaceae	Dioscorea	–	–	–	–	–	–	LE	–	–	–	LE	–	–	2						
Ebenaceae	Diospyros	–	–	–	–	–	–	–	–	–	–	FR	FR	–	2	32	20	Yellow	Yes		
Elaeocarpaceae	Muntingia	–	–	FR	–	–	–	–	–	–	–	–	–	–	1			Red	No		
Elaeocarpaceae	Sloanea	–	–	FR	–	–	–	–	–	–	–	FR	FR	–	3			Red	No		

Family	Genus	1	2	3	4	5	6	7	8	9	10	n
Euphorbiaceae	Alchornea	–	–	FR	–	–	–	–	–	–	–	1
Euphorbiaceae	Alchorneopsis	–	–	–	FR	–	–	–	–	–	–	1
Euphorbiaceae	Drypetes	–	–	–	FR	FR	–	–	LE	–	–	3
Euphorbiaceae	Hyeronima	–	–	FR	FR	–	FR	FR	FR	FR	–	7
Euphorbiaceae	Micrandra	–	–	FR	–	–	–	–	–	–	–	1
Euphorbiaceae	Paradrypetes	–	–	–	–	–	–	–	FR	FR	–	1
Euphorbiaceae	Richeria	–	–	–	–	–	–	–	FR	FR	–	1
Euphorbiaceae	Sapium	–	–	LE, FR	–	–	–	–	–	FR	FR	3
Fabaceae	Acacia	FL, LE	–	–	–	–	–	–	–	–	–	1
Fabaceae	Bauhinia	–	–	–	–	–	FL	–	LE	–	–	2
Fabaceae	Brownea	–	–	–	–	–	LE	–	–	–	–	1
Fabaceae	Cassia	–	–	–	–	–	–	–	FR	–	–	1
Fabaceae	Cedrelinga	–	–	–	SE	–	–	–	–	–	–	1
Fabaceae	Centrolobium	–	LE	–	–	LE	–	–	–	–	–	1
Fabaceae	Copaifera	–	–	–	FR	–	–	–	FR	–	–	2
Fabaceae	Cynometra	–	–	–	LE	–	–	–	–	–	–	1
Fabaceae	Dalbergia	–	–	–	–	–	LE, FL	–	LE	–	–	2
Fabaceae	Dialium	–	FR	FR	FR	FR	LE	–	–	–	–	4
Fabaceae	Dimorphandra	–	–	–	FR, WD	–	–	–	–	–	–	1
Fabaceae	Dioclea	–	–	–	SE	–	–	–	–	–	–	1
Fabaceae	Dipteryx	–	FR, FR	FR, FL	LE	–	FR	–	–	FR	–	5
Fabaceae	Enterolobium	LE, FR	–	FL	FL	–	FR	–	–	–	–	3
Fabaceae	Eperua	–	–	FL, LE	–	–	–	–	–	–	–	1
Fabaceae	Erythrina	–	–	–	–	–	–	–	FL, LE	–	–	1
Fabaceae	Fissicalyx	–	–	–	FL	–	FL	–	–	–	–	1

Red	No			**68.4**	21.7
Green	No	21			
Blue/Purple	No	6	2	6.9	42.7
Orange	No	30	28	1.0	47.6
Mixed	No		4		
Brown	Yes	20	8	3.9	65.5
Yellow	No	50	11	5.2	54.4
Black	Yes				

(cont.)

Table 4.3 (*cont.*)

Family	Genus [2]	Punta Laguna [3]	Tikal [4]	Punta Santa Rosa [5]	Rio Claro [6]	BCI [6]	Rio Tawadu [6]	Voltzberg	Ilha de Maracá	Tinigua	La Macarena [6]	Yasuní [7]	Cocha Cashu [7]	Lago Caiman [8]	# Sites [9]	Fruit size, max diam (mm) [9]	Seed size, max diam (mm) [9]	Fruit color [10]	Fruit protection [10]	% Lipids [11]	% CHO [11]
Fabaceae	Hymenaea	–	–	LE	–	–	–	–	LE, FL	–	–	FR	–	–	**3**	120+	32	Brown	Yes		
Fabaceae	Hymenolobium	–	–	–	–	–	–	LE	–	–	–	–	–	–	**1**						
Fabaceae	Inga	–	–	–	FR	FR	FR	FR, WD	FR, SE	FR	FR	FR	FR	FR	**10**	80+	28	Green	Yes	5.3	77.3
Fabaceae	Lecointea	–	–	–	–	–	–	–	–	–	–	–	FR	–	**1**						
Fabaceae	Machaerium	–	–	–	–	LE	–	–	–	–	–	LE	–	–	**2**						
Fabaceae	Myroxylum	–	–	–	–	–	–	–	–	–	–	FR, LE	–	–	**1**						
Fabaceae	Ormosia	–	–	–	–	–	–	SE	–	–	–	–	–	–	**1**						
Fabaceae	Parkia	–	–	–	–	–	–	FL, FR	–	–	–	–	–	–	**1**	340+	22	Black	Yes	9.0	
Fabaceae	Peltogyne	–	–	–	–	–	–	FR	–	–	–	–	–	–	**1**						
Fabaceae	Piptadenia	–	–	–	–	–	–	FL	–	SE	–	–	–	–	**2**						
Fabaceae	Pithecellobium	–	–	–	–	FR	–	FL, WD	–	–	–	–	–	–	**3**						
Fabaceae	Platymiscium	–	–	–	–	–	–	FL	–	–	–	–	–	–	**1**						
Fabaceae	Platypodium	–	–	–	–	LE, FR	–	–	–	FL, LE	–	–	–	–	**2**						
Fabaceae	Pterocarpus	–	–	–	–	–	–	LE	–	–	–	–	–	–	**1**						
Fabaceae	Swartzia	–	–	–	–	–	–	–	–	LE	–	–	–	–	**1**						
Fabaceae	Vataireopsis	–	–	–	–	–	–	LE	–	–	–	–	–	–	**1**		20	Green	Yes		
Fabaceae	Zollernia	–	–	–	–	–	–	–	FR	–	–	–	–	–	**1**						
Flacourtiaceae	Casearia	–	–	–	–	–	–	–	–	–	–	FR	FR	–	**2**			Red	No		
Flacourtiaceae	Hasseltia	–	–	–	FR	–	–	–	–	–	–	FR	–	–	**2**	5	2	Mixed	No		
Flacourtiaceae	Homalium	–	–	–	–	–	–	–	–	–	FR	–	–	–	**1**						
Flacourtiaceae	Laetia	–	–	–	–	–	FR	FR	FR, SE	FR	–	–	–	–	**4**	25	3	Mixed	No	**21.4**	**22.1**

96

Table (rotated 90° on the page; a portion of a larger continued table). Codes in the data columns: FR = fruit, FL = flower, LE = leaf, SE = seed, WD = wood; "–" indicates no record.

Family	Genus	Plant-part records (read across)	N	A	B	Colour	Cons.	%	%
Flacourtiaceae	Lindackeria	FR	1	3	–	Mixed	No	–	–
Flacourtiaceae	Pleuranthrodendron	FR, FL	1	2	–	Red	No	–	–
Flacourtiaceae	Tetrathylacium	FR, FL	1	–	–	–	–	–	–
Flacourtiaceae	Xylosma	LE, FR	1	–	–	–	–	–	–
Gesneriaceae	INDET	–	1	–	–	–	–	–	–
Gnetaceae	Gnetum	FR, SE	4	–	–	Red	No	1.1	53.6
Goupiaceae	Goupia	FR, LE	2	–	–	–	–	–	–
Gramineae	Chusquea	LE, FR	1	–	–	–	–	–	–
Hippocrataceae	Anthodon	FR	1	–	–	–	–	–	–
Hippocrataceae	Cuervea	FR	1	–	–	–	–	–	–
Hippocrataceae	Salacia	FR	2	10	–	Yellow	Yes	–	–
Humiriaceae	Humariastrum	FR	1	–	–	–	–	–	–
Humiriaceae	Sacoglottis	FR, WD	2	–	–	–	–	3.1	69.8
Humiriaceae	Vantanea	FR	1	–	–	–	–	–	–
Icacinaceae	Calatola	FR	1	–	–	–	–	–	–
Icacinaceae	Leretia	FR	2	–	–	–	–	–	–
Lauraceae	Aniba	FR	1	–	–	Mixed	No	–	–
Lauraceae	Beilschmiedia	FR	1	–	–	–	–	–	–
Lauraceae	Endlicheria	FR	2	14	–	Mixed	No	–	–
Lauraceae	INDET (12 morphotypes)	FR	1	–	–	–	–	–	–
Lauraceae	INDET (6 morphotypes)	FR	1	–	–	–	–	–	–
Lauraceae	Licaria	LE, FR	4	9	–	Green	No	–	–
Lauraceae	Nectandra	WD, FR, FL	3	–	–	–	–	–	–
Lauraceae	Ocotea	FR	7	31	14	Variable	No	**23.3**	12.0
Lauraceae	Persea	FR	1	–	–	–	–	–	–
Lauraceae	Rhodostemonodaphne	SE, WD, FR	2	15	–	Mixed	No	–	–
Lecythidaceae	Couratari	SE, WD	1	–	–	–	–	–	–

(cont.)

Table 4.3 (cont.)

Family	Genus [2]	Punta Laguna [3]	Tikal [4]	Punta Santa Rosa [5]	Rio Claro [6]	BCI [6]	Rio Tawadu [6]	Voltzberg	Ilha de Maracá	Tinigua	La Macarena [6]	Yasuní [7]	Cocha Cashu [7]	Lago Caiman [8]	# Sites	Fruit size, max diam (mm) [9]	Seed size, max diam (mm) [9]	Fruit color [10]	Fruit protection [10]	% Lipids [11]	% CHO [11]
Lecythidaceae	Eschweilera	–	–	–	–	–	–	SE, FR	SE	–	–	–	–	–	2	56	18	Brown	No		
Lecythidaceae	Gustavia	–	–	–	–	FR, LE	–	FR	–	FR	–	–	–	–	3	60	18	Variable	Yes		
Lecythidaceae	Lecythis	–	–	–	–	–	–	FR	SE	–	–	–	–	–	2						
Loganiaceae	Strychnos	–	–	–	–	–	–	FR	FR	–	–	FR	FR	–	4	29	18	Yellow	Yes	8.0	33.8
Loranthaceae	INDET	–	–	–	–	–	–	LE	–	–	–	–	–	–	1						
Loranthaceae (Santalaceae)	Phoradendron	–	–	–	–	–	–	–	–	FR	–	–	–	–	1			Red	No		
Lythraceae	Lafoensia	–	–	–	–	–	–	–	–	–	FR	–	–	–	1						
Malpighiaceae	Bunchosia	–	–	–	–	–	–	–	–	–	–	FR	–	–	1						
Malpighiaceae	Byrsonima	–	–	–	FR	FR	–	–	FR	FR	–	FR	–	FR	6	15	5	Yellow	No		
Marcgraviaceae	INDET	–	–	–	–	–	–	–	–	–	FR	–	–	–	1						
Marcgraviaceae	Marcgravia	–	–	–	–	FR	–	–	–	–	–	–	FR	–	2			Mixed	No		
Marcgraviaceae	Norantea	–	–	–	–	–	–	FL	–	–	–	–	–	–	1			Mixed	No		
Marcgraviaceae	Souroubea	–	–	–	FR	–	–	FR	–	–	–	FR	–	–	3		2	Mixed	No		
Melastomataceae	Bellucia	–	–	–	–	–	–	–	–	FR	–	–	–	FR	2	6	1	Green	No		
Melastomataceae	Henrietella	–	–	–	–	–	–	–	–	FR	–	–	–	–	1			Green	No		
Melastomataceae	INDET	–	–	FR	–	–	–	–	–	–	–	–	–	–	1						
Melastomataceae	Miconia	–	–	–	FR	–	–	–	–	FR	–	–	–	–	2		1	Variable	No		
Melastomataceae	Mouriri	–	–	–	FR	–	–	–	–	–	–	FR	–	–	2	23	12				
Meliaceae	Carapa	–	–	–	–	–	–	FL, SE, LE	–	–	–	–	–	–	1						

Family	Genus														Color			
Meliaceae	Cedrela	–	–	–	FR	–	SE	–	–	LE, SE	–	**3**					17.5	
Meliaceae	Guarea	–	FR	FR	FR, LE, FL	FR	FR, FL	–	FR	–	–	**4**	59	24	Mixed	No	4.1	50.2
Meliaceae	Trichilia	–	FR	FR	FR	FR	FR	–	FR	FR	–	**6**	25	22	Mixed	No		
Menispermaceae	Abuta	–	FR	FR	FR	FR	FR	–	FR, LE	–	–	**5**			Yellow	Yes		
Menispermaceae	Anomospermum	–	–	FR	–	–	–	–	–	FR, LE	–	**2**					1.5	48.3
Menispermaceae	Borismene	–	–	–	–	–	–	–	–	FR	–	**1**						
Menispermaceae	INDET	–	–	–	–	–	–	FR	–	FR	–	**1**						
Menispermaceae	Odontocarya	–	–	–	–	–	–	–	–	FR	–	**1**			Yellow	No		
Menispermaceae	Orthomene	–	–	–	–	–	–	–	–	FR	–	**1**						
Menispermaceae	Telitoxicum	–	–	–	–	–	–	FR	–	–	–	**1**						
Moraceae	Bagassa	–	–	–	–	FR	FR	–	FR	–	–	**2**						23.7
Moraceae	Batocarpus	–	–	–	–	–	–	–	FR	LE, FR	–	**2**		9	Variable	Variable		
Moraceae	Brosimum	FR, LE	FR	FR, LE, FL	FR	FR	FR, SE	FR	FR	FL, LE, FR, SE	FR, FL	**12**	30	13	Variable	No	0.6	75.5
Moraceae	Castilla	–	–	–	–	FR	–	–	–	–	–	**1**	70	13	Yellow	Yes	0.8	11.5
Moraceae	Clarisia	–	FR	FR	FR	FR	FR	FR	FR	FL, LE, FR	–	**7**	46	31	Variable	No	8.4	70.4
Moraceae	Ficus	FR	FR, LE	FR, LE	FR	FR	FR, LE	FR	FR	FR, LE	–	**13**	45	1	Variable	No	2.9	29.6
Moraceae	Helicostylis	–	–	–	FR	FR	FR	–	FR	–	–	**4**	30	8	Yellow	No	5.9	50.4
Moraceae	INDET	–	–	–	–	–	LE	–	–	–	–	**1**						
Moraceae	Maclura	–	–	–	–	–	–	–	–	LE, FR	–	**1**						
Moraceae	Maquira	–	LE	LE	–	–	LE	–	FR	LE	–	**4**			Yellow	Yes		

(cont.)

Table 4.3 (*cont.*)

Family	Genus [2]	Punta Laguna [3]	Tikal [4]	Santa Rosa [5]	Punta Rio Claro [6]	BCI [6]	Rio Tawadu [6]	Voltzberg	Ilha de Maracá	Tinigua	La Macarena [6]	Yasuní [7]	Cocha Cashu [7]	Lago Caiman [8]	# Sites [9]	Fruit size, max diam (mm) [9]	Seed size, max diam (mm) [9]	Fruit color [10]	Fruit protection [10]	% Lipids [11]	% CHO [11]
Moraceae	Naucleopsis	–	–	–	–	–	–	–	–	–	–	FR, LE	–	–	1	82	16	Green	Yes		
Moraceae	Olmedia	–	–	–	–	FR	–	–	–	–	–	–	–	–	1						
Moraceae	Perebea	–	–	–	–	–	FR	–	–	FR, SE	–	FR	FR	–	4	49	12	Variable	No		
Moraceae	Poulsenia	–	–	–	–	LE, FR	–	–	–	–	–	LE	LE	–	3						
Moraceae	Pseudolmedia	–	–	–	FR	–	FR	–	–	FR, SE	FR	FR	LE, FR, SE	FR	7	25	14	Variable	No	5.5	4.2
Moraceae	Sorocea	–	–	–	–	FR, LE	–	–	–	–	–	–	LE, FR	–	2	9	6	Blue/ Purple	No		
Moraceae	Trymatococcus	–	–	–	–	–	–	FR	–	–	–	–	–	–	1						
Myristicaceae	Compsoneura	–	–	–	–	–	–	–	–	–	–	FR	–	–	1						
Myristicaceae	Iryanthera	–	–	–	FR	–	–	–	–	–	–	FR	–	–	2	58	33	Mixed	No	23.5	46.9
Myristicaceae	Otoba	–	–	–	–	–	–	–	–	–	–	FR	FR	–	2					6.7	49.8
Myristicaceae	Virola	–	–	–	FR	FR	FR	FR	FR, SE	FR	FR	FR	FR	–	9	48	27	Mixed	No	43.4	18.7
Myrtaceae	Campomanesia	–	–	–	–	–	–	FR	FR	–	–	–	–	–	1			Brown	Yes		
Myrtaceae	Eugenia	–	–	–	–	FR	–	FR	–	–	–	FR	FR	–	6	26	15	Variable	No	5.2	85.5
Myrtaceae	INDET	–	–	–	–	–	–	–	–	–	–	FR	–	–	1						
Myrtaceae	Myrcia	–	–	–	–	–	–	–	FR	–	–	FR	FR	–	3						
Myrtaceae	Pimienta	–	–	FR	–	–	–	–	–	–	–	–	–	–	1						
Myrtaceae	Plinia	–	–	–	–	–	–	–	–	–	–	FR	–	–	1	49	28	Blue/ Purple	Yes		

100

Table (continued). Plant families and genera with food-part records and fruit characteristics. (Columns are sparse; "–" indicates no record; FR = fruit, LE = leaf, FL = flower.)

Family	Genus	N												n₁	n₂	Colour	Eaten	%
Myrtaceae	Psidium	2	–	–	–	–	–	–	FR	–	–	–	–					
Myrtaceae	Syzygium	1	–	–	–	–	–	–	FR	–	–	–	–					
Nyctaginaceae	Guapira	1	–	–	–	–	–	–	FR	FR	–	–	–	23	18	Variable	No	
Nyctaginaceae	Neea	2	–	–	–	–	–	–	–	–	–	–	–			Variable	No	
Ochnaceae	Ouratea	1	–	–	–	FR	–	–	–	FR	FR	–	–			Mixed	No	**38.6**
Olacaceae	Heisteria	4	–	–	–	FR	–	–	FR	FR	FR	FR	–	18	9	Mixed	No	
Olacaceae	Minquartia	3	–	–	–	–	–	–	–	FR	FR	FR	–					
Opiliaceae	Agonandra	2	–	–	–	–	–	–	–	FR	FR	FR	–					
Passifloraceae	Passiflora	3	–	–	–	FR	–	FR	FR	FR	FR	FR	–			Variable	Yes	
Phytolaccaceae	Trichostigma	3	–	–	–	–	–	LE	FR	FR	FR	FR	–	5	3	Mixed	No	
Piperaceae	Peperomia	3	–	–	–	LE	–	LE	LE	LE	LE	–	–			Green	No	
Poaceae	Guadua	1	–	–	–	LE	–	LE	LE	LE	–	–	–					
Polygalaceae	Moutabea	1	–	FR	–	FR	–	–	–	–	–	–	–					36.2
Polygonaceae	Coccoloba	5	–	FR	–	FR	–	–	FR	FR	LE	LE	–	13	9	Variable	No	3.4
Polygonaceae	Triplaris	1	–	–	LE	–	–	–	–	–	LE	LE	–					
Polypodiaceae	Polypodium	1	–	–	–	–	–	–	–	–	–	–	–					
Rhamnaceae	Rhamnidium	1	–	–	–	–	–	–	FR	FR	FR	–	–					
Rhamnaceae	Ziziphus	2	–	–	–	–	–	–	FR	FR	FR	LE, FR	–					
Rosaceae	Prunus	4	FR	–	–	–	–	–	FR	FR	FR	FR	–	37	10	Blue/Purple	No	12.0
Rubiaceae	Alibertia	2	–	–	–	FR	–	–	FR	FR	FR	–	–			Black	No	
Rubiaceae	Coussarea	2	–	FR	–	FR	–	–	–	–	–	–	–					
Rubiaceae	Duroia	1	–	–	FR	–	–	–	–	FR	–	–	–			Yellow	Yes	
Rubiaceae	Faramea	2	–	FR	–	FR	–	–	–	FR	–	–	–					
Rubiaceae	Genipa	4	–	FR	–	FR	–	–	FR	FR	FR	FR	–	120	14	Brown	Yes	
Rubiaceae	Guettarda	3	–	FR	–	FR	–	–	FR	FR	–	–	–			Black	No	
Rubiaceae	Hillia	1	–	FL	–	–	–	–	–	–	–	–	–					
Rubiaceae	INDET	1	FR	–	–	–	–	–	–	–	–	–	–					

(cont.)

Table 4.3 (*cont.*)

Family	Genus [2]	Punta Laguna [3]	Tikal [4]	Punta Santa Rosa [5]	Rio Claro [6]	BCI [6]	Rio Tawadu [6]	Voltzberg	Ilha de Maracá	Tinigua	La Macarena [6]	Yasuní [7]	Cocha Cashu [7]	Lago Caiman [8]	# Sites [9]	Fruit size, max diam (mm) [9]	Seed size, max diam (mm) [9]	Fruit color [10]	Fruit protection [10]	% Lipids [11]	% CHO [11]
Rubiaceae	Palicourea	–	–	–	–	–	–	–	–	–	–	FR	–	–	1	19	13				
Rubiaceae	Pentagonia	–	–	–	–	–	–	–	–	–	–	FR	–	–	1	42	2	Brown	No		
Rubiaceae	Posoqueria	–	–	–	FR	FR	–	–	–	–	–	–	–	–	2			Yellow	Yes		
Rubiaceae	Psychotria	–	–	–	–	FR	–	–	–	–	–	–	–	–	1			Variable	No		
Rubiaceae	Randia	–	–	–	–	FR	–	–	–	–	–	–	–	–	1			Yellow	Yes		
Rubiaceae	Tocoyena	–	–	–	–	FR	–	–	–	–	–	–	–	–	1						
Rutaceae	Zanthoxylum	–	–	–	–	FR	–	LE	–	–	–	–	–	–	2						
Sapindaceae	Allophylus	–	–	–	–	FR	–	–	–	–	–	FR	FR	–	3	11	8	Variable	No	7.0	
Sapindaceae	Cupania	–	FR	–	–	–	–	–	–	LE, FR	–	FR	FR	–	4		9	Mixed	No		
Sapindaceae	Dildendron	–	–	–	FR	–	–	–	–	–	–	–	–	–	1						
Sapindaceae	Dipterodendron	–	–	FR	–	–	–	–	–	–	–	–	–	–	1						
Sapindaceae	Matayba	–	–	–	–	–	–	–	–	–	–	FR	–	–	1						
Sapindaceae	Paullinia	–	–	–	–	FR	FR	FR, LE, SE	FR	FR	–	FR	FR	–	7	44	22	Mixed	No	10.0	
Sapindaceae	Serjania	–	–	–	–	LE	–	–	–	–	–	–	–	–	1						
Sapindaceae	Talisia	–	FR	–	–	FR	FR	FR	–	FR	–	FR	–	–	6	28	17	Variable	Yes		
Sapotaceae	Achrouteria	–	–	–	–	–	FR	FR	–	–	–	–	–	–	2						
Sapotaceae	Chrysophyllum (Prieurella)	–	–	–	–	FR	–	FR	FR	FR	FR	FR	FR	–	7	55	30	Variable	No		45.3
Sapotaceae	Ecclinusa	–	–	–	–	–	–	FR	FR	–	–	FR	–	–	3						
Sapotaceae	Elaeoluma	–	–	FR	–	–	FR	–	–	–	–	FR	FR	–	2						
Sapotaceae	Manilkara	FR	FR, LE	–	–	–	–	–	–	–	–	FR	FR	–	5						

Family	Genus	1	2	3	4	5	6	7	8	9	Sp			Color			
Sapotaceae	Mastichodendron	–	FR	–	–	–	–	–	–	–	**1**						
Sapotaceae	Micropholis	–	–	FR	–	FR	–	–	FR	–	**3**					3.5	48.8
Sapotaceae	Pouteria	–	FR	FR	–	FR, SE	FR	–	FR	FR	**9**	65	41	Variable	Yes	15.0	53.1
Sapotaceae	Pradosia	–	–	–	–	FR, SE, LE	–	FR	–	–	**1**						
Sapotaceae	Prieurella	–	–	–	–	FR, SE	–	–	–	–	**1**						
Sapotaceae	Pseudocladia	–	–	FR	–	–	–	–	–	–	**1**						
Sapotaceae	Sarcaulus	–	–	–	–	–	FR	–	FR	–	**2**	25	20	Yellow	Yes	2.5	21.2
Sapotoceae	Sideroxylon	FR	–	–	–	–	–	–	–	–	**1**						
Simaroubaceae	Quassia	–	–	–	–	FR, WD	–	–	–	–	**1**						
Simaroubaceae	Simarouba	–	–	FR	–	–	FR	–	FR	–	**3**	18	16	Black	No	0.2	78.4
Smilacaceae	Smilax	–	–	LE, FL	–	–	–	–	–	–	**1**			Orange	No		
Solanaceae	Cestrum	–	–	–	–	–	FL, FR, LE	–	–	–	**1**			Black	No		
Solanaceae	INDET	–	–	–	–	–	–	–	FL	–	**1**						
Solanaceae	Solanum	–	FR	FR	–	FR	FR	–	–	–	**3**	37	2	Variable	No		24.1
Sterculiaceae	Byttneria	–	–	–	–	–	–	–	SE	SE	**1**						
Sterculiaceae	Guazuma	FR	–	–	FR	FR	–	–	FR	FR	**5**	21	2	Black	No		
Sterculiaceae	INDET	–	–	–	–	LE	–	–	LE	–	**1**						
Sterculiaceae	Sterculia	–	–	FR	FL	FR, FL	–	–	FL	FL	**4**	59	17	Variable	No		
Sterculiaceae	Theobroma	–	–	FR	–	FR	FR	–	LE	LE	**3**	130	25	Variable	Yes		
Styracaceae	Styrax	–	–	–	FR	FR	–	–	–	–	**1**						
Tiliaceae	Apeiba	–	–	LE, FR	FR, FL	FR, FL	–	–	FR	–	**5**	80	4	Black	Yes		
Tiliaceae	Mortoniodendron	–	FR	–	–	–	–	–	–	–	**1**						
Tiliaceae	Pentaplaris	–	–	–	–	–	–	–	LE, FR	–	**1**						
Ulmaceae	Ampelocera	–	–	–	LE, FL, FR	–	–	–	LE, FR	LE, FR	**4**	22	13	Yellow	No		
Ulmaceae	Celtis	–	–	FR, LE	–	FR	FR	–	FR, SE	–	**4**	18	7	Variable	No	4.0	31.8

(cont.)

Table 4.3 (cont.)

Family	Genus [2]	Punta Laguna [3]	Tikal [4]	Punta Santa Rosa [5]	Rio Claro [6]	BCI [5]	Rio Tawadu [6]	Voltzberg [7]	Ilha de Maracá [8]	Tinigua [9]	La Macarena [6]	Yasuní [7]	Cocha Cashu [7]	Lago Caiman [8]	# Sites [9]	Fruit size, max diam (mm) [9]	Seed size, max diam (mm) [9]	Fruit color [10]	Fruit protection [10]	% Lipids [11]	% CHO [11]
Ulmaceae	Trema	–	–	–	–	–	–	–	–	FR	–	–	–	–	1			Red	No		
Urticaceae	Urera	–	–	–	–	–	–	–	–	–	–	–	FR	–	1			Variable	No		
Verbenaceae	Aegiphila	–	–	–	–	FR	–	–	–	–	–	–	–	–	1			Blue/Purple	No		
Verbenaceae	Petrea	–	–	–	–	–	–	–	–	–	–	LE	–	–	1						
Verbenaceae	Vitex	–	FR	–	–	–	–	–	FR	FR	–	FR	–	–	4	16		Green	No		
Violaceae	Leonia	–	–	–	–	–	–	FR	FR	FR	–	FR	FR	–	5	70	18	Brown	Yes	2.2	7.6
Vitaceae	Cissus	–	–	–	–	–	–	–	FR	FR	–	FR	FR	–	4	11	4	Variable	No	5.0	
Vitaceae	Vitis	–	–	–	FR	–	FR	–	–	–	–	–	–	–	2		4				
Vochysiaceae	Qualea	–	–	–	–	–	–	FL	–	–	–	–	–	–	1						
	Min # Genera [12]	9	24	12	27	103	58	134	56	103	23	151	95	17							
	Source	[1]	[2]	[3]	[4]	[5]	[6]	[7]	[8]	[9]	[10]	[11]	[12]	[13]							

[1] Parts eaten are scored as 'FR' for fruits (both ripe and immature, though for all genera the vast majority of fruits are consumed ripe), 'FL' for flowers (including flower buds, nectar, and inflorescences), 'WD' for wood and bark, 'SE' for seeds, and 'LE' for leaves (both young and mature, though for all genera the vast majority of leaves are consumed when young). The category 'LE' also includes other plant vegetative parts such as leaf buds, meristems, and petioles. Where more than one dietary item is noted per genus, those items are listed in order of relative importance as best as can be determined from the data in the original sources. Sites are presented in order of decreasing latitude.

[2] Names in parentheses are either accepted synonymns or subsumed generic names for the noted genus, which are used in one or more of the original sources (*Maximiliana* = *Attalea*, *Prieurella* = *Chrysophyllum*, *Didymopanax* = *Schefflera*) or reflect the generic name used in an original source for a species more conventionally assigned to the listed genus. For example, van Roosmalen (1985) recognized '*Iriartea exorrhiza*' rather than '*Socratea exorrhiza*' recognized at other sites.

[3] The original source lists only the top ten species in the diet, but these account for >85% of feeding records. One unidentified morphospecies accounting for ~2% of feeding records is not listed here.

[4] The diet list for this site excludes several items that could not be identified or for whom only a local name was recorded. We have scored the genus for one item for which the original source reported only an easily recognized and widely used local name – i.e. *Prunus* (Rosaceae) for "capulin."

[5] A complete diet list is not given in the original source, only items that comprise ≥1.0% of the annual diet, thus other genera may contibute minimally to the *Ateles* diet at this site.

[6] The diet list for this site included in the original source lists only fruit species so a number of genera providing leaves and flowers are undoubtedly not reported.

[7] The original source does not allow relative proportions of different food items from the same genus to be evaluated, so these are presented in the order given in the original source.

[8] A complete diet list is not given in the original source, only items that comprise ≥0.5% of the annual diet, thus other genera may contibute minimally to the *Ateles* diet at this site.

[9] For each genus, the maximum recorded dimension, length or breadth for any morphospecies within the genus is noted, combining data from Stevenson *et al.* (2002) and Di Fiore and Link, unpublished data. Note that this combines data from multiple study sites. For seed dimensions, only genera whose seeds are routinely swallowed and passed intact are included.

[10] Fruit color and fruit protection data for each genus are compiled from Link and Stevenson (2004), supplemented by Di Fiore and Link (unpublished data). For color, 'mixed' means that mature fruits may be more than one color, while 'variable' signifies that fruits of different species within the genus are different colors. For protection, 'variable' signifies that at least one different species within the genus is protected while others are not.

[11] Nutritional data are compiled from Castellanos and Chanin (1996), Simmen and Sabatier (1996), Machado and Pizo (2000), L. Arnedo (unpublished data), A. Derby (unpublished data), and Di Fiore (unpublished data). Where multiple species per genus were examined in a study, the average value across species is noted. Where more than one study provides data on a genus, the midrange of values reported across studies is noted.

[12] Scored as the number of genera plus INDET morphotypes noted at the site, thus value may exceed the total number of nonblank cells.

Sources: [1] Ramos-Fernández and Ayala-Orozco (2003); [2] Cant (1977); [3] Chapman (1989); [4] Riba-Hernández *et al.* (2003); [5] Campbell (2000); [6] Castellanos (1997); [7] van Roosmalen (1985); [8] Nunes (1998); [9] Stevenson *et al.* (2002), Link, unpublished data; [10] Klein and Klein (1977); [11] Dew (2005), Suarez (2006), Di Fiore and Link, unpublished data; [12] Symington (1987), [13] Wallace (2005).

Table 4.4 *Correlations between dietary diversity measures and habitat variables*

	Log min # fruit morphospecies in diet	Log min # plant morphospecies in diet	Log min # fruit morphospecies comprising ≥1% of total diet	Log min # fruit genera comprising ≥1% of total diet	% of total diet from top 5 morphospecies	% of total diet from top 5 genera	# Genera comprising ≥50% of total diet
Log absolute latitude *Pearson Correlation*	−0.65	−0.60	−0.47	−0.51	0.70	0.64	−0.72
Sig. (2-tailed)	0.06	0.12	0.20	0.13	0.04	0.07	0.02
# sites	9	8	9	10	9	9	10
Log annual rainfall *Pearson Correlation*	**0.67**	**0.79**	**0.90**	**0.90**	**−0.64**	**−0.70**	**0.57**
Sig. (2-tailed)	0.05	0.02	0.00	0.00	0.06	0.05	0.11
# sites	9	8	8	9	9	8	9

Data used come from Table 4.2. For sites where data from multiple studies were available (i.e. Tinigua, Yasuní), data from the longest study (which typically yielded the highest diversity measures) were used. Thus, for Tinigua, data were taken from Link (unpublished data) and for Yasuní, data were taken from Suarez (2006). Significant correlations are identified in bold text, but note that all nonsignificant correlations are large and in the same, expected direction (i.e. greater diversity at lower latitudes and in sites with greater rainfall).

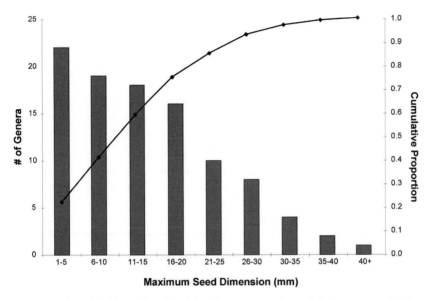

Figure 4.2 Distribution of seed sizes (average or midrange of the largest seed width) for genera consumed by spider monkeys from across their geographic range. More than 25% of genera consumed have a seed size of >20 mm.

endosperm per fruit, $n = 20$). Seed spitting in Yasuní is likewise rare; spider monkeys are known to systematically spit the seeds of only two plant species, both of which are palms (*Iriartea deltoidea* and *Socratea exorrhiza*) from which large seeds can be easily separated from the pulp and discarded (Link and Di Fiore, 2006). This pattern of ripe fruit preference and rare seed predation and spitting is common across populations and species of spider monkeys (van Roosmalen and Klein, 1988; Russo *et al.*, 2005; Wallace, 2005; Di Fiore and Campbell, 2007).

Data on additional physical and chemical characteristics of the fruits consumed by spider monkeys, where available, are compiled in Table 4.3. Fruits eaten by spider monkeys range from being very small (e.g. *Hyeronima*) to very long (e.g. *Inga, Parkia*) or large (e.g. *Cheiloclinium*), and spider monkeys routinely swallow and defecate seeds that range in size from ~1 mm (e.g. *Ficus*) to over 30 mm (e.g. *Cheiloclinium, Iryanthera, Oenocarpus, Pouteria*) (Dew, 2005; Russo *et al.*, 2005; A. Di Fiore and A. Link, unpublished data). In fact, 25% of genera whose seeds are commonly swallowed by spider monkeys have maximal seed dimensions of more than 20 mm (Figure 4.2). Fruits whose seeds are in this size range are generally not dispersed by other mammalian frugivores, thus spider monkeys are likely to be among the most significant dispersers of many species of large-seeded Neotropical plants (Stevenson, 2002; Link and Di

Fiore, 2006). Indeed, Dew (2005) reports that spider monkeys in Yasuní swallowed and dispersed larger seeds than did sympatric woolly monkeys (*Lagothrix lagotricha poeppigii*), which are similar in body size and likewise highly frugivorous (for a more in-depth discussion of seed dispersal by *Ateles*, see Dew, this volume).

Many of the fleshy fruits consumed by *Ateles* tend to be relatively rich in lipids (Table 4.3), which may help to explain the coexistence of spider monkeys, at some sites, with other heavily frugivorous primates such as woolly monkeys (Di Fiore, 1997; Stevenson *et al.*, 2000; Dew, 2005). For example, despite gross similarity in the ripe fruit proportion of the spider and woolly monkey diet in Yasuní (87% and 73%, respectively), spider monkeys are four times more likely to consume lipid-rich fruits (e.g. from the families Arecaeae, Lauraceae, Meliaceae, and Myristicaceae) than are woolly monkeys (Dew, 2005).

Leaves and flowers

Although leaves and flowers make up a relatively small proportion (~10%: range ~7% to ~20%) of the overall diet of well-studied spider monkey populations, they can be seasonally important feeding sources, in some months representing the bulk of the food consumed, particularly during periods of ripe fruit scarcity. For example, in Tinigua National Park (La Macarena), Colombia, young leaves and flowers represent more than half of the diet of *Ateles belzebuth belzebuth* during the end of the rainy season, a period characterized by low fruit availability at this site (A. Link, unpublished data). Similarly, more than 36% of the diet of *A. b. chamek* in northeastern Bolivia consists of leaves during the middle of the dry season, a time of ripe fruit scarcity (Wallace, 2005), and a comparable pattern is seen for *A. paniscus* in Suriname (Van Roosmalen, 1985).

The diversity of leaves included in the diet is generally much lower that that of fruits. Van Roosmalen and Klein (1988) note that spider monkeys in Suriname eat leaves from 28 plant species while consuming fruits more than six times that number of species. In Tinigua, *A. b. belzebuth* eat the leaves of 33 species of plants (in the 71% of feeding bouts on leaves where taxonomic identification was possible), while they consume the fruits of at least 117 different species (A. Link, unpublished data). Together, only three genera (*Dalbergia, Brosimum* and *Peperomia*) account for more than 39% of total leaf consumption at this site. The diversity of leaves eaten by spider monkeys is significantly greater during periods of fruit scarcity, suggesting that they are indeed used as fallback items as phenological data demonstrate that leaf flush is not particularly common during this period (Stevenson, 2002). Most of the leaves consumed in Tinigua come from large canopy trees (e.g. from the families Moraceae, Bombacaceae), but epiphytes (Araceae, Piperaceae) and lianas (Bignonianceae) are also important leaf sources. Similarly, over a roughly 1-year period in Yasuní National Park, Ecuador, *A. b. belzebuth* fed on leaves from only five species of plants but on fruits from 238 different species (Suarez, 2006). It is likely that leaf consumption is underestimated or underreported in many studies, as many bouts of leaf feeding tend to be short and leaves are more difficult to identify taxonomically. Nonetheless, the pattern of much lower diversity in the foliage versus fruit component of the diet is robust.

Although flowers represent only a small proportion of the overall diet of spider monkeys at all sites (~5%: range ~0% to ~14%), they, too, can be seasonally important items. In Tinigua, spider monkeys eat flowers from at least 17 different species (A. Link, unpublished data). Moreover, during the period of fruit scarcity at Tinigua, flowers are heavily consumed (comprising up to 35% of the monthly diet), and just five species make up ~65% of the total time annually that spider monkeys devoted to eating flowers. The flowers consumed come mostly from large canopy trees (i.e. *Platypodium elegans*) and from several lianas in the families Bignoniaceae and Fabaceae. In both Tinigua and Yasuní, spider monkeys also feed extensively on immature flowers of several palm species (e.g. *Astrocaryum chambira*, *Iriartea deltoidea*, *Socratea exorrhiza*), some of which they extract from the tough sheaths surrounding developing inflorescences (A. Di Fiore and A. Link, unpublished data).

Animal prey

In terms of both time spent eating them and diversity of taxa consumed, insects and other animal prey constitute a very small proportion of the annual diet of *Ateles*. In fact, prey consumption by spider monkeys is limited mainly to a few species of caterpillars, meliponid bees and termites (Link, 2003). Across sites, the only consistency seen in spider monkey faunivory is the occasional consumption of a few species of caterpillars that "massively" bloom in specific host trees. Consumption of caterpillars has been reported in several study populations (Cant, 1977; van Roosmalen, 1985; Chapman, 1987; Symington, 1987; Campbell, 2000; Link, 2003), and all observations have been reported to be limited to within a short time frame, typically only one to two weeks each year. Feeding bouts on caterpillars can be very long (~ 50 min), during which time spider monkeys actively search for larvae within a single tree crown. Larvae are first unrolled from the leaves; the supporting branch is then pulled to the mouth and the larvae removed directly with the mouth. Meliponid bees are eaten by spider monkeys at both Tinigua and Yasuní (Link, 2003; Pozo Rivera, 2004). At Tinigua, spider monkeys have been observed feeding on bees from two different nests, although the majority of observations were restricted to a single nest. To capture bees, the monkeys position themselves within arm's reach of the nest and actively disturb it with their hands. The monkeys then capture the bees as they emerge and become entangled in the monkeys' hair. An average feeding bout on bees lasts 6.8 ± 3.6 min ($n = 8$ bouts), and feeding rates are estimated at 18.0 ± 6.2 bees/min ($n = 9$ bouts), thus the monkeys could consume well over 100 bees in a typical bout (Link, 2003). Multiple monkeys, up to three, have been observed feeding simultaneously on bees, and most individuals in a subgroup wait for a turn to access this resource. Occasionally during these feeding bouts, spider monkeys would actively disturb the bee nest, reactivating the bees' activity. Termite consumption by *Ateles* was first reported by van Roosmalen (1985), who suggested that spider monkeys might preferentially select certain types of termites. In other studies reporting the consumption of termite nests, monkeys have not been observed eating termites directly, and this behavior is most commonly interpreted as a form of geophagy (Klein and Klein, 1977; Castellanos and Chanin, 1996; Link, 2003; Dew, 2005).

Geophagy

Spider monkeys in some populations in western Amazonia regularly visit terrestrial mineral licks or "saladeros" where they eat soil and occasionally drink water from small streams (Izawa, 1993; Dew, 2005; Link et al., 2006; Suarez, 2006). As noted by Campbell et al. (2005), it is still uncertain why this behavior in spider monkeys is almost entirely restricted to South American sites. Groups of spider monkeys that regularly visit mineral licks are extremely vigilant and cautious before coming down to the ground to visit the mineral licks, and will both scan for predators and rest for extended periods (up to four hours) around mineral lick sites before accessing them for very short time periods (generally ~ 1 min per individual) (Suarez, 2003; Link et al., 2006). Although several nonexclusive hypotheses have been suggested for the adaptive and ecological significance of geophagy in primates, mineral supplementation, pH buffering and toxin absorption have received most support (see Krishnamani and Mahaney, 2000 and references therein). Izawa (1993) and Dew (2005) have analyzed the chemical composition of soils from mineral licks and termitaries used by spider monkeys and compared them with control soil samples taken from other parts of the monkeys' range. In general, no marked differences in mineral concentrations between soil samples from mineral licks and termitaria compared with control sites have been found, though Dew (2005) notes that soils eaten by spider monkeys had somewhat higher phosphorus (K) concentrations and Izawa (1993) reports higher sodium (Na) concentration in water collected at saladeros when compared with control samples. To date, ecologists' understanding of the significance of geophagy by vertebrates is still unclear, even though many taxa in addition to spider monkeys (e.g. tapirs, peccaries, deers, armadillos, rodents, parrots, guans and howler monkeys) regularly visit mineral licks in the Neotropics.

Decayed wood

Decayed wood represents an additional important item in the diet of at least some spider monkey groups in northwestern Amazonian rain forests (Klein and Klein, 1977; Suarez, 2003; A. Di Fiore, A. Link and S. Spehar, unpublished data) and at other sites (van Roosmalen, 1985). For example, Klein and Klein (1977) report that wood from several specific dead or decaying trees made up ~9% of the diet of a group of *Ateles belzebuth belzebuth* in La Macarena, Colombia, and they inferred that some of these trees might have been used for several years. However, several other groups of spider monkeys at a nearby site in Tinigua National Park have not been observed eating large amounts of decayed wood (A. Link unpublished data). Decayed wood has also been observed to be a major item in the diet of one group of *A. b. belzebuth* studied in Yasuní National Park, Ecuador, since the mid 1990s (Suarez, 2003; A. Di Fiore, A. Link and S. Spehar, unpublished data). Interestingly, the first studies on this group did not report spider monkeys eating decayed wood (Pozo Rivera, 2004), although Dew (2001) started to observe a few decayed wood feeding events in 1995. By 1999, however, decayed wood was the second most important item in the diet of this group (after ripe fruits), when it accounted for 9.6% of total feeding time (Suarez, 2003, 2006). Nearly all dead wood consumption took place in a single "dead standing" tree, particularly from one of its low, lateral branches. By 2002 to 2003, spider monkeys had literally eaten most of the branches of the tree, and only the remaining tree trunk (at the time, ~15 m in height) was still standing. During that year, decayed wood accounted for 3.6% of the spider monkeys' feeding time, and spider monkeys frequently visited that single tree and spent large amounts of time feeding on it. By 2005, the dead tree trunk was only about 8 m high and was still being used by spider monkeys, although access to it was becoming more difficult as it was starting to be disconnected from the surrounding canopy trees (A. Di Fiore and A. Link, unpublished data).

Diet, foraging ecology, and the social organization of spider monkeys

The social organization of spider monkeys is tied tightly to their diet and foraging strategies. As described elsewhere in this volume, spider monkeys tend to travel and feed in small parties that comprise a flexible subset of the members of a larger community (Klein, 1972; Cant, 1977; van Roosmalen, 1985; White, 1986; Symington, 1988b; Ahumada, 1989; Chapman, 1990; Chapman *et al.*, 1995; Shimooka, 2003). This pattern of *fission–fusion sociality* is typically interpreted as an adaptation that allows spider monkeys to focus their diets on scarce and patchily distributed high-quality food resources, particularly ripe fruits (Klein and Klein, 1977; Wrangham, 1980, 1987; Symington, 1990; Di Fiore and Campbell, 2007). Consistent with this interpretation, several studies of spider monkeys have demonstrated that the size of foraging parties is positively correlated with the habitat-wide availability of ripe fruits (Symington, 1987, 1988b; Chapman *et al.*, 1995). Below, we review some aspects of the foraging behavior and diet of spider monkeys that are relevant to understanding the fission–fusion sociality that characterizes the genus.

Characteristics of feeding patches and feeding bouts

Despite the small average size of their foraging parties, spider monkeys tend to feed from large forest trees or from epiphytic plants supported by large trunks. Nunes (1998) compared the size distribution of *Ateles belzebuth belzebuth* feeding trees in Roraima, Brazil, with that based on a floristic sample at the same site and found that spider monkey feeding trees were significantly larger. A similar pattern was seen for *A. b. belzebuth* in Yasuní where the average diameter at breast height (DBH) of spider monkey feeding trees was more that twice that of trees located in a series of botanical transects (47 ± 28 cm versus 22 ± 22 cm) (A. Di Fiore and A. Link, unpublished data).

Spider monkeys also generally use upper levels of the forest for feeding. In a long-term study of *Ateles paniscus* in Suriname, for example, van Roosmalen (1985) found that 77% of fruits eaten were obtained in the higher levels of the canopy (>25 m), while 22% of fruits consumed came from the subcanopy level (15 to 25 m), and less than 1% of the fruits were obtained from the understory (<15 m). Similarly, in Yasuní, *A. b. belzebuth* feed primarily from trees and lianas in the subcanopy, canopy and emergent levels of the forest (mean $= 16.6 \pm 4.9$ m: Dew, 2001; mean $= 23 \pm 6.9$ m: Suarez, 2003; mean $= 21 \pm 2.5$ m: Pozo Rivera, 2004).

Long-term studies on the ecological strategies of spider monkeys in Yasuní provide additional data on the characteristics of feeding patches and feeding bouts. First, feeding patches used by spider monkeys in Yasuní tend to be located far from one another. Suarez (2006) found that the distance between successively visited spider monkey food patches in Yasuní averages 420 m, which is far greater than the average distance between successive patches fed in by

Figure 4.3 An adult female *Ateles belzebuth belzebuth* gorging on *Oenocarpus* palm fruits at the Tiputini Biodiversity Station in Yasuní National Park, Ecuador. (Photo: A. Link.)

groups of sympatric (and also highly frugivorous) woolly monkeys (*Lagothrix lagotricha poeppigii*; A. Di Fiore, unpublished data). However, the estimated size of the fruit crops of trees fed in by spider monkeys is not statistically different from those fed in by woolly monkeys (A. Di Fiore and A. Link, unpublished data).

Second, although *Ateles* feeding bouts in Yasuní ranged from less than one to 45 minutes in length, most are very short in duration. During two separate study periods of ~1 year, for example, average feeding bout length was 8.1 ± 10.6 min (median = 5 min) (1998 to 1999: Suarez, 2006) and 6.6 ± 6.0 min (median = 5 min) (2002 to 2003: A. Di Fiore and A. Link, unpublished data). Despite the short length of most feeding bouts, spider monkeys can nevertheless ingest large numbers and large volumes of fruits very rapidly (Figure 4.3). For example, Klein and Klein (1977) report the example of a feeding bout in which a single spider monkey ate more than 100 *Pouteria* fruits (which have seeds that average 20–25 mm in size) in less than seven minutes. Similarly, Link and De Luna (2004) estimate that spider monkeys at Tinigua can ingest up to 1.35 kg of *Oenocarpus bataua* fruits in a single feeding bout (average 0.29 kg/bout,

$n = 90$ bouts), which translates to more than 15% of the body weight of a typical animal with much of that weight comprising indigestible seeds. Dew (2005) also notes a similar bout of *Oenocarpus* feeding by spider monkeys in Yasuní, in which one individual consumed more than 70 fruits – and the equivalent of ~10% of its body weight in indigestible seeds – in a single 21-minute feeding bout.

Morphological adaptations to frugivory

Spider monkeys show a number of anatomical and physiological characteristics associated with their highly frugivorous diets. Perhaps most obvious of these are the skeletal modifications associated with their habitual semibrachiating pattern of locomotion – i.e. reduction of the thumb; elongation of the tail, limbs, and phalanges; more dorsal position and vertical orientation of the scapula (Hill, 1962; Erikson, 1963; see also Rosenberger *et al.*, this volume) – which are presumably adaptations that allow them to travel rapidly and efficiently between widely dispersed feeding patches (Cant, 1977; Cant *et al.*, 2003). *Ateles* also have broad, spatulate incisors, which are used in picking and tasting fruits and in peeling away the thick husks that protect the mesocarps or arils of many consumed species, and small molars with low, rounded cusps, which are a hallmark of primate species with highly frugivorous diets.

In addition, spider monkeys possess a simple, unelongated digestive tract and have fast gut passage times relative to their body size (Milton, 1981, 1984; Lambert, 1998). This allows them to follow a foraging strategy of rapidly ingesting large numbers of individual fruits with minimal postincisive processing, extracting the easily absorbed carbohydrates and lipids, and then clearing the gut of its "ballast" of indigestible seeds (Link and Di Fiore, 2006). Dew (2005), in fact, argues that the more gracile jaws and teeth of spider monkeys relative to those of closely related woolly monkeys (*Lagothrix* spp.) can be seen as adaptation "to a specialization on ripe fruits swallowed whole with a minimum of processing" (p. 1129). Such a "high-throughput" fruit foraging strategy essentially requires that spider monkeys specialize in fruits of high nutritional quality.

Seasonal and daily variation in diet

Seasonal changes in the availability of high-quality resources influence the composition and diversity of spider monkey diets. For example, in two different populations of *Ateles belzebuth chamek*, the proportion of the monthly diet consisting of ripe fruit was positively related to indices describing the habitat-wide availability of this resource ($r = 0.75, p < 0.01$: Symington, 1990; $r = 0.70$, $p < 0.05$: Wallace, 2005). In addition, leaf consumption was neg-atively correlated with the availability of ripe fruit in these two populations, suggesting that the monkeys include lower-quality plant vegetative parts in the diet when preferred higher-quality resources are scarce. Stevenson *et al.* (2000) found similar patterns between fruit availability on the one hand and both ripe fruit and leaf consumption for *A. b. belzebuth* in Colombia (ripe fruit: $r = 0.89$, $p < 0.01$; leaves and other vegetative parts: $r = -0.71$: $p < 0.01$). For *A. b. belzebuth* in Brazil, too, Nunes (1998) found a similar relationship between ripe fruit consumption and the availability of this resource ($r = 0.484, p < 0.01$), and also noted that the taxonomic breadth of the fruit diet likewise correlated positively with fruit availability ($r = 0.751, p < 0.01$).

Spider monkey diets also vary over the course of the day. Typically, there are several peaks in fruit consumption during the day, in the morning and afternoon, with less activity around midday (van Roosmalen, 1985; Castellanos and Chanin, 1996). In several populations, leaf consumption tends to peak late in the afternoon (van Roosmalen, 1985; Chapman and Chapman, 1991; Castellanos and Chanin, 1996). Chapman and Chapman (1991) argue that spider monkeys may opt to focus on leaves – which are the primary source of protein in the diet but require long retention times to process (Lambert, 1998) – at the end of the day to take advantage of the long, "obligated" period of overnight resting when animals cannot devote time to searching for more readily digested, higher quality fruits. Such a strategy also would help reduce the costs of transporting a "ballast" of partially digested leaves during diurnal travel, when the gut could more efficiently be filled with fruits.

Cross-site comparisons

Recently, Russo *et al.* (2005) compared the dietary preferences of spider mon-keys at four Neotropical sites from across the geographic range of the genus: Barro Colorado Island (Panama), Tinigua National Park (Colombia), Yasuní National Park (Ecuador), and Voltzberg Nature Reserve (Suriname). Both across sites and within the same site over time, only a few plant genera – *Brosimum* (Moraceae), *Ficus* (Moraceae), *Cecropia* (Cecropiaceae), and *Virola* (Myris-ticaceae) – are consistently preferred and/or consumed in large proportions. The authors speculate that intersite differences in the consumption of particular plant taxa can be attributed to (1) variability across sites in the composition and abundance of different plant taxa, (2) differences between spider monkey species in dietary preferences, and (3) differences in the set of sympatric and competing frugivores at each site. In addition, they suggest that interannual variation in the diet of spider monkeys at one site, Yasuní, was attributable to superannual phenological cycles for particular plant taxa.

Here, we broaden and revise these between- and within-site comparisons of *Ateles* diets by expanding the data set used by Russo *et al.* (2005) to include additional sites. Thus, Table 4.3 compares the taxonomic composition of *Ateles* diets at 13 different sites from across the geographic range of the genus (Figure 4.1), and Table 4.5 summarizes data on the relative importance of different plant genera and families for eight different sites where *Ateles* foraging ecology has been studied long-term. Table 4.5 also provides a perspective on interannual variation in the diet at two of these sites, Tinigua and Yasuní. A review of Table 4.3 highlights some of the points made above. First, sites differ greatly in dietary diversity (as measured in terms of the number of genera included in the diet), with diets being much more diverse at sites located close to the Equator. Second, very few genera are consumed by spider monkeys at more than a handful of sites. For example, more than 50% of the genera listed in Table 4.3 are consumed at only one site, and almost 90% of the genera are consumed at four or fewer sites. Only 15 genera (4.3% of those listed) from 11 families are consumed at more than half of the sites, with only *Ficus* (Moraceae), *Brosimum* (Moraceae), *Spondias* (Anacardiaceae), and *Inga* (Fabaceae) being consumed in more than 75% of sites (Figure 4.4). These results echo those of Russo *et al.* (2005) concerning the broad regional importance of several of these genera in the *Ateles* diet.

To crudely examine dietary similarity across the set of sites, we calculated Jaccard's coefficient based on the lists of genera consumed for every pairwise combination of sites, where

$$S_{ij} = \text{(number of genera consumed in both sites i and j)/(total number of genera consumed in either site i or j).}$$

We then used a neighbor-joining clustering algorithm to link sites where the genera of plants consumed by spider monkeys were more similar (Figure 4.5a). Not unexpectedly, this analysis links sites in the western Amazon (Yasuní, Tinigua, Cocha Cashu), two sites on the Guianan shield (Rio Tawadu and Voltzberg), and three sites from Central America (Punta Laguna, Tikal, and Santa Rosa), all regions within which forests are likely to be similar in terms of floristic composition. The only odd grouping in this analysis joins Punta Rio Claro, La Macarena, and Lago Caiman, and, for the first two of these sites, only genera contributing to the fruit diet rather than the over-all diet were available for inclusion in Table 4.3. Nonetheless, this analysis also reveals that diets in even the most similar sites are not really very similar in terms of their generic composition. For example, Jaccard's coefficient for even the two most similar sites, Yasuní and Tinigua, was only 0.363: that is, only ~36% of the genera recorded in the diets of spider monkeys

Table 4.5 *Genera and families of plants comprising at least 1% of the total diet for spider monkeys at eight sites in Central and South America*

Tikal [1]

Genus	Family	# Species	% of diet	Parts eaten	Family	# Genera	# Species	% of diet
Brosimum	(Moraceae)	1	56.3	FR, SE, LE	Moraceae	2	4	64
Ficus	(Moraceae)	3	7.8	FR, LE	Anacardiaceae	1	1	7
Spondias	(Anacardiaceae)	1	7.2	FR	Araliaceae	1	1	6
Dendropanax	(Araliaceae)	1	5.7	FR	Sapotaceae	2	2	5
Manilkara	(Sapotaceae)	1	4.1	FR, LE	Rosaceae	1	1	2
INDET 1	–	1	2.0	FR	Sapindaceae	2	2	1
Prunus	(Rosaceae)	1	1.7	FR				
Cupania	(Sapindaceae)	1	1.0	FR				
INDET 3	–	1	1.0	LE				
TOTAL		11	86.8			9	11	84.6

(*cont.*)

117

Table 4.5 (*cont.*)

Punta Rio Claro		# Species	% of diet [3]	Part eaten
Mortoniodendron	(Tiliaceae)	1	23.7	FR
Elaeoluma	(Sapotaceae)	1	22.5	FR
Inga	(Fabaceae)	1	3.8	FR
Byrsonima	(Malpighiaceae)	1	2.4	FR
Spondias	(Anacardiaceae)	1	2.3	FR
Virola	(Myristicaceae)	1	2.3	FR
Souroubea	(Marcgraviaceae)	1	1.8	FR
Lacmellea	(Apocynaceae)	1	1.8	FR
Jacaratia	(Caricaceae)	1	1.5	FR
INDET	(Convolvulaceae)	1	1.4	FR
Humiriastrum	(Humiriaceae)	1	1.4	FR
Pseudolmedia	(Moraceae)	1	1.3	FR
Pouteria	(Sapotaceae)	1	1.2	FR
Dendropanax	(Araliaceae)	1	1.2	FR
Clusia	(Clusiaceae)	1	1.2	FR
Brosimum	(Moraceae)	1	1.2	FR
INDET	(Convolvulaceae)	1	1.1	FR
TOTAL		17	72.1	

(*cont.*)

Table 4.5 (*cont.*)

Barro Colorado Island

Genus	Family	# Species	% of diet	Parts eaten	Family	# Genera	# Species	% of diet
Quararibea	(Bombacaceae)	1	10.6	FR	Moraceae	6	12	16.2
Spondias	(Anacardiaceae)	2	8.9	FR	Bombacaceae	3	4	10.9
Virola	(Myristicaceae)	1	7.3	FR, LE	Anacardiaceae	2	3	8.9
Poulsenia	(Moraceae)	1	7.3	FR	Myristicaceae	1	1	7.3
Hyeronima	(Euphorbiaceae)	1	5.7	FR	Euphorbiaceae	3	3	6.3
Doliocarpus	(Dillenaceae)	2	4.8	FR	Arecaceae	3	3	5.7
Ficus	(Moraceae)	7	6.1	FR	Rubiaceae	6	6	5.2
Astrocaryum	(Arecaceae)	1	4.0	FR	Dillenaceae	1	2	4.8
Brosimum	(Moraceae)	1	3.7	FR, LE, FL	Burseraceae	2	3	4.6
Chrysophyllum	(Sapotaceae)	2	3.5	FR	Sapotaceae	2	3	4.0
Tetragastris	(Burseraceae)	1	3.4	FR	Fabaceae	3	4	3.4
Abuta	(Menispermaceae)	1	3.2	FR	Menispermaceae	1	1	3.2
Trichilia	(Meliaceae)	1	2.8	FR	Araceae	4	7	3.1
Philodendron	(Araceae)	4	2.5	FR, LE, RT	Meliaceae	1	1	2.8
Coussarea	(Rubiaceae)	1	2.3	FR	Myrtaceae	3	4	2.3
Cecropia	(Cecropiaceae)	2	2.2	FR, LE	Cecropiaceae	1.0	2	2.2
Dipteryx	(Fabaceae)	1	2.1	FR, FL	Lecythidaceae	1	1	1.6
Faramea	(Rubiaceae)	1	2.0	FR	Lauraceae	1	1	1.4
Oenocarpus	(Arecaceae)	1	1.6	FR				
Gustavia	(Lecythidaceae)	1	1.6	FR, LE				
Belschmiedia	(Lauraceae)	1	1.4	FR				
Protium	(Burseraceae)	2	1.2	FR				
Syzygium	(Myrtaceae)	1	1.1	FR				
Eugenia	(Myrtaceae)	2	1.1	FR				
Platypodium	(Fabaceae)	1	1.1	FR, LE				
TOTAL		40	91.7			44	61	93.9

(*cont.*)

Table 4.5 (cont.)

Voltzberg genus	Surinam family	# Species	% of diet	Parts eaten
Virola	(Myristicaceae)	2	10.8	FR
Inga	(Fabaceae)	12	9.5	FR, WD
Guarea	(Meliaceae)	2	4.3	FR, FL
Tetragastris	(Burseraceae)	2	3.7	FR
Ecclinusa	(Sapotaceae)	1	3.3	FR
Cecropia	(Cecropiaceae)	2	3.2	FR, LE
Dimorphandra	(Fabaceae)	1	3.1	FR, WD
Philodendron	(Araceae)	2	2.8	LE, FR
Bagassa	(Moraceae)	1	2.8	FR
Achrouteria	(Sapotaceae)	1	2.6	FR
Chrysophyllum (Prieurella)	(Sapotaceae)	2	2.2	FR
Laetia	(Flacourtiaceae)	1	2.2	FR
Protium	(Burseraceae)	2	1.8	FR
Cordia	(Boraginaceae)	3	1.6	FR
Vataireopsis	(Fabaceae)	1	1.6	LE
Ephedranthus	(Annonaceae)	1	1.5	FR
Guatteria	(Annonaceae)	1	1.4	FR
Brosimum	(Moraceae)	2	1.4	FR
Licania	(Chrysobalanaceae)	3	1.3	FR, LE, WD
Cayaponia	(Cucurbitaceae)	2	1.2	FR
Clarisia	(Moraceae)	1	1.0	FR
Couratari	(Lecythidaceae)	2	1.0	SE
Apeiba	(Tiliaceae)	4	1.0	FR, FL
Rheedia	(Clusiaceae)	2	1.0	FR
Clusia	(Clusiaceae)	5	1.0	FR, FL
TOTAL		58	67.5	

Family	# Genera	# Species	% of diet
Fabaceae	18	32	16.3
Myristicaceae	1	2	10.8
Sapotaceae	5	7	8.5
Moraceae	8	21	7.1
Meliaceae	4	6	5.8
Burseraceae	2	4	5.6
Cecropiaceae	3	8	4.6
Annonaceae	5	5	3.6
Araceae	3	4	3.1
Flacourtiaceae	1	1	2.2
Bignoniaceae	≥ 7	≥ 11	2.1
Clusiaceae	4	9	2.1
Chrysobalanaceae	3	6	1.8
Bombacaceae	2	2	1.8
Lecythidaceae	4	7	1.7
Boraginaceae	1	3	1.6
Cucurbitaceae	1	2	1.2
Convolvulaceae	3	4	1.1
Celastraceae	3	6	1.1
Tiliaceae	1	4	1.0
Cactaceae	2	3	1.0
TOTAL	39	47	95.8

(cont.)

Table 4.5 (cont.)

Ilha de Maraca

Genus	Family	# Species	% of diet	Parts eaten	Family	# Genera	# Species	% of diet
Pradosia	(Sapotaceae)	1	28.9	FR, LE	Sapotaceae	5	6	38.9
Ocotea	(Lauraceae)	2	9.8	FR	Moraceae	4	8	13.5
Ficus	(Moraceae)	3	9.2	FR	Lauraceae	1	2	9.8
Tetragastris	(Burseraceae)	1	7.6	FR, SE	Burseraceae	5	6	8.2
Eugenia	(Myrtaceae)	1	7.5	FR	Myrtaceae	3	3	8.1
Pouteria	(Sapotaceae)	2	5.3	FR, SE	Anacardiaceae	3	3	4.1
Ecclinusa	(Sapotaceae)	1	4.4	FR	Rubiaceae	4	4	3.0
Spondias	(Anacardiaceae)	1	3.7	FR	Malpighiaceae	1	1	3.0
Brosimum	(Moraceae)	3	3.1	FR, LE	Fabaceae	5	6	1.6
Byrsonima	(Malpighiaceae)	1	3.0	FR	Arecaceae	4	4	1.6
Cordia	(Boraginaceae)	1	1.5	FR	Boraginaceae	1	1	1.5
Guettarda	(Rubiaceae)	1	1.3	FR	Chrysobalanaceae	2	2	1.4
Licania	(Chrysobalanaceae)	1	1.2	FR	Celastraceae	1	1	1.1
Attalea (Maximiliana)	(Arecaceae)	1	1.2	FR				
Maytenus	(Celastraceae)	1	1.1	FR, SE				
Genipa	(Rubiaceae)	1	1.1	FR, SE				
TOTAL		22	89.9			39	47	95.8

(cont.)

Table 4.5 (*cont.*)

Timigua (1990–1991)

Genus	Family	# Species	% of diet [3]	Parts eaten
Ficus	(Moraceae)	6	16.5	FR
Oenocarpus	(Arecaceae)	2	10.3	FR
Virola	(Myristicaceae)	2	7.4	FR
Gustavia	(Lecythidaceae)	1	7.4	FR
Protium	(Burseraceae)	4	6.2	FR
Sarcaulus	(Sapotaceae)	1	5.9	FR
Brosimum	(Moraceae)	4	5.5	FR, SE
Cecropia	(Cecropiaceae)	2	4.7	FR
Pourouma	(Cecropiaceae)	3	4.3	FR
Pseudolmedia	(Moraceae)	3	4.1	FR, SE
Spondias	(Anacardiaceae)	2	2.9	FR
Coussapoa	(Cecropiaceae)	1	2.5	FR
Trichilia	(Meliaceae)	2	2.3	FR
Astrocaryum	(Arecaceae)	1	2.0	SE
Inga	(Fabaceae)	6	1.9	FR
Socratea	(Arecaceae)	1	1.4	FR, SE
Ocotea	(Lauraceae)	2	1.4	FR
Laetia	(Flacourtiaceae)	1	1.4	FR
Leonia	(Violaceae)	1	1.3	FR
Enterolobium	(Fabaceae)	1	1.3	FR
Crepidospermum	(Burseraceae)	1	1.2	FR
TOTAL		47	92.0	

Family	# Genera	# Species	% of diet
Moraceae	6	16	27.0
Arecaceae	5	6	15.4
Cecropiaceae	3	6	11.5
Burseraceae	2	5	7.4
Myristicaceae	1	2	7.4
Lecythidaceae	1	1	7.4
Sapotaceae	2	2	6.4
Fabaceae	2	7	3.2
Anacardiaceae	1	2	2.9
Meliaceae	1	2	2.3
Lauraceae	1	2	1.4
Flacourtiaceae	1	1	1.4
Violaceae	1	1	1.3
TOTAL	27	53	95.04

(*cont.*)

Table 4.5 *(cont.)*

Tinigua (2001–2002)

Genus	Family	# Species	% of diet	Parts eaten
Ficus	(Moraceae)	9	8.1	FR, LE
Cecropia	(Cecropiaceae)	2	6.8	FR, LE, FL
Gustavia	(Lecythidaceae)	1	6.8	FR
Oenocarpus	(Arecaceae)	2	6.2	FR
Trichilia	(Meliaceae)	1	5.4	FR
Brosimum	(Moraceae)	4	4.4	FR
Pseudolmedia	(Moraceae)	3	4.0	FR
Spondias	(Anacardiaceae)	2	3.1	FR
Dalbergia	(Fabaceae)	1	2.7	LE, FL
Jacaratia	(Caricaceae)	1	2.5	FR, FL, LE
Bursera	(Burseraceae)	1	2.4	FR
Fissicalyx	(Fabaceae)	1	2.4	FL
Pourouma	(Cecropiaceae)	3	2.4	FR
Ocotea	(Lauraceae)	2	2.3	FR
Virola	(Myristicaceae)	4	2.1	FR
Astrocaryum	(Arecaceae)	1	1.8	FL, FR
Phryganocydia	(Bignoniaceae)	1	1.7	FL, LE
Coccoloba	(Polygonaceae)	2	1.4	FR
Talisia	(Sapindaceae)	1	1.3	FR
Pouteria	(Sapotaceae)	2	1.2	FR
Syagrus	(Arecaceae)	1	1.2	FR, FL
Peperomia	(Piperaceae)	1	1.1	LE
TOTAL		46	71.3	

Family	# Genera	# Species	% of diet
Moraceae	≥7	≥20	18.3
Arecaceae	≥8	≥9	10.5
Cecropiaceae	3	6	9.3
Lecythidaceae	1	1	6.8
Meliaceae	1	1	5.4
Fabaceae	6	11	4.7
Anacardiaceae	2	3	3.5
Burseraceae	4	7	3.1
Lauraceae	3	4	2.6
Caricaceae	1	1	2.5
Myristicaceae	1	4	2.1
Sapotaceae	3	4	2.0
Sapindaceae	2	3	1.7
Polygonaceae	1	2	1.4
Piperaceae	1	1	1.1
Annonaceae	5	5	1.0
TOTAL	≥49	≥82	76.0

(cont.)

Table 4.5 (*cont.*)

Yasuní (1995–1996)

Genus	Species	# Species	% of diet	Parts eaten	Family	# Genera	# Species	% of diet
Spondias	(Anacardiaceae)	1	14.0	FR	Anacardiaceae	1	1	14.0
Virola	(Myristicaceae)	4	9.4	FR	Myristicaceae	2	7	12.1
Iriartea	(Arecaceae)	1	5.9	FR, IMM SE	Arecaceae	≥3	≥3	9.4
Persea/Ocotea	(Lauraceae)	3	5.9	FR	Lauraceae	≥2	5	5.9
Matisia	(Bombacaceae)	3	5.2	FR	Bombacaceae	≥2	≥3	5.3
Guarea	(Meliaceae)	1	4.1	FR	Meliaceae	2	3	5.1
Oenocarpus	(Arecaceae)	1	3.2	FR	Cecropiaceae	2	3	4.0
Cecropia	(Cecropiaceae)	1	2.7	FR	Moraceae	3	5	2.6
Iryanthera	(Myristicaceae)	1	2.7	FR	Fabaceae	1	5	1.5
Naucleopsis	(Moraceae)	1	1.8	FR	Annonaceae	≥2	≥3	1.5
Guatteria/ Oxandra	(Annonaceae)	4	1.5	FR	Caricaceae	1	1	1.3
Inga	(Fabaceae)	9	1.5	FR	Sapindaceae	2	3	1.0
Pourouma	(Moraceae)	3	1.3	FR				
TOTAL		33	59.2			≥23	≥42	63.7

(*cont.*)

Table 4.5 (*cont.*)

Yasuní (1998–1999)

Genus	Family	# Species	% of diet	Parts eaten	Family	# Genera	# Species	% of diet
Virola	(Myristicaceae)	8	11.1	FR	Moraceae	8	26	16.3
Coccoloba	(Polygonaceae)	3	6.0	FR	Myristicaceae	3	10	14.2
Naucleopsis	(Moraceae)	2	5.3	FR	Polygonaceae	1	3	6.7
Matisia	(Bombacaceae)	3	4.8	FR	Lauraceae	≥1	13	5.0
Byrsonima	(Malpighiaceae)	1	4.0	FR	Bombacaceae	2	3	4.8
Hyeronima	(Euphorbiaceae)	1	3.5	FR	Cecropiaceae	3	9	4.5
Brosimum	(Moraceae)	3	3.4	FR	Anacardiaceae	2	2	4.5
Tapirira	(Anacardiaceae)	1	3.3	FR	Arecaceae	6	6	4.3
Inga	(Fabaceae)	10	3.3	FR, LE	Malpighiaceae	1	1	4.0
Cecropia	(Cecropiaceae)	3	3.3	FR	Fabaceae	4	13	3.8
Pseudolmedia	(Moraceae)	4	3.1	FR	Euphorbiaceae	3	3	3.7
Iryanthera	(Myristicaceae)	1	3.0	FR	Ulmaceae	2	3	2.7
Ampelocera	(Ulmaceae)	1	2.6	FR, LE	Meliaceae	2	≥9	2.6
Clarisia	(Moraceae)	1	2.2	FR	Sapotaceae	7	22	2.4
Iriartea	(Arecaceae)	1	2.2	FR, INFL	Annonaceae	7	15	2.0
Schefflera	(Araliaceae)	1	1.8	FR	Araliaceae	1	1	1.8
Simarouba	(Simaroubaceae)	1	1.6	FR	Simaroubaceae	1	1	1.6
Trichilia	(Meliaceae)	3	1.5	FR	Rubiaceae	2	2	1.5
Alibertia	(Rubiaceae)	1	1.5	FR	Sterculiaceae	3	4	1.1
Pouteria	(Sapotaceae)	12	1.4	FR				
Guatteria	(Annonaceae)	7	1.3	FR				
Batocarpus	(Moraceae)	1	1.2	FR				
Spondias	(Anacardiaceae)	1	1.2	FR				
Guarea	(Meliaceae)	≥3	1.1	FR				
Oenocarpus	(Arecaceae)	1	1.0	FR				
Pourouma	(Cecropiaceae)	4	1.0	FR				
TOTAL		≥78	75.7			≥59	≥146	87.5

(*cont.*)

125

Table 4.5 (*cont.*)

Yasuní (2002–2003)

Genus	Family	# Species	% of diet	Parts eaten	Family	# Genera	# Species	% of diet
Spondias	(Anacardiaceae)	1	9.6	FR	Anacardiaceae	2	2	11.0
Virola	(Myristicaceae)	6	7.9	FR	Cecropiaceae	3	10	8.9
Cecropia	(Cecropiaceae)	4	7.5	FR	Moraceae	8	13	8.9
Iriartea	(Arecaceae)	1	6.3	FR, IMM FL	Myristicaceae	3	9	8.3
Guarea	(Meliaceae)	1	4.3	FR	Arecaceae	≥4	≥4	7.2
Cissus	(Vitaceae)	1	4.1	FR	Fabaceae	4	11	5.6
Quararibea	(Bombacaceae)	1	4.0	FR, FL	Meliaceae	2	3	5.2
Pentaplaris	(Tiliaceae)	1	3.1	LE	Bombacaceae	≥3	≥4	5.0
Bauhinia	(Fabaceae)	2	2.7	LE	Vitaceae	1	1	4.1
Inga	(Fabaceae)	7	2.5	FR	Tiliaceae	2	2	3.2
Ficus	(Moraceae)	1	2.0	FR, LE	Annonaceae	≥5	≥7	2.7
Pseudolmedia	(Moraceae)	2	1.8	FR	Sapindaceae	4	4	2.1
Cupania	(Sapindaceae)	1	1.6	FR	Sapotaceae	3	4	1.5
Naucleopsis	(Moraceae)	2	1.4	FR	Lauraceae	≥3	4	1.4
Tapirira	(Anacardiaceae)	1	1.4	FR	Verbenaceae	1	1	1.2
Pourouma	(Cecropiaceae)	4	1.3	FR, LE	Polygonaceae	1	2	1.2
Pouteria	(Sapotaceae)	2	1.2	FR	Myrtaceae	3	3	1.2
Vitex	(Verbenaceae)	1	1.2	FR	Rosaceae	1	1	1.1
Coccoloba	(Polygonaceae)	2	1.2	FR				
Porcelia	(Annonaceae)	1	1.2	FR				
INDET	(Lauraceae)	1	1.1	FR				
Guatteria	(Annonaceae)	2	1.1	FR				
Prunus	(Rosaceae)	1	1.1	FR				
Poulsenia	(Moraceae)	1	1.0	LE				
Clarisia	(Moraceae)	1	1.0	FR				
TOTAL		48	71.8			≥53	≥85	79.9

(*cont.*)

Table 4.5 (*cont.*)

Lago Caiman		# Species	% of diet	Parts eaten
Ampelocera	(Ulmaceae)	1	16.4	FR
Brosimum	(Moraceae)	2	13.1	FR
Spondias	(Anacardiaceae)	1	12.7	FR
Ficus	(Moraceae)	2	10.7	FR, FL
Euterpe	(Arecaceae)	1	5.2	FR
Byrsonima	(Malpighiaceae)	1	4.5	FR
Huberodendron	(Bombacaceae)	1	4.0	FR, FL
Pseudolmedia	(Moraceae)	2	2.7	FR
Bellucia	(Melastomaceae)	1	2.4	FR
Sapium	(Euphorbiaceae)	1	1.8	FR
Inga	(Fabaceae)	1	1.6	FR
Attalea (Maximiliana)	(Arecaceae)	1	1.5	FR
Schefflera (Didymopanax)	(Araliaceae)	1	1.3	FR
Clarisia	(Moraceae)	1	1.1	FR
TOTAL		17	79.1	

[1] Parts eaten are scored as 'FR' for fruits (both ripe and immature, though for all genera the vast majority of fruits are consumed ripe), 'FL' for flowers (including flower buds, nectar, and inflorescences), 'WD' for wood and bark, 'SE' for seeds, and 'LE' for leaves (both young and mature, though for all genera the vast majority of leaves are consumed when young). The category 'LE' also includes other plant vegetative parts such as leaf buds, meristems, and petioles. 'IMM' signifies that the part is eaten when immature or not fully developed. Where more than one dietary item is noted per genus, those items are listed in order of relative importance as best as can be determined from the data in the original sources. Note that "parts eaten" may not match exactly between this table and Table 4.3 because studies at the same site in different years may have recorded different parts eaten for particular genera, and these have been lumped together in Table 4.3.
[2] Proportions refer to the fruit component of the diet only, as data concerning the proportional contribution of specific genera of leaves to the diet are not noted in the original source.

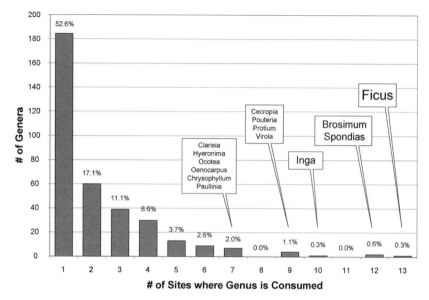

Figure 4.4 Distribution of plant genera consumed across the set of 13 sites where long-term studies of spider monkey diets have been conducted. More than 50% of genera consumed by *Ateles* are reported for only one site. Only four genera are part of the diet at more than 75% of sites.

in these two locales were consumed at both sites. Repeating this analysis, but focusing only on those genera that make up more 1% or more of the annual diet (and considering data from multiple study years at Tinigua and Yasuní) yields a rather different pattern (Figure 4.5b). Here, diets from both Tinigua and Yasuní in different years do cluster together, but, the similarity from year to year is not terribly impressive. Tinigua and Yasuní cluster with Voltzberg and Barro Colorado Island, to the exclusion of two Central American sites (Punta Rio Claro and Tikal) plus Lago Caiman, as they did before, but the position of the other Guianan shield site (Ilha de Maracá) has changed dramatically. While crude, this analysis underscores the fact that specific diets can differ widely between regions, between sites, and within the same sites over time.

Part of the variation noted in the diets of spider monkeys at different study sites within the same broad geographic region is undoubtedly associated with the different local habitats these study groups occupy. For example, in Tinigua one main study group uses both terra firme and seasonally flooded forests (A. Link, unpublished data) while the entire home range of the principal group studied in Yasuní National Park, Ecuador, comprises upland terra firme forest (Pozo Rivera, 2004; Dew, 2005; Suarez, 2006; A. Di Fiore and A. Link, unpublished data). Not surprisingly, figs (*Ficus* spp.), which are much more common in

(a)

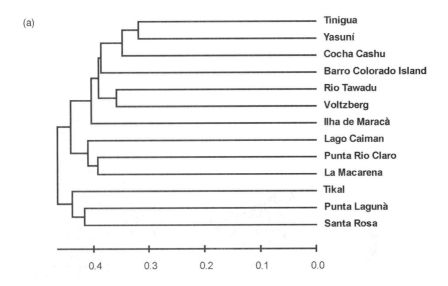

(b)

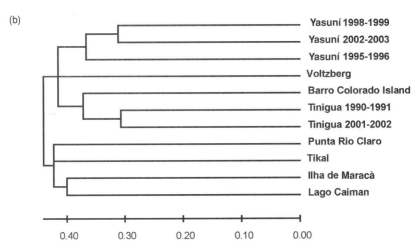

Figure 4.5 Linearized neighbor-joining trees of dietary similarity across sites and study periods for (a) the complete list of genera consumed by *Ateles* across all sites (see Table 4.3) and (b) the combined set of genera contributing to greater than 1% of the total diet at each site (see Table 4.5). Note that panel (b) incorporates separate diet lists recorded at Tinigua and Yasuní for two and three different study periods, respectively. The axis for each figure represents Jaccard's distance, calculated as $(1 - S_{ij})$ (see details in the text).

seasonally flooded forest than in terra firme, formed a much more significant part of the diets of spider monkeys at Tinigua than Yasuní. Differences in diet attributed to the different habitats may even be seen within a single population. For example, members of a spider monkey community studied over eight years in upland terra firme in Yasuní were never observed feeding on the "morete" palm, *Mauritia flexuosa* (Arecaceae), even though local indigenous informants consider it a major food resource for spider monkeys and even target large "moretals" ("morete" swamps) when hunting spider monkeys (Wampi "Humberto" Ahua, personal communication). This palm is typically associated with flat, low-lying, and semi-indundated patches of forest, and was thus absent from the long-term terra firme study site. More recent observations on a newly habituated study group in Yasuní that uses both terra firme and semi-inundated swampy habitats reveal that Yasuní *Ateles* do feed frequently on *Mauritia* palms (A. Link, unpublished data).

One clear pattern that does emerge from this cross-site comparison is that, while spider monkey diets are usually very diverse, only a few plant taxa typically make up the bulk of the diet. For example, more than 50% of the annual diet derives from 12 or fewer genera in all of the studies summarized in Tables 4.3 and 4.5. Even more striking, at some sites only a handful of species together comprise a substantial proportion of the annual diet. For example, in Tikal, a single species (*Brosimum alicastrum*) constituted more than 50% of the diet of *Ateles geoffroyi*, and at Ilha de Maracá, ~29% of the annual diet of *A. b. belzebuth* came from a single plant species, *Pradosia surinamensis* (Nunes, 1998). South of the Equator, at Lago Caiman, only seven species from five genera (*Ampelocera*, *Spondias*, *Ficus*, *Brosimum* and *Euterpe*) together made up more than 58% of the diet of *A. b. chamek* (Wallace, 2005).

This concentration on a relatively small number of taxa characterizes monthly diets as well, even in hyperdiverse western Amazonian forests like those at Tinigua and Yasuní. At Tinigua, for example, an average of 55.1% of the diet each month came from only the five top fruit species (SD = 14.5, range = 36.4–84.7, $n = 13$ months: A. Link, unpublished data), and at Yasuní the top five fruit species each month made up an average of 51.1% of the diet (SD = 7.6, range = 40.4–63.2, $n = 10$ months: A. Di Fiore and A. Link, unpublished data). This pattern was even more pronounced at Lago Caiman, where just the three top fruit species in the diet each month made up between 50 and 95% of foods consumed that month (Wallace, 2005).

Across sites, too, the set of top taxa can change dramatically from month to month, reflecting the fact that most of the important genera in the *Ateles* diet at each site fruit synchronously and do not have extended fruiting periods (Stevenson, 2002; Wallace, 2005; A. Di Fiore, unpublished data). Interestingly, many of the genera that are important at multiple sites from across the geographic range of *Ateles* (Figure 4.4) are exceptions to this rule. Seven of the eight top genera in terms of cross-site importance – *Ficus, Brosimum, Inga, Cecropia, Pouteria, Protium* and *Virola* – are all speciose genera, and, at most sites, multiple species from these genera are included in the diet, not all of which fruit synchronously. It is thus perhaps not surprising that this set of genera features so prominently in the spider monkey diet at multiple sites. The remaining genus in this group, *Spondias*, tends to fruit at a time of year when other ripe, fleshy fruits are rare, at least in Yasuní (Di Fiore, 1997, 2004) and at Lago Caiman (Wallace, 2005), although individual *Spondias* trees may not fruit every year in Yasuní, leading to dramatic interannual variation in *Spondias* availability (A. Di Fiore, unpublished data). A number of additional genera that are important at some sites (e.g. *Iriartea* and *Oenocarpus* in Yasuní: Di Fiore, 1997; Dew, 2001) and *Bursera, Gustavia, Oenocarpus* and *Iriartea* in Tinigua: (Stevenson, 2002; Stevenson *et al.*, 2005) also provide resources during periods of fruit scarcity, and some of these have extended phenologies, providing fruit for multiple months each year. As such, they may represent reliable resources that animals can turn to when other food items are scarce.

Conclusions

Summary of spider monkey socioecology

What emerges from this review is a picture of *Ateles* foraging ecology in which animals rely primarily on high-quality foods (specifically, ripe fruits rich in easily digested carbohydrates and/or lipids) that are presented in ephemeral and widely dispersed patches in the canopies of large trees. Spider monkeys move quickly and efficiently between these patches, where they then ingest large numbers of food items rapidly, without devoting much time to preingestion processing. Usually, spider monkeys focus most of their feeding bouts on a relatively small number of key fruiting plant taxa each month, and the composition of this set of taxa changes following the annual or supraannual phenological cycles of preferred species. After feeding, spider monkeys often rest for long periods of time, usually defecating large numbers of indigestible seeds from their bowels during or close to the end of these resting bouts (Dew, 2001; Link and Di Fiore, 2006). Their high-quality fruit diet is supplemented throughout the year by small amounts of various other items (e.g. young leaves, decaying wood, soils) that are likely to be sources of important nutrients (protein, key minerals) not provided by their fruit diets.

Throughout the year, spider monkeys reduce competition over high-quality food resources from other members of their communities by breaking into flexible foraging parties, and the average size of these foraging parties is related to the habitat-wide abundance of ripe fruits (Symington, 1988b; Chapman *et al.*, 1995). During periods when high-quality fruits are particularly scarce, spider monkeys turn to other resources (van Roosmalen, 1985; Chapman, 1987; Symington, 1988b; Castellanos, 1995; Nunes, 1998), sometimes with a loss of body condition (Wallace, 2005). They also adjust both their foraging and grouping behavior, typically traveling less and visiting fewer feeding sources each day (Nunes, 1995; Wallace, 2005) and associating in smaller parties, presumably to further reduce intragroup feeding competition (Symington, 1988b; Chapman *et al.*, 1995; Shimooka, 2003).

Implications for tropical forests

Their diverse diets – coupled with their propensity to swallow and pass seeds intact (van Roosmalen, 1985; Dew, 2005; Link and Di Fiore, 2006) and their wide daily ranging patterns (Symington, 1988a; Nunes, 1995; Shimooka, 2005; Suarez, 2006) – make spider monkeys especially effective seed dispersers for an enormous number of plant species in most Neotropical forests (Chapman, 1989; Zhang and Wang, 1995; Pacheco and Simonetti, 2000; Stevenson *et al.*, 2002; Russo, 2003; Russo and Augspurger, 2004; Dew, 2005; Link and Di Fiore, 2006; Dew, this volume). Spider monkeys are also likely to be key dispersers for plants that produce large-seeded fruits, as they are the only arboreal frugivores capable of routinely swallowing and passing those seeds (Dew and Wright, 1998; Link and Di Fiore, 2006). Species such as the palms *Oenocarpus bataua* and *Iriartea deltoidea*, as well as members of such large-seeded genera as *Virola* and *Iryanthera* (Myristicaceae), *Guarea* (Meliaceae), *Quararibea* (Bombacaceae), and *Pouteria* (Sapotaceae) may well rely on spider monkeys as their primary dispersal agents. Moreover, given their heavy dietary focus in most sites on a relatively small number of plant species, it is likely that spider monkeys are particularly important dispersers for those taxa. Thus, spider monkeys play a very significant role in Neotropical forest dynamics across their wide geographic range.

In addition, across sites, many of the plant genera that feature heavily in the spider monkey diet (e.g. *Ficus*, *Brosimum*, *Virola*) tend to be rare, occurring at low population density. Thus, the area needed to encompass enough individuals of these plant taxa to support a viable population of spider monkey is typically quite large. Not surprisingly, spider monkeys have larger home ranges than any other Neotropical primate, with the exception of some populations of woolly monkeys (*Lagothrix* spp.) (Peres, 1989; Defler, 1996; Di Fiore, 2003; Di Fiore and Campbell, 2007), a pattern undoubtedly tied to their dietary strategy of focusing on ripe fruits distributed in widely dispersed patches. Clearly, if effective strategies for managing or conserving Neotropical forests are to be devised, they must take into account both the ecological role that spider monkeys play as seed dispersers and the area requirements prescribed by their dietary strategy.

References

Ahumada, J. A. (1989). Behavior and social structure of free ranging spider monkeys (*Ateles belzebuth*) in La Macarena. *Field Studies of New World Monkeys, La Macarena, Colombia*, **2**, 7–31.

Campbell, C. J. (2000). The reproductive biology of black-handed spider monkeys (*Ateles geoffroyi*): integrating behavior and endocrinology. Unpublished Ph.D. thesis, University of California, Berkeley.

Campbell, C. J., Aureli, F., Chapman, C. A., *et al.* (2005). Terrestrial behavior of *Ateles* spp. *Int. J. Primatol.*, **26**, 1039–1051.

Cant, J. G. H. (1977). Ecology, locomotion, and social organization of spider monkeys (*Ateles geoffroyi*). Unpublished Ph.D. thesis, University of California, Davis.

Cant, J. G. H. (1990). Feeding ecology of spider monkeys (*Ateles geoffroyi*) at Tikal, Guatemala. *Hum. Evol.*, **5**, 269–281.

Cant, J. G. H., Youlatos, D. and Rose, M. D. (2003). Suspensory locomotion of *Lagothrix lagothricha* and *Ateles belzebuth* in Yasuní National Park, Ecuador. *J. Hum. Evol.*, **44**, 685–699.

Carpenter, C. R. (1935). Behavior of red spider monkeys in Panama. *J. Mammal.*, **16**, 171–180.

Castellanos, H. G. (1995). Feeding behavior of *Ateles belzebuth* E. Geoffroy 1806 (Cebidae: Atelinae) in Tawadu Forest Southern Venezuela. Unpublished Ph.D. thesis, University of Exeter, UK.

Castellanos, H. G. (1997). Ecología del comportamiento alimentario del marimona (*Ateles belzebuth belzebuth* Geoffroy, 1806) en el Río Tawadu, Reserva Forestal "El Caura". *Scientia Guiainæ*, **7**, 309–341.

Castellanos, H. G. and Chanin, P. (1996). Seasonal differences in food choice and patch preference of long-haired spider monkeys (*Ateles belzebuth*). In *Adaptive Radiations of Neotropical Primates*, ed. M. A. Norconk, A. L. Rosenberger and P. A. Garber, New York: Plenum Press, pp. 451–466.

Chapman, C. (1987). Flexibility in diets of three species of Costa Rican primates. *Folia Primatol.*, **49**, 90–105.

Chapman, C. A. (1989). Primate seed dispersal: the fate of dispersed seeds. *Biotropica*, **21**, 148–154.

Chapman, C. A. (1990). Association patterns of spider monkeys: the influence of ecology and sex on social organization. *Behav. Ecol. Sociobiol.*, **26**, 409–414.

Chapman, C. A. and Chapman, L. J. (1991). The foraging itinerary of spider monkeys: when to eat leaves? *Folia Primatol.*, **56**, 162–166.

Chapman, C. A., Wrangham, R. W. and Chapman, L. J. (1995). Ecological constraints on group size: an analysis of spider monkey and chimpanzee subgroups. *Behav. Ecol. Sociobiol.*, **36**, 59–70.

Defler, T. R. (1996). Aspects of the ranging pattern in a group of wild woolly monkeys (*Lagothrix lagothricha*). *Am. J. Primatol.*, **38**, 289–302.

Dew, J. L. (2001). Synecology and seed dispersal in woolly monkeys (*Lagothrix lagotricha peoppigii*) and spider monkeys (*Ateles belzebuth belzebuth*) in Parque Nacional Yasuní, Ecuador. Unpublished Ph.D. thesis, University of California, Davis.

Dew, J. L. (2005). Foraging, food choice, and food processing by sympatric ripe-fruit specialists: *Lagothrix lagotricha poeppigii* and *Ateles belzebuth belzebuth*. *Int. J. Primatol.*, **26**, 1107–1135.

Dew, J. L. and Wright, P. (1998). Frugivory and seed dispersal by four species of primates in Madagascar's eastern rain forest. *Biotropica*, 425–437.

Di Fiore, A. (1997). Ecology and behavior of lowland woolly monkeys (*Lagothrix lagotricha poeppigii*, Atelinae) in Eastern Ecuador. Unpublished Ph.D. thesis, University of California, Davis.

Di Fiore, A. (2003). Ranging behavior and foraging ecology of lowland woolly monkeys (*Lagothrix lagotricha poeppigii*) in Yasuní National Park, Ecuador. *Am. J. Primatol.*, **59**, 47–66.

Di Fiore, A. (2004). Diet and feeding ecology of woolly monkeys in a western Amazonian rainforest. *Int. J. Primatol.*, **24**, 767–801.

Di Fiore, A. and Campbell, C. J. (2007). The atelines: variation in ecology, behavior, and social organization. In *Primates in Perspective*, ed. C. J. Campbell, A. Fuentes, K. C. MacKinnon, M. Panger and S. K. Beader, New York: Oxford University Press, pp. 155–185.

Erikson, G. E. (1963). Brachiation in New World monkeys and in anthropoid apes. *Symp. Zool. Soc. Lond.*, **10**, 135–164.

Hill, W. C. O. (1962). *Primates: Comparative Taxonomy and Anatomy*. Vol. V: *Cebidae B*. Edinburgh: Edinburgh University Publications.

Iwanaga, S. and Ferrari, S. F. (2001). Party size and diet of syntopic atelids (*Ateles chamek* and *Lagothrix cana*) in southwestern Brazilian Amazonia. *Folia Primatol.*, **72**, 217–227.

Izawa, K. (1993). Soil-eating by *Alouatta* and *Ateles*. *Int. J. Primatol.*, **14**, 229–242.

Klein, L. L. (1972). The ecology and social behavior of the spider monkey, *Ateles belzebuth*. Unpublished Ph.D. thesis, University of California, Berkeley.

Klein, L. L. and Klein, D. J. (1977). Feeding behavior of the Colombian spider monkey, *Ateles belzebuth*. In *Primate Ecology: Studies of Feeding and Ranging Behaviour in Lemurs, Monkeys, and Apes*, ed. T. H. Clutton-Brock, London: Academic Press, pp. 153–181.

Krishnamani, R. and Mahaney, W. C. (2000). Geophagy among primates: adaptive significance and ecological consequences. *Anim. Behav.*, **59**, 899–915.

Lambert, J. E. (1998). Primate digestion: interactions among anatomy, physiology and feeding ecology. *Evol. Anthropol.*, **7**, 8–20.

Link, A. (2003). Insect-eating by spider monkeys. *Neotrop. Primates*, **11**, 104–107.

Link, A. and De Luna, A. G. (2004). The importance of *Oenocarpus bataua* (Arecaceae) in the diet of spider monkeys at Tinigua National Park, Colombia. *Folia Primatol.*, **75**(S1), 391.

Link, A., Di Fiore, A. and Spehar, S. N. (2006). Predation risk affects subgroup size in spider monkeys (*Ateles belzebuth*) at Yasuní National Park, Ecuador. *Folia Primatol.*, **77**, 318.

Link, A. and Di Fiore, A. (2006). Seed dispersal by spider monkeys and its importance in the maintenance of neotropical rain-forest diversity. *J. Trop. Ecol.*, **22**, 335–346.

Link, A. and Stevenson, P. R. (2004). Fruit dispersal syndromes in animal disseminated plants at Tinigua National Park, Colombia. *Revista Chilena de Historia Natural*, **77**, 319–334.

Machado, G. and Pizo, M. A. (2000). The use of fruits by the neotropical harvestman *Neosadocus variabilis* (Opiliones, Laniatores, Gonyleptidae). *J. Arachnol.*, **28**, 357–360.

Milton, K. (1981). Food choice and digestive strategies of two sympatric primate species. *Am. Nat.*, **117**, 496–505.

Milton, K. (1984). The role of food-processing factors in primate food choice. In *Adaptations for Foraging in Non-Human Primates*, ed. P. S. Rodman and J. G. H. Cant, New York: Columbia University Press, pp. 249–274.

Nunes, A. (1995). Foraging and ranging patterns of white-bellied spider monkeys. *Folia Primatol.*, **65**, 85–99.

Nunes, A. (1998). Diet and feeding ecology of *Ateles belzebuth belzebuth* at Maracá Ecological Station, Roraima, Brazil. *Folia Primatol.*, **69**, 61–76.

Pacheco, L. F. and Simonetti, J. A. (2000). Genetic structure of a mimosoid tree deprived of its seed disperser, the spider monkey. *Conserv. Biol.*, **14**, 1766–1775.

Peres, C. A. (1989). Costs and benefits of territory defense in wild golden lion tamarins, *Leontopithecus rosalia*. *Behav. Ecol. Sociobiol.*, **25**, 227–233.

Pozo Rivera, W. E. (2004). Agrupación y dieta de *Ateles belzebuth belzebuth* en el Parque Nacional Yasuní, Ecuador. *Anuar Investigación Científica*, **2**, 77–102.

Ramos-Fernández, G. and Ayala-Orozco, B. (2003). Population size and habitat use of spider monkeys at Punta Laguna, Mexico. In *Primates in Fragments: Ecology and Conservation*, ed. L. K. Marsh, New York: Kluwer Academic/Plenum Publishers, pp. 191–209.

Riba-Hernández, P., Stoner, K. E. and Lucas P. W. (2003). The sugar composition of fruits in the diet of spider monkeys (*Ateles geoffroyi*) in tropical humid forest in Costa Rica. *J. Trop. Ecol.*, **19**, 709–716.

Russo, S. E. (2003). Linking spatial patterns of seed dispersal and plant recruitment in a neotropical tree, *Virola calophylla* (Myristicaceae). Unpublished Ph.D. thesis, University of Illinois, Urbana-Champaign.

Russo, S. E. and Augspurger, C. K. (2004). Aggregated seed dispersal by spider monkeys limits recruitment to clumped patterns in *Virola calophylla*. *Ecol. Lett.*, **7**, 1058–1067.

Russo, S. E., Campbell, C. J., Dew, J. L., Stevenson, P. R. and Suarez, S. A. (2005). A multi-forest comparison of dietary preferences and seed dispersal by *Ateles* spp. *Int. J. Primatol.*, **26**, 1017–1037.

Shimooka, Y. (2003). Seasonal variation in association patterns of wild spider monkeys (*Ateles belzebuth belzebuth*) at La Macarena, Colombia. *Primates*, **44**, 83–90.

Shimooka, Y. (2005). Sexual differences in ranging of *Ateles belzebuth belzebuth* at La Macarena, Colombia. *Int. J. Primatol.*, **26**, 385–406.

Simmen, B. and Sabatier, D. (1996). Diets of some French Guianan primates: food composition and food choices. *Int. J. Primatol.*, **17**, 661–693.

Stevenson, P. R. (2002). Frugivory and seed dispersal by woolly monkeys at Tinigua National Park, Colombia. Unpublished Ph.D. thesis, State University of New York, Stony Brook.

Stevenson, P. R., Castellanos, M. C., Pizarro, J. C. and Garavito, M. (2002). Effects of seed dispersal by three ateline monkey species on seed germination at Tinigua National Park, Colombia. *Int. J. Primatol.*, **23**, 1187–1204.

Stevenson, P. R., Link, A. and Ramírez, B. H. (2005). Frugivory and seed fate in *Bursera inversa* (Burseraceae) at Tinigua Park, Colombia: implications for primate conservation. *Biotropica*, **37**, 431–438.

Stevenson, P. R., Quiñones, M. J. and Ahumada, J. A. (2000). Influence of fruit availability on ecological overlap among four neotropical primates at Tinigua National Park, Colombia. *Biotropica*, **32**, 533–544.

Suarez, S. A. (2003). Spatio-temporal foraging skills of white-bellied spider monkeys (*Ateles belzebuth belzebuth*) in the Yasuní National Park, Ecuador. Unpublished Ph.D. thesis, State University of New York, Stony Brook.

Suarez, S. A. (2006). Diet and travel costs for spider monkeys in a nonseasonal, hyperdiverse environment. *Int. J. Primatol.*, **27**, 411–436.

Symington, M. M. (1987). Ecological and social correlates of party size in the black spider monkey, *Ateles paniscus chamek*. Unpublished Ph.D. thesis, Princeton University, Princeton, NJ.

Symington, M. M. (1988a). Demography, ranging patterns, and activity budgets of black spider monkeys (*Ateles paniscus chamek*) in the Manu National Park, Peru. *Am. J. Primatol.*, **15**, 45–67.

Symington, M. M. (1988b). Food competition and foraging party size in the black spider monkey (*Ateles paniscus chamek*). *Behaviour*, **105**, 117–132.

Symington, M. M. (1990). Fission-fusion social organization in *Ateles* and *Pan*. *Int. J. Primatol.*, **11**, 47–61.

van Roosmalen, M. G. M. (1985). Habitat preferences, diet, feeding strategy and social organization of the black spider monkey (*Ateles paniscus paniscus* Linnaeus 1758) in Surinam. *Acta Amazonica*, **15**, 1–238.

van Roosmalen, M. G. M. and Klein, L. L. (1988). The spider monkeys, genus *Ateles*. In *Ecology and Behavior of Neotropical Primates*, ed. R. A. Mittermeier, A. B. Rylands, A. F. Coimbra-Filho and G. A. B. da Fonseca, Washington, DC: World Wildlife Fund, pp. 455–537.

Wallace, R. B. (2005). Seasonal variations in diet and foraging behavior of *Ateles chamek* in a southern Amazonian tropical forest. *Int. J. Primatol.*, **26**, 1053–1075.

White, F. (1986). Census and preliminary observations on the ecology of the black-faced black spider monkey (*Ateles paniscus chamek*) in Manu National Park, Peru. *Am. J. Primatol.*, **11**, 125–132.

Wrangham, R. W. (1980). An ecological model of female-bonded primate groups. *Behaviour*, **75**, 262–300.

Wrangham, R. W. (1987). Evolution of social structure. In *Primate Societies*, ed. B. B. Smuts, D. L. Cheney, R. M. Seyfarth, R. W. Wrangham and T. T. Struhsaker, Chicago: University of Chicago Press, pp. 282–296.

Zhang, S.-Y. and Wang, L.-X. (1995). Consumption and seed dispersal of *Ziziphus cinnamomum* (Rhamnaceae) by two sympatric primates (*Cebus apella* and *Ateles paniscus*) in French Guiana. *Biotropica*, **27**, 397–401.

5 Factors influencing spider monkey habitat use and ranging patterns

ROBERT B. WALLACE

Introduction

The main influence on primate ranging behavior is food abundance and distribution (Clutton-Brock, 1977; Bennett, 1986). Temporal variations in the availability and distribution of preferred resources shape primates' ranging patterns (Zhang and Wang, 1995; Agetsuma and Noma, 1995; Defler, 1996; Olupot *et al.*, 1997), and can affect the size and shape of home ranges (Harvey and Clutton-Brock, 1981). Other factors known to affect primate ranging behavior include the position of water resources (Altmann and Altmann, 1970), location of sleeping sites (Rasmussen, 1979; Chapman *et al.*, 1989), climatic extremes (Chivers, 1974), the need to patrol boundary areas of the home range (Goodall, 1986; Watts and Mitani, 2001; Williams *et al.*, 2002), and variation in the perceived predation risk of differing habitats (Cowlishaw, 1997).

As large ripe fruit specialists, spider monkeys (*Ateles* spp.) are known for their relatively wide ranging behavior (van Roosmalen and Klein, 1988; Symington, 1988; Chapman, 1990; Castellanos, 1995; Nunes, 1995; Shimooka, 2005; Wallace, 2006), with only *Lagothrix*, *Cacajao*, *Chiropotes*, *Brachyteles* and possibly *Oreonax* displaying comparable or larger ranges within the New World monkeys. Published accounts of *Ateles* home range sizes vary between 95 and 390 hectares in continuous forests (Table 5.1), and the Barro Colorado Island group ranges more than 900 hectares (Campbell, 2000). As such several distinct habitat types are found in most spider monkey territories (Table 5.1). Here I briefly review general patterns of habitat use and ranging patterns across their distributional range before going on to assess the different factors that are known to affect spider monkeys' use of space.

Habitat use

Spider monkeys are generally associated with relatively tall evergreen and semideciduous tropical forest types throughout their geographic range (van

Spider Monkeys: Behavior, Ecology and Evolution of the Genus Ateles, ed. Christina J. Campbell. Published by Cambridge University Press. © Cambridge University Press 2008.

Table 5.1 *Ranging parameters from long-term* Ateles *study sites*

Taxa	Study site	Home range (ha)	Female ind. 100% (ha)	Female ind. 80% core (ha)	Male ind. 100% (ha)	Male ind. 80% core (ha)	Mean daily range (ha)	Mean day journey length and range (m)	# of distinct habitat types	Source
A. paniscus	Voltzberg, Surinam	220						2300 (500–5000)	3	van Roosmalen, 1985; Norconk & Kinzey, 1994
A. belzebuth belzebuth	Macarena, Colombia	260–390						(500–4000)	3	Klein and Klein, 1977
A. belzebuth belzebuth	Macarena (Tinigua), Colombia	169							4	Shimooka, 2005
A. belzebuth belzebuth	Maracá, Brazil	316	87.5		140			1750	5	Nunes, 1995; Mendes Pontes, 1997
A. belzebuth belzebuth	Venezuela	240						ca. 3100	5	Castellanos, 1995
A. belzebuth belzebuth	Yasuní, Ecuador	314	245.8	196.4	253.6	228.2		3311 (723–6039)	1	Suarez, 2006; Spehar, 2006
A. belzebuth chamek	Cocha Cashu (Manu), Peru	153–231	92	49				1977 (465–4070)	3	Symington, 1988; Terborgh, 1983
A. belzebuth chamek	Lago Caiman, Bolivia	295					13.3	2338 (460–5690)	5	Wallace, 2006
A. belzebuth chamek	La Chonta, Bolivia	340							4	A. Felton, pers. comm.
A. geoffroyi	Yucatan, Mexico	95–166		10		18		2302 (1182–3872)	2	Ramos Fernández & Ayala-Orozco, 2002
A. geoffroyi	Borro Colorado Island, Panama	962						2055 (500–4500)	2	Campbell, 2000
A. geoffroyi	Santa Rosa, Costa Rica	170	54.9		81.4				2	Chapman, 1990
A. geoffroyi	Santa Rosa, Costa Rica	158								A. Campbell, pers. comm.

Roosmalen and Klein, 1988). Nevertheless, previous studies and short-term surveys have revealed that *Ateles* is found, at least temporarily, in other forest types, for example, igapo and varzea forest (Klein and Klein, 1976; Branch, 1983; Peres, 1994), upland forest types including the lower tropical slopes of the Andes (Anderson, 1997; Apaza-Quevedo, 2002, Gómez *et al.*, in press), upland Amazonian forests including the edges of cerrado forest (Braza and Garcia, 1988; Wallace *et al.*, 1998), deciduous forests (Freese, 1976; Chapman, 1988), and even mangrove forest (Eisenberg and Kuehn, 1966).

Although spider monkeys have been observed in a wide variety of habitats, in longer-term studies they usually show a preference for taller forest types (van Roosmalen, 1985; Mendes Pontes, 1997; Cant *et al.*, 2001; Ramos-Fernández and Ayala-Orozco, 2002). In Bolivia this relationship is more complicated with spider monkeys preferring certain taller forest habitats, and spending long periods in the abundant tall forest although actually statistically avoiding this habitat and low vine forest (Wallace, 2006). Most study sites have detailed multiple habitat types within spider monkey home ranges (Table 5.1) and at those sites where habitat use has been studied in detail clear seasonal variations in habitat use have been demonstrated (van Roosmalen, 1985; Castellanos, 1995; Ramos-Fernández and Ayala-Orozco, 2002; Wallace, 2006). Indeed, I have suggested that local habitat diversity influences local densities of *Ateles* with greater local diversity permitting greater densities and carrying capacity because of greater and more consistent year-round fruit availability (Wallace, 1998, 2006; Wallace *et al.*, 2000).

Early studies demonstrated a clear preference for the upper levels of forest strata; for example, in Suriname *A. paniscus* were recorded over 70% of time in the upper canopy or emergent tree forest strata, around 20% in the middle levels of the forest and less than 10% in lower forest strata (van Roosmalen, 1985). These general patterns have also been documented on Maracá Island in Brazil where *A. belzebuth belzebuth* were encountered in upper canopy or emergents in over 90% of sightings (Mendes Pontes, 1997). More recently little has been published on forest strata preferences, although a recent cross-site comparative study focused exclusively on the relatively rare terrestrial behavior of *Ateles* (Campbell *et al.*, 2005). Virtually nothing has been published on the distribution of observations in terms of vertical height in the forest except observations of foraging heights by Dew (2005) in Ecuador that recorded a mean foraging height of 16.4 m for *A. belzebuth belzebuth* in the middle to upper canopy strata. Only van Roosmalen (1985) examined preferences for edge versus nonedge habitat for *A. paniscus* indicating that in general nonedge habitats are preferred, but that in nonhunted populations spider monkeys do not completely avoid edge habitats.

Ranging patterns

Studies have varied in the way ranging behavior is presented and this has limited the number of cross-site analyses that can be made in this review. Indeed, a clear recommendation for the future would be for researchers to present standard ranging parameters to facilitate a better understanding of how these ranging parameters relate to other habitat and ecological variables. The most important recommended variables would be: day journey lengths in meters, overall home ranges, 80% and 100% individual core areas, and mean overall and monthly day range areas. All of these variables would be more useful if broken down into male and female values, as well as examined for seasonal variations.

Table 5.1 summarizes the ranging variables for published studies to date on *Ateles*. Home range size is fairly consistent, usually lying between 150 and 350 hectares. Relatively few studies have included more than one study group, but Symington (1988) reports a 20–25% overlap between two *A. belzebuth chamek* groups at Cocha Cashu (Manu), Peru, and maps published in Ramos Fernández and Ayala-Oruzco (2003) suggest overlap of around 15 hectares (*c.* 10–15%) at Otoch Ma'ax Yetel Koh Reserve (Punta Laguna), Mexico between two groups of *A. geoffroyi yucatanensis*. Finally, at La Macarena (Tinigua), Colombia home range overlap of less than 10% between three study groups of *A. belzebuth belzebuth* is reported (Izawa, 2000, cited in Shimooka, 2005).

The most commonly reported variable for ranging data is day journey length (Symington, 1988; Norconk and Kinzey, 1994; Castellanos, 1995; Nunes, 1995; Campbell, 2000; Ramos Fernández and Ayala-Orozco, 2002; Wallace, 2006). Strikingly, day journey length data are remarkably similar across studies, with site-specific ranges usually varying between 500 and 4500 m with means between 1750 and 3311 m. Five of eight mean estimates for day journey lengths fall between 2000 and 2300 m. In addition, several studies report significant gender differences in day journey lengths, with adult males consistently ranging further than adult females: at Lago Caiman, Bolivia males average 2546 m versus 1972 m for adult females (Wallace, 1998, 2007); at Otoch Ma'ax Yetel Koh Reserve, Mexico, across two study groups daily ranges average 2777 m for males versus 2237 m for females (Ramos-Fernández and Ayala-Orozco, 2002); and at La Macarena (Tinigua), Colombia, for follow days with more than six hours data, the ranges are 1960 m for males versus 980 m for females (Shimooka, 2005).

Overall home range sizes are important but the figures in Table 5.1 hide another important pattern in published ranging studies; spider monkeys do not use their home ranges uniformly. Nunes (1995) reports that 40% of *A. belzebuth*

belzebuth time is spent in an area of just 30 hectares and 80% in 115 ha of the 316 ha home range. I have demonstrated a similar pattern for *A. belzebuth chamek* with 50% of time in just 28 ha and 80% of time in 83 ha of a 295 ha home range (Wallace, 2006). Similarly, Spehar (2006) reports that spider monkeys in Yasuní spent 80% of their time in 217 ha of the 314 ha home range.

Patterns of spider monkey ranging also vary seasonally, with studies demonstrating significant differences in ranging data including monthly values for day journey lengths (Nunes, 1995; Wallace, 2006) and seasonal differences in overall group range areas (Castellanos, 1995; Nunes, 1995; Wallace, 2006). However, it is important to highlight that seasonal ranging patterns also vary in the degree of clumping within the community home range. At Lago Caiman in Bolivia, the degree of clumping in monthly ranging patterns varies dramatically, with the entire spider monkey group even abandoning a large portion of their territory for almost four months (Wallace, 2006). Although few other studies have examined ranging in as much detail, in Brazil only 12% of range cells were visited during five or more months, suggesting a similar temporarily shifting ranging pattern (Nunes, 1995).

A less consistent ranging phenomenon is that of individual core ranging areas. Many authors have referred to individual core areas within the community home range (Table 5.1) and have also detailed that typically female individual core areas are smaller than male core areas. Although van Roosmalen (1985) introduced the concept of individual core areas, Symington (1988) was the first to detail individual core areas for adult female *A. b. chamek*; eight females had 100% core areas of between 52 and 118 ha and 80% areas of between 34 and 66 ha, and overlap between individual female core areas was substantial, suggesting that core areas are not exclusive. Minimum convex polygon core areas for males were significantly larger than females (males mean 64 ha, range 49–85 ha; females mean 34 ha, range 18–43 ha). In Costa Rica (Chapman, 1990), on average male spider monkeys had significantly larger 100% individual core areas than adult females (81.4 ha for males versus 54.9 ha for females). Dew (2005) details one adult female using 80 ha, with a 50 ha core area, whereas males extended ranging further than the 350 ha survey area. However, other studies have indicated that in general individual core areas are less relevant with individual ranging patterns occurring across the majority of the overall community range (Campbell, 2000; Spehar, 2006; Wallace, 2006).

Finally, recent studies regarding spider monkey travel paths at the Otoch Ma'ax Yetel Koh Reserve in the Yucatan region of Mexico have demonstrated travel path segments are extremely linear, usually ending at a resource, and that linear path segments ending at resources are much longer than spider monkey visibility in the canopy, suggesting that spider monkeys use spatial memory when planning and traveling through the forest (Valero and Byrne, 2007).

Factors affecting habitat use and ranging patterns

Fruit resource availability and distribution

The most important factor that influences spider monkey ranging behavior and habitat use is the availability and distribution of fruit resources in the forest. Resource availability, particularly ripe fruit, can affect spider monkey ranging behavior in a number of ways. Firstly, where multiple habitats are present, observed seasonal variation and shifting patterns of habitat use have been related to monthly fruit production of primary spider monkey fruit resources (Wallace, 2006) with spider monkeys spending more time in habitats that are temporarily richer in important fruit resources. Unfortunately, few studies have examined habitat use quantitatively but several have observed temporal shifting patterns of habitat use and related this to spatial variations in fruit abundance (van Roosmalen, 1985; Castellanos, 1995; Ramos-Fernández and Ayala-Orozco, 2002).

Secondly, overall spider monkey ranging patterns within a given home range are far from uniform and can generally be described as clumped and inconsistent (Rasmussen, 1980) with seasonal shifts in overall core areas responding to patterns of fruit distribution and clumped patterns also concentrating around specific fruiting trees (Nunes, 1995; Shimooka, 2005; Wallace, 2006). For example, Nunes (1995) demonstrated that feeding resource trees are not uniformly distributed across the study area on the Maracá Island in Brazil, and that *c.* 5% of registered individual feeding trees contribute to over 30% of spider monkey feeding records. Likewise, in Colombia spider monkey range use concentrates around specific grid cells associated with large fruiting trees such as *Spondias mombin*, *Protium glablescens*, *Trichilia pallida*, *Pseudolmedia laevis* and *Pourouma bicolor* (Shimooka, 2005).

At Lago Caiman in Bolivia spider monkeys shift ranging patterns and habitat use according to changing patterns of ripe fruit distribution within the focal community home range (Wallace, 2006). Between May 1996 and April 1997, for long periods of the dry season (June to October) spider monkey ranging behavior was distributed across the range, but range use was concentrated in piedmont forest from November through February with spider monkeys feeding firstly almost exclusively on *Ampelocera ruizii*, and then *Spondias mombin*. In March and April ranging shifted to sartenejal or swamp forest with the diet dominated by *Ficus* spp. trees and *Euterpe precatoria* palms (Wallace, 2006).

Campbell (2000) suggests that the lack of observed individual female core areas for the reintroduced study group on Barro Colorado Island is a behavioral response to spatially explicit fruiting patterns and the lack of neighboring groups on the island which removes an important ranging limitation (Campbell, 2000, 2002), present in differing degrees for all other groups studied (Wallace, 2007). Intriguingly, although generally individual core areas were irrelevant at the Lago Caiman site, at times of fruit scarcity spider monkeys appeared to limit ranging behavior, preferring to restrict themselves to smaller areas, and fed on alternative food resources such as flowers, unripe fruit, leaves and leaf buds (Wallace, 1998, 2006).

Indeed, theoretically the relevance of *Ateles* female individual core areas might also be expected to be influenced by the broad distribution patterns of fruit resources at a given site, with more uniform overall distribution patterns favoring the adoption of core areas (van Roosmalen, 1985; Symington, 1988; Chapman, 1990). Nevertheless, individual core areas are large and overlap almost totally at the Yasuní study site (Spehar, 2006) and this was hypothesized as possibly because populations are below carrying capacity due to historical hunting activities and/or because as an aseasonal and relatively productive site the uniform distribution of fruit encouraged wider overall ranging patterns. Perhaps then, individual and more exclusive core areas are only relevant at times of fruit scarcity (van Roosmalen, 1985; Wallace, 2006), which may be an overall situation if for long portions of the year core areas are worth defending, or a temporal response to specific conditions at sites where fruit abundance is usually not a limiting factor.

Seasonal variations in day journey lengths are also affected by the availability and distribution of fruit resources with 81% of monthly variation explained by these variables at the Lago Caiman study site (Wallace, 2006) where monthly means vary between 1460 and 3541 m. In contrast to many other primate studies (Dunbar, 1988), spider monkeys travel further as fruit availability increases. This phenomenon has been demonstrated at the Maracá Island site, with day journey lengths greater during the wet season than in the dry season and increasing as the availability of fruiting trees increased (Nunes, 1995). Spider monkeys appear to respond to fruit scarcity by reducing the time spent moving and shifting to a more folivorous diet, typically during the dry season (Wallace, 2001, 2005). Indeed, the site with greatest average day journey lengths is the aseasonal Yasuní study site where spider monkeys consistently travel great distances, apparently to maximize energetic intake (Suarez, 2006).

Finally, given the major influence of fruit distribution and availability on overall broader scale ranging behavior it seems probable that at a finer scale the size and distribution pattern of individual fruit patches may also be expected to influence ranging behavior, particularly as they have been shown to influence other aspects of spider monkey ecology such as subgroup size and feeding bout length (Wallace, 2008).

Water and mineral licks

At La Macarena (Tinigua) in Colombia, mineral licks and drinking sources were noted as an important resource for *A. belzebuth belzebuth* (Izawa *et al.*, 1979) and 25 years later the influence of these factors on individual and community ranging patterns is still evident with salt and mineral lick sites having concentrated spider monkey range use (Shimooka, 2005). Campbell *et al.* (2005) conclude from a review of *ad libitum* observations that access to drinking water is likely to be an important driving force behind terrestrial behavior in *Ateles* in Central America, but not in South America. Seasonality of rainfall in South America is not as great, thus meaning that water collected in tree holes is likely to be available year round. In Central America, however, there are months of the year where tree holes dry up and the only source of water is from ground streams and creeks. Wallace (2006) indicates that *A. belzebuth chamek* visits cliffs at a seasonal Bolivian study site in southern Amazonia and uses a probable mineral lick and drinks water from caves on the cliff. The influence of drinking sites may therefore be relevant at seasonal sites at the northern and southern frontiers of the distributional range of *Ateles*.

Territorial defense

Recently several studies have highlighted the importance of territoriality in interpreting spider monkey behavior in general, especially the ranging behavior of adult and subadult males (Shimooka, 2005; Aureli *et al.*, 2006; Wallace, 2007). Most studied spider monkey groups have been shown to at least partially defend a defined home range or territory (Fedigan and Baxter, 1984; van Roosmalen, 1985; Symington, 1988; Chapman, 1990; Nunes, 1995; Wallace, 1998; Shimooka, 2005). Apart from clearly defined home ranges and observed territorial disputes between neighboring groups, male-dominated patrolling behavior along territorial boundaries has been observed in several published accounts of long-term research projects (Symington, 1990; Shimooka, 2005; Wallace, 1998, 2007).

Shimooka (2005) demonstrated that *A. b. belzebuth* males travel further, faster and in larger parties than females, and also frequently visit boundary areas of the community range. Larger overall subgroup size in community border areas has also been reported in previous studies (Symington, 1988; Chapman, 1990; Wallace, 1998). Indeed, I have suggested (Wallace, 1998, 2007) that the number of males in a spider monkey group can be predicted by the length of risky boundary, or boundary that directly borders adjacent groups, in a given community territory. This argument is further supported by recent observations of male raiding parties at Otoch Ma'ax Yetel Koh Reserve, Punta Laguna in Mexico (Aureli *et al.*, 2006), with male parties traveling terrestrially and in silence up to 1950 m in total within neighboring groups' home ranges, and chasing and attacking females in neighboring groups. This revelation, along with observed boundary disputes (van Roosmalen, 1985; Symington, 1988; Wallace, 1998) demonstrates that intergroup male aggression is indeed a risk for spider monkeys and boundary defense clearly significantly influences male ranging behavior. It is worth highlighting that in some cases the shape, and hence boundary length, of spider monkey home ranges is probably influenced by the overall distribution pattern of fruit resources (Wallace, 2006, 2007).

Accident risk

The risk of falling for arboreal primates has been documented for several genera including *Ateles* (Schultz, 1956; Goodall, 1986; Bulstrode 1987; Chapman and Chapman, 1987), and health evaluations of immobilized spider monkeys at the Lago Caiman site revealed healed fractures on the limbs of three of nine sampled individuals (Karesh *et al.*, 1998). Frequently used travel paths have been reported in various studies (Milton, 1988; Wallace, 1998; Di Fiore and Suarez, 2007) and may serve as a source of social and/or ecological information for a fission–fusion community-living primate, but they may also represent a relatively safe travel route. Field data regarding frequency and characteristics of primate falls, including spider monkeys, are understandably scarce and future studies at long-term and intensive study sites might focus on accident rates along well-traveled paths versus other paths and rates for youngsters versus adults.

Predation risk

Despite the difficulty in documenting predation under field conditions, variations across habitat types in primate predation risk have been shown to influence primate habitat use and ranging behavior. Cowlishaw (1997) showed that for baboons (*Papio cynocephalus ursinus*) the key factors regarding the predation risk of a given habitat are differences in visibility from the baboon's perspective, as well as ambush opportunities for large felid predators.

Being relatively large-bodied there is a limited list of known predators for *Ateles*, including the two largest forest raptors – harpy eagle (*Harpia harpyja*;

L. Painter, pers. comm. to R. Wallace, 1997) and crested eagle (*Morphnus guianensis*; Julliot, 1994) – and the jaguar (*Panthera onca*; Emmons, 1987; Matsuda and Izawa, 2007), and puma (Castellanos, 1995; Matsuda and Izawa, 2007). Vocalizing and/or aggressive branch shaking behavior has been documented for all of these species when encountered by habituated spider monkey communities, as well as slightly smaller ornate hawk eagles (*Spizaetus ornatus*), tayras (*Eira barbara*), ocelots (*Leopardus pardalis*), large boa constrictors, and caiman, perhaps because of a risk to juveniles (Castellanos, 1995; Wallace, 1998; C. Campbell, unpublished data).

Anecdotal observations at a number of sites suggest that predation risk and subsequent avoidance strategies may well influence spider monkey behavior including habitat use and ranging patterns. Izawa and colleagues (Izawa *et al.*, 1979) noted that large subgroups of up to 30 *A. belzebuth belzebuth* are associated with the use of terrestrial natural salt licks, with one or two individuals from each large subgroup simultaneously using the salt lick whilst others remain vigilant. Campbell *et al.* (2005) also suggest differences in predation risk as the most likely explanation of variable rates of observed terrestrial behavior across several *Ateles* sites, and increased vigilance by other members of the subgroup in association with this terrestrial behavior. Specifically, during a game of "chase" played on the ground by juveniles, older infants and younger subadults, adult males were always located nearby, watching, but not participating, apparently acting as sentries (Campbell *et al.*, 2005).

At Lago Caiman on several occasions subgroups traveling towards piedmont forest were observed to stop on the edge of a broad swathe of low vine forest and appear to wait until other individuals arrived from the west before traveling east across the strip of low vine forest (Wallace, 1998). The community almost always used exactly the same travel route across this habitat and individuals often risked jumping considerable distances between large emergents in low vine forest rather than taking a less precarious route through lower lying dense vegetation. This low vine forest is characterized by relatively low forest interspersed with stretches of particularly low, dense and liana-infested vegetation where visibility is extremely poor. Predation risk and ambush opportunities are presumably greater than in taller continuous canopy habitats and spider monkeys appeared to actively avoid descending to the dense liana tangles to travel. It is possible that liana tangles are not a desirable travel medium for other reasons, for example, thorns, stinging invertebrates and/or arboreal venomous snakes. However, larger subgroup sizes than would be predicted by resource availability for both low vine forest and cerrado forest fits with a perceived increase in predation risk, and spider monkeys appeared particularly nervous in low vine forest during habituation.

Sleeping sites

Finally, the location of sleeping sites might also influence spider monkey ranging behavior. Regularly used sleeping sites tend to be large, usually in the form
of emergent trees (van Roosmalen and Klein, 1988; Chapman, 1989; Wallace,
unpublished data), and this presumably allows resident sleeping subgroups to
be relatively large (Chapman, 1989). Thus, the threat of nocturnal predation
has been proposed as a possible selective pressure explaining this tendency
amongst spider monkey populations (van Roosmalen, 1985; Chapman, 1989).
An alternative explanation is that regular sleeping sites permit multiple central
place foraging (Chapman *et al.*, 1989) and act as centers of resource information exchange and/or social bonding for members of the community (Chapman,
1989).

At the Lago Caiman study site, the spider monkey community used a 200–
300 m longitudinal section of the Huanchaca escarpment cliffs as a preferred
site for the entire community to sleep in close visual proximity, occurring on
most evenings between November 1996 and February 1997 (Wallace, 2006).
Before assessing the potential reasons for this preference it is important to
recognize the risks involved in climbing these cliffs with a potential fall of up
to 200 m. In line with the above predation risk arguments, the last 100–200 m of
piedmont forest before reaching the actual cliffs was usually very low (<10 m)
transitional forest, and the focal community almost exclusively used two routes
across this habitat to reach the cliffs. Other possible explanations for this cliff
sleeping site preference include the existence of a salt lick and easy access to
water, and that the rocks of the escarpment may have a significant heat-retaining
capacity.

Conclusions

Spider monkey habitat use and ranging behavior is a dynamic phenomenon with
variations across study sites apparent as well as clear seasonal variations at the
majority of long-term study sites to date. Curiously the exception to seasonal
variation seems to be the only site that is described as aseasonal in Yasuní,
Ecuador (Suarez, 2006). Sites with multiple habitat types have demonstrated
multiple habitat use that usually varies over time and seems primarily linked to
resource distribution and availability, usually in the form of a few important ripe
fruit resources. As such it seems reasonable to assume that although a number of
factors have been shown or are strongly suspected to influence habitat use and
ranging behavior, patterns of range use are primarily driven by fruit distribution

and availability. Spider monkeys specialize on ripe fleshy fruit, an ephemeral and patchily distributed resource within tropical forest environments. As such spider monkeys respond to a dynamic food resource availability and distribution situation through corresponding variability in ranging behavior.

Nevertheless, other factors are important in interpreting spider monkey ranging behavior and habitat use. These can be divided into two types of factors: those linked to variation in access to another nonfood-related type of resource such as drinking water, mineral licks or sleeping sites, all of which might be expected to vary according to the specific habitat configuration, soil types and seasonality of the study site; and those that constitute risk limitations of different types, such as predation risk, accident risk, and in the case of territory defense, the risk of invasion and physical injury from neighboring community individuals.

Biologically it seems reasonable to conclude that most of these additional factors, apart from drinking water, are subordinate to the principal influence of food distribution and availability on ranging behavior. It also seems likely that both sexes are equally influenced by some of these factors such as drinking water and mineral licks, but not by others. For example, territorial defense and patrolling of home range borders seems to be an activity that primarily influences male behavior, suggesting that males are indeed defending access to females, although an analysis of whether females avoid boundary areas has yet to be conducted. Conversely predation risk may be more relevant to female spider monkeys given that they are often accompanied by infants and/or juveniles and several of the documented and potential spider monkey predators have more relevance to younger animals. Juvenile animals may also be more prone to accident risk. Indeed, female spider monkeys may be more limited in the overall distances traveled due to being held back by younger animals and/or compromised energetically by the need to carry infants. Taken together these explanations go some way in explaining sexual differences in ranging patterns, with males less constrained and able to range further than females and in so doing able to defend the community range.

A surprising conclusion to this review is that few studies have quantitatively examined habitat use in terms of overall habitat preferences, the vertical use of forest strata within habitats, and edge versus nonedge habitat use. Future studies and/or collaborative reviews might benefit from examining these variables in more detail and tying them to some of the factors detailed that are known to influence habitat use and ranging behavior. Site-specific variations in other spider monkey behaviors detailed in other chapters in this volume are likely to be better interpreted when a more complete understanding of the ecological factors affecting spider monkey foraging and ranging behavior is available.

References

Agetsuma, N. and Noma, N. (1995). Rapid shifting of foraging pattern by Yakushima macaques (*Macaca fuscata yakui*) in response to heavy fruiting of *Myrcia rubra*. *Int. J. Primatol.*, **16**, 247–260.

Altmann, S. A. and Altmann, J. (1970). *Baboon Ecology*. Chicago: University of Chicago Press.

Anderson, S. (1997). Mammals of Bolivia, taxonomy and distribution. *Bull. Amer. Mus. Nat. Hist.*, **231**, 1–652.

Apaza-Quevedo, A. E. (2002). Comportamiento alimentario de *Ateles chamek* (Cebidae) y disponibilidad de frutos en época húmeda en los Yungas del PN-ANMI Cotapata, La Paz. Undergraduate thesis, Universidad Mayor San Andres, La Paz, Bolivia.

Aureli, F., Schaffner C. M., Verpooten, J., Slater, K. and Ramos-Fernández, G. (2006). Raiding parties of male spider monkeys: insights into human warfare? *Am. J. Phys. Anthropol.*, **131**, 486–497.

Bennett, E. L. (1986). Environmental correlates of ranging behaviour in the banded langur, *Presbytis melalophus*. *Folia Primatol.*, **47**, 26–38.

Branch, L. C. (1983). Seasonal and habitat differences in the abundances of primates in the Amazon (Tapajos) National Park, Brazil. *Primates*, **24**, 424–431.

Braza, F. and Garcia, J. E. (1988). Rapport préliminaire sur les singes de la région montagneuse de Huanchaca, Bolivie. *Folia Primatol.*, **49**, 182–186.

Bulstrode, C. (1987). What happens to wild animals with broken bones. *Spec. Sci. Tech.*, **10**, 245–253.

Campbell, C. J. (2000). The reproductive biology of black-handed spider monkeys (*Ateles geoffroyi*): integrating behavior and endocrinology. Unpublished Ph.D. thesis, University of California, Berkeley.

Campbell, C. J. (2002). The influence of a large home range on the social structure of free ranging spider monkeys (*Ateles geoffroyi*) on Barro Colorado Island, Panama. *Am. J. Phys. Anthropol. Suppl.*, **34**, 51–52.

Campbell, C. J., Aureli, F., Chapman, C. A., *et al.* (2005). Terrestrial behaviour of *Ateles* spp. *Int. J. Primatol.*, **26**, 1039–1051.

Cant, J. G. H., Youlatos, D. and Rose, M. D. (2001). Locomotor behavior of *Lagothrix lagotricha* and *Ateles belzebuth* in Yasuní National Park, Ecuador: general patterns and nonsuspensory modes. *J. Human Evol.* **41**, 141–166.

Castellanos, H. G. (1995). Feeding behaviour of *Ateles belzebuth* E. Geoffroy 1806 (Cebidae: Atelinae) in Tawadu Forest, southern Venezuela. Unpublished Ph.D. thesis, University of Exeter, UK.

Chapman, C. (1988). Patterns of foraging and range use by three species of Neotropical primates. *Primates*, **29**, 177–194.

Chapman, C. (1989). Spider monkey sleeping sites: use and availability. *Am. J. Primatol.*, **18**, 53–60.

Chapman, C. (1990). Association patterns of spider monkeys: the influence of ecology and sex on social organization. *Behav. Ecol. Sociobiol.*, **26**, 409–414.

Chapman, C. and Chapman, L. J. (1987). Social responses to the traumatic injury of a juvenile spider monkey (*Ateles geoffroyi*). *Primates*, **28**, 271–275.

Chapman, C. A., Chapman, L. J. and McLaughlin, R. L. (1989). Multiple central place foraging by spider monkeys: travel consequences of using many sleeping sites. *Oecologia*, **79**, 506–511.

Chivers, D. J. (1974). The siamang in Malaya: a field study of a primate in tropical rain forest. *Contr. Primatol.*, **4**, 1–335.

Clutton-Brock, T. H. (1977). Some aspects of intraspecific variation in feeding and ranging behaviour in primates. In *Primate Ecology: Studies of Feeding and Ranging Behaviour in Lemurs, Monkeys, and Apes*, ed. T. H. Clutton-Brock, London: Academic Press, pp. 539–556.

Cowlishaw, G. (1997). Trade-offs between foraging and predation risk determine habitat use in a desert baboon population. *Anim. Behav.*, **53**, 667–686.

Defler, T. R. (1996). Aspects of the ranging pattern in a group of wild woolly monkeys (*Lagothrix lagotricha*). *Am. J. Primatol.*, **38**, 289–302.

Dew, J. L. (2005). Foraging, food choice, and food processing by sympatric ripe-fruit specialists: *Lagothrix lagotricha poeppigii* and *Ateles belzebuth belzebuth*. *Int. J. Primatol.*, **26**, 1107–1135.

Di Fiore, A. and Suarez, S. A. (2007). Route-based travel and shared routes in sympatric spider and woolly monkeys: cognitive evolutionary implications. *Anim. Cogn.* (Online First 10.1107/s10071-007-0067-x).

Dunbar, R. I. M. (1988). *Primate Social Systems*. London and Sydney: Croom Helm.

Eisenberg, J. F. and R. E. Kuehn. (1966). The behaviour of *Ateles geoffroyi* and related species. *Smithson. Misc. Coll.*, **151**, 1–63.

Emmons, L. H. (1987). Comparative feeding ecology of felids in a Neotropical rainforest. *Behav. Ecol. Sociobiol.*, **20**, 271–283.

Fedigan, L. M. and Baxter, M. J. (1984). Sex differences and social organization in free-ranging spider monkeys (*Ateles geoffroyi*). *Primates*, **25**, 279–294.

Freese, C. (1976). Censusing *Alouatta palliata*, *Ateles geoffroyi* and *Cebus capucinus* in the Costa Rican dry forest. In *Neotropical Primates: Field Studies and Conservation*, ed. R. W. Thorington and P. G. Heltne, Washington, DC: National Academy of Sciences, pp. 4–9.

Gómez, H., Ayala, G. and Wallace. R. B. (in press). Biomasa de primates y ungulados en bosques amazónicos preandinos en el Parque Nacional y Área Natural de Manejo Integrado Madidi (La Paz, Bolivia). *Mastozoologia Neotropical*.

Goodall, J. (1986). *The Chimpanzees of Gomb: Patterns of Behaviour*. London: Belknap Harvard Press.

Harvey, P. H. and Clutton-Brock, T. H. (1981). Primate home-range size and metabolic needs. *Behav. Ecol. Sociobiol.*, **8**, 151–155.

Izawa, K., Kimura, K. and Samper Nieto, A. (1979). Grouping of the wild spider monkey. *Primates*, **20**, 503–512.

Julliot, C. (1994). Predation of a young spider monkey (*Ateles paniscus*) by a crested eagle (*Morphnus guianensis*). *Folia Primatol.*, **63**, 75–77.

Karesh, W. B., Wallace, R. B., Painter, R. L. E., *et al.* (1998). Immobilization and health assessment of free-ranging black spider monkeys (*Ateles paniscus chamek*). *Am. J. Primatol.*, **44**, 107–123.

Klein, L. L. and Klein, D. J. (1976). Neotropical primates, aspects of habitat usage, population density, and regional distribution in La Macarena, Columbia. In

Neotropical Primates: Field Studies and Conservation, ed. R. W. Thorington and P. G. Heltne, Washington, DC: National Academy of Sciences, pp. 70–78.

Klein, L. L. and Klein, D. J. (1977). Feeding behavior of the Colombian spider monkey, *Ateles belzebuth*. In *Primate Ecology: Studies of Feeding and Ranging Behaviour in Lemurs, Monkeys, and Apes*, ed. T. H. Clutton-Brock, London: Academic Press, pp. 153–181.

Laska, M., Scheuber, H.-P., Carrera Sanchez, E. and Rodriguez Luna, E. (1999). Taste difference thresholds for sucrose in two species of nonhuman primates. *Am. J. Primatol.*, **48**, 153–160.

Matsuda, I. and Izawa, K. (2007). Predation of wild spider monkeys at La Macarena, Colombia. *Primates* (Online First 10.1007/s10329-007-0042-5).

Mendes Pontes, A. R. (1997). Habitat partitioning among primates in Maracá Island, Roraima, Northern Brazilian Amazonia. *Int. J. Primatol.*, **18**, 131–157.

Milton, K. (1988). Foraging behaviour and the evolution of primate intelligence. In *Machiavellian Intelligence: Social Expertise and the Evolution of Intellect in Monkeys, Apes and Humans*, ed. R. Byrne and A. Whiten, Oxford: Clarendon Press, pp. 285–306.

Norconk, M. A. and Kinzey, W. G. (1994). Challenge of neotropical frugivory: travel patterns of spider monkeys and bearded sakis. *Am. J. Primatol.*, **34**, 171–183.

Nunes, A. (1995). Foraging and ranging patterns in white-bellied spider monkeys. *Folia Primatol.*, **65**, 85–99.

Olupot, W., Chapman, C. A., Waser, P. M. and Isabirye-Basuta, G. (1997). Mangabey (*Cercocebus albigena*) ranging patterns in relation to fruit availability and the risk of parasite infection in Kibale National Park, Uganda. *Am. J. Primatol.*, **43**, 65–78.

Peres, C. A. (1994). Primate responses to phenological changes in an Amazonian terra firme forest. *Biotropica*, **26**, 98–112.

Ramos-Fernández, G. and Ayala-Orozco, B. (2003). Population size and habitat use of spider monkeys in Punta Laguna, Mexico. In *Primates in Fragments: Ecology and Conservation*, ed. L. K. Marsh, New York: Kluwer Plenum Press, pp. 191–209.

Rasmussen, D. R. (1979). Correlates of patterns of range use of a troop of yellow baboons (*Papio cynocephalus*). I. Sleeping sites, impregnable females, births, and male emigrations and immigrations. *Anim. Behav.*, **27**, 1098–1112.

Rasmussen, D. R. (1980). Clumping and consistency in primates' patterns of range use: definitions, sampling, assessment and applications. *Folia Primatol.*, **34**, 111–139.

Rondinelli, R. and Klein, L. L. (1976). Analysis of adult social spacing tendencies and related social interactions in a colony of spider monkeys (*Ateles geoffroyi*) at San Francisco Zoo. *Folia Primatologica*, **25**, 122–142.

Schultz, A. H. (1956). The occurrence and frequency of pathological and teratological conditions and of twinning among non-human primates. *Primatologia*, **1**, 965–1014.

Shimooka, Y. (2005). Sexual differences in ranging of *Ateles belzebuth belzebuth* at La Macarena, Colombia. *Int. J. Primatol.*, **26**, 385–406.

Spehar, S. N. A. (2006). The function of the long call in white-bellied spider monkeys (*Ateles belzebuth*) in Yasuní National Park, Ecuador. Unpublished Ph.D. thesis, New York University, NY.

Suarez, S. (2006). Diet and travel costs for spider monkeys in a nonseasonal, hyperdiverse environment. *Int. J. Primatol.*, **27**, 411–436.

Symington, M. (1988). Demography, ranging patterns, and activity budgets of black spider monkeys (*Ateles paniscus chamek*) in the Manu National Park, Peru. *Am. J. Primatol.*, **15**, 45–67.

Symington, M. (1990). Fission-fusion social organization in *Ateles* and *Pan*. *Int. J. Primatol.* **11**, 47–61.

Terborgh, J. (1983). *Five New World Primates: A Study in Comparative Ecology*. Princeton, NJ: Princeton University Press.

Valero, A. and Byrne, R. W. (2007). Spider monkey ranging patterns in Mexican subtropical forest: do travel routes reflect planning? *Anim. Cogn.* (Online First 10.1007/s10071-006-0066-z).

van Roosmalen, M. G. M. (1985). Habitat preferences, diet, feeding strategy, and social organization of the black spider monkey (*Ateles p. paniscus* Linnaeus 1758) in Surinam. *Acta Amazonica*, **15**, 1–238.

van Roosmalen, M. G. M. and Klein, L. L. (1988). The spider monkeys, genus *Ateles*. In *Ecology and Behavior of Neotropical Primates*, ed. R. A. Mittermeier, A. B. Rylands, A. Coimbra-Filho and G. A. B. Fonseca, Washington DC: World Wildlife Fund, pp. 455–539.

Wallace, R. B. (1998). The behavioural ecology of black spider monkeys in north-eastern Bolivia. Unpublished Ph.D. thesis, University of Liverpool, UK.

Wallace, R. B. (2001). Diurnal activity budgets of black spider monkeys, *Ateles chamek*, in a southern Amazonian tropical forest. *Neotrop. Primates*, **9**, 101–107.

Wallace, R. B. (2005). Seasonal variations in the diet and foraging behavior of the black spider monkey, *Ateles chamek*, in a southern Amazonian tropical forest. *Int. J. Primatol.*, **26**, 1053–1075.

Wallace, R. B. (2006). Seasonal variations in black spider monkey, *Ateles chamek*, habitat use and ranging behavior in a southern Amazonian tropical forest. *Am. J. Primatol.*, **68**, 313–332.

Wallace, R. B. (2007). Towing the party line: territoriality, risky boundaries and male group size in spider monkey fission-fusion societies. *Am. J. Primatol.* DOI 10.1002/ajp.20484.

Wallace, R. B. (2008). The influence of feeding patch size and relative fruit availability on black-faced black spider monkey, *Ateles chamek*, foraging behavior. *Biotropica* doi: 10.1111/j.1744-7429.2007.00392.x.

Wallace, R. B., Painter, R. L. E., Rumiz, D. I. and Taber, A. B. (2000). Primate diversity, distribution and relative abundances in the Rios Blanco y Negro Wildlife Reserve, Santa Cruz Department, Bolivia. *Neotrop. Primates*, **8**, 24–28.

Wallace, R. B., Painter, R. L. E. and Taber, A. B. (1998). Primate diversity, habitat preferences and population density estimates in Noel Kempff Mercado National Park, Santa Cruz, Bolivia. *Am. J. Primatol.*, **46**, 197–211.

Watts, D. P. and Mitani, J. C. (2001). Boundary patrols and intergroup encounters in wild chimpanzees. *Behaviour*, **138**, 299–327.

Williams, J. M., Pusey, A. E., Carlis, J. V., Farms, B. P. and Goodall, J. (2002). Female competition and male territorial behaviour influence female chimpanzee ranging patterns. *Anim. Behav.*, **63**, 347–360.

Zhang, S.-Y. and Wang, L.-X. (1995). Consumption and seed dispersal of *Ziziphus cinnamomum* (Rhamnaceae) by two sympatric primates (*Cebus apella* and *Ateles paniscus*) in French Guiana. *Biotropica*, **27**, 397–401.

6   *Spider monkeys as seed dispersers*

J. LAWRENCE DEW

Introduction

Plants, which cannot move on their own, must rely on external forces such as wind, water, or interactions with motile organisms for pollination and seed dispersal. Dispersal away from the parent plant is vital as seeds are unlikely to survive predation or disease if they are not moved away from the crown of the parent. Additionally, the layered canopy of the tropical forest prevents as much as 98% of sunlight from reaching the ground (Gentry, 1983). Tropical canopy plants are thus under selective pressure to produce relatively large seeds, as their seedlings often require considerable energy reserves to survive in the forest understory. Large seeds are not easily dispersed by wind or water, and most tropical tree species use the vertebrate gut as a vector for dispersal (Gentry, 1983). These seeds are encased in fleshy, nutritious fruits that are swallowed by animals and defecated elsewhere, a dispersal strategy known as endozoochory (van der Pijl, 1957).

Frugivorous primates play important roles in this interaction. With their relatively large body sizes (compared with most birds, for example) and their long day ranges, primates are among the most effective dispersers of the large seeds produced by most tropical trees (Gautier-Hion *et al.*, 1983; Bourliere, 1985; Garber, 1986; Leiberman and Lieberman, 1986; White, 1986; Gautier-Hion and Michaloud, 1989; Chapman, 1989, 1995). As a result, there could be serious community-level effects should these seed dispersers be removed from the ecosystem (Howe, 1977, 1984; Gilbert, 1980; Howe and Smallwood, 1982; Janzen, 1985; Terborgh, 1986; Cox *et al.*, 1991; Chapman, 1995; Given, 1995; Chapman and Onderdonk, 1998; Lambert and Garber, 1998; Birkinshaw, 1999; Willson and Traveset, 2000).

Of all Neotropical primates, the spider monkey specializes more intensely than any other on a diet of ripe fruits (Richard, 1970; Cant, 1977; Klein and Klein, 1977; Milton, 1981; Mittermeier and van Roosmalen, 1981; van Roosmalen, 1985a; White, 1986; Chapman, 1987, 1989; Ahumada, 1989; Cant,

Spider Monkeys: Behavior, Ecology and Evolution of the Genus Ateles, ed. Christina J. Campbell. Published by Cambridge University Press. © Cambridge University Press 2008.

1990; Castellanos and Chanin, 1996; Nunes, 1998; Symington, 1988; Di Fiore and Campbell, 2007). Being among the largest bodied of the New World primates, and with exceedingly long day ranges (see Di Fiore and Campbell, 2007), spider monkeys are among the most important seed-dispersing animals in the New World. This has been documented across the range of *Ateles*. In the northern part of their distribution in Tikal, Guatemala, Cant (1977) and Muskin and Fishgrund (1981) described numerous fruits that they disperse there. In dry forests in western Costa Rica Colin Chapman (1989) found similar results. In the Guianas Marc van Roosmalen conducted a detailed study of spider monkeys' interactions with plants (1985a) and described hundreds of dispersed plant species in his definitive book *Fruits of the Guianan Flora* (1985b). Years later Zhang and Wang (1995) examined French Guianan spider monkeys' seed dispersal abilities. The role of Amazonian spider monkeys as seed dispersers in Colombia has been described by Klein and Klein (1977), van Roosmalen and Klein (1988), and Stevenson *et al.* (2002, 2005). In Ecuador, their role in this activity has been studied by Dew (2001, 2005), and Link and Di Fiore (2005). In Cocha Cashu, Peru, the importance of spider monkeys as seed dispersers was first described by Frances White (1986) and detailed examinations of their ecological roles were conducted by Sabrina Russo (2003) and her collaborators (Russo *et al.*, 2005). Spider monkeys have demonstrated their importance in terms of numbers of species dispersed (van Roosmalen, 1985a), numbers of seeds removed (Russo, 2003), size ranges of seeds swallowed (Dew, 2001, 2005; Russo *et al.*, 2005), distances dispersed (Dew, 2001; Link and Di Fiore, 2005), and survivorship of dispersed seeds (Dew, 2001).

Spider monkeys are large-bodied, vocal, easily shot, and commonly targeted by hunters. Their slow birth rates and large home ranges make them particularly sensitive to disturbance, and they are often the first Neotropical primates to disappear from a disturbed area (Yost and Kelley, 1983; Mittermeier, 1987; Mittermeier *et al.*, 1989; Peres, 1990, 1991; Redford, 1992). This reality provides an additional compelling reason to understand their ecological importance in the forests they inhabit.

My research has examined the seed dispersal efficacy of the Amazonian white-bellied spider monkey, *Ateles belzebuth belzebuth*, compared with that of other Neotropical primates, in particular with the ecologically similar Humboldt's woolly monkey, *Lagothrix lagothricha*. In Amazonian Ecuador I followed focal animals, recorded feeding behaviors, identified food plants, and measured the seed shadows they disseminated. I looked for relationships in which particular plant species might rely on particular primate species for seed dispersal.

Table 6.1 lists the 10 species of primates found at our field site, the Proyecto Primates research site in Yasuní National Park, Ecuador (Figure 6.1). The table

Table 6.1 *Primate species on the Proyecto Primates trail system at Yasuní*

Genus and species	Common name	Encounter rank	Weight (kg)
Lagothrix lagothricha poeppigii	Humboldt's woolly monkey	1	8.0
Ateles belzebuth belzebuth	White-bellied spider monkey	2	8.0
Saimiri sciureus macrodon	Squirrel monkey	3	0.9
Cebus albifrons aequatorialis	White-fronted capuchin	4	2.8
Pithecia monachus monachus	Monk saki	5	1.5
Saguinus tripartitus	Golden-mantled tamarin	6	0.4
Callicebus cupreus* discolor	Dusky titi	7	0.8
Allouatta seniculus seniculus	Red howler	8	8.0
Aotus vociferans	Owl monkey	9	0.8
Cebuella pygmaea	Pygmy marmoset	10	0.1

Woolly and spider monkeys are the largest and most populous primates in Yasuní.
*Also called *Callicebus moloch cupreus*.
Sources: Emmons (1990); Di Fiore (1997).

illustrates why woolly monkeys provide a particularly interesting ecological comparison with spider monkeys. Both species are diurnal canopy dwellers approximately 6–8 kg in weight. Apart from the primarily folivorous red howler, *Alouatta seniculus*, these two are the largest primate species in Amazonia by several kilograms. Second, both woolly and spider monkeys are obligate frugivores of the same subfamily (Atelinae), and both are known dispersers of many fleshy-fruited plants (Peres, 1994, 1996; Defler and Defler, 1996; Stevenson, 2000; Dew, 2001, 2005; Russo, 2003; Stevenson *et al.*, 2005; Link and Di Fiore, 2006). If there are particularly large-seeded plant species that rely primarily on primates for seed dispersal, then these monkeys are their most likely dispersers. Finally, these are the most populous monkey species in the area. Together they account for a majority of the region's entire primate biomass (Peres, 1996, 2000a, 2000b; Peres and Dolman, 2000). They are thus capable of dispersing not only the largest seeds, but also the greatest numbers of seeds overall. In doing so, they are likely to have large impacts across a wide variety of plant species.

Study site and study species

This research took place in the Napo region of Eastern Ecuador, in Yasuní National Park, a UNESCO Man and the Biosphere Reserve in Eastern Ecuador of approximately 900 000 ha (75°28′W, 0°42′S; Figure 6.1). It is a region of

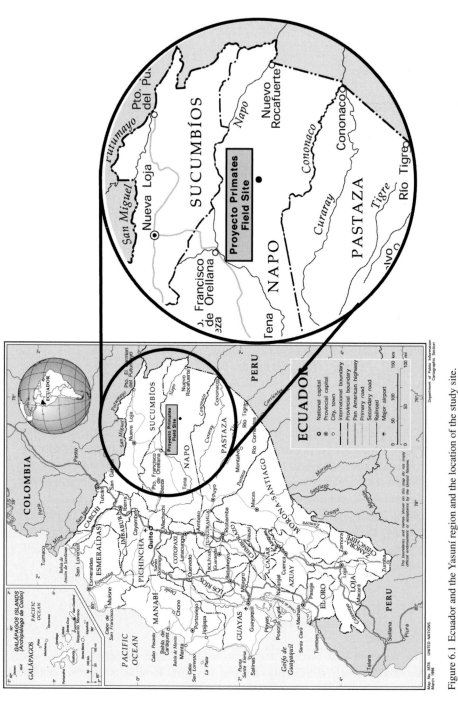

Figure 6.1 Ecuador and the Yasuní region and the location of the study site.

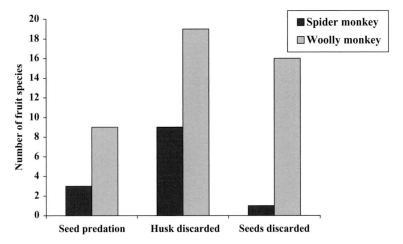

Figure 6.2 A comparison of fruit processing by spider monkeys and woolly monkeys. Spider monkeys prey on fewer seeds, and discard fewer fruit husks and seeds than do woolly monkeys. Instead, spider monkeys typically swallow fruits whole.

extensive primary rain forest threatened by development for petroleum extraction, bushmeat hunting, and slash-and-burn cultivation. In 1994 researchers from the University of California, Davis, established the Proyecto Primates trail system as a community reserve in the area (Di Fiore, 1997).

Yasuní lies within the "tropical Andes" biodiversity hot spot (Myers *et al.*, 2000). The area is characterized by primary *terra firme* rain forest with possibly the highest tree species diversity on Earth (Gentry, 1988; Pitman *et al.*, 1999) and abundant wildlife. The trail system covers an area of approximately 350 hectares. The climate is ever wet, without the pronounced dry season seen at some other Amazonian sites (Pitman *et al.*, 1999; Terborgh, 1983).

Data collection and analysis

The study began with a six-month period of habituation followed by a 12-month period of regular focal animal time sampling (Altmann, 1974; Martin and Bateson, 1993). In order to track seed ingestion and dispersal, a focal animal was followed for alternating periods of one to three consecutive days whenever possible, at which point one or two days of botanical investigation would commence before a new focal animal was chosen.

Each fruit-bearing food source was measured, tagged, and its location was recorded for mapping. Swallowed seeds were followed from ingestion to deposition whenever possible. Site spread and seed dispersion was measured with

Table 6.2 *Composition of the spider monkey community*

Age-sex class	# of animals	Names
Reproductive adult females	10–11	Mara, Blondie, Grace, Jane, Karen, Maya, Freckles, Iman, Kali, Mary?
Nulliparous adult or subadult females	3	Venus, Paloma, Charlotte
Juvenile and/or infant females	8–9	Fluffy *et al.*
Adult males	3	Lefty, Chimpo, Odysseus
Juvenile and/or infant males	2	Boy, Jesus
Total	26–28	

The study population was composed of 21 to 23 females and 5 males. This ratio of 4.4 females per male is one of the highest known. Three additional females appear to have been removed by hunters during the early months of the study.

a tape measure, and seeds were collected for identification and measurement. Gut-passed seeds were replaced on the forest floor at 766 simulated deposition sites marked with plastic surveyor's flags. Seed survivorship, germination, and seedling growth were monitored over a period of 10 months.

Results

Grouping, ranging, and population density

As elsewhere, spider monkeys at Yasuní live in a "fission–fusion" community in which multiple adult females' overlapping home ranges are cooperatively defended by groups of several adult males. Animals may range up to several kilometers per day. The spider monkey community at our site consisted of 26 to 28 animals or more, including 10 or 11 adult females with offspring and three adult males (Table 6.2). This is a fairly typical community size for Amazonian spider monkeys (Di Fiore and Campbell, 2007).

The home range of the most completely documented adult female spider monkey was estimated to be approximately 80 hectares in size, though her activities were typically concentrated within a core area of roughly 50 ha. In contrast, the ranges of the community's three adult males exceeded the boundaries of the entire 350-ha main trail system. These home ranges are comparable to or larger than others described in Amazonia (Di Fiore and Campbell, 2007), and with these data I estimate that spider monkey density at Yasuní was approximately 11.5 animals/km^2. This density is in the middle range of those found at other South American sites (Muckenhirn *et al.*, 1975; Bernstein *et al.*, 1976; Klein and Klein, 1976; van Roosmalen and Klein, 1988; P. Stevenson, personal communication), and considerably lower than those of some Central American spider monkey populations (e.g. 26–27 animals/ha, Cant, 1977; Gonzalez-Kirchner, 1999).

Fruit diet, dietary diversity and overlap

Frugivory was consistently high throughout the year. Ripe fruits made up between 64% and 100% of spider monkeys' monthly diets (average: 87%). If these data represent a typical annual pattern they suggest that spider monkeys at Yasuní are rarely driven to rely heavily on "fallback" resources such as insects, flowers or leaves. Spider monkeys fed on 349 individual fruiting plants during the study period. These plants came from at least 29 families and 44 different genera. At least 73 fruit morphospecies were consumed in all. The fruits chosen by spider monkeys included a high proportion of large-seeded lipid-rich fruits from the families Arecaceae, Lauraceae, Meliaceae and Myristicaceae (see Castellanos and Chanin, 1996). Spider monkeys spend 27% of all fruit feeding records eating lipid-rich fruits.

Seed spitting and swallowing

Of all the fruits in the diet, only a single species had its flesh swallowed and seeds discarded by spider monkeys. This was the palm *Astrocaryum chambira*, which has large yellow fleshy fruits with a single huge, pointed seed roughly 10 cm in length. In contrast, woolly monkeys were observed removing the flesh and spitting out some or all of the seeds of at least 16 fruit species, the seeds of virtually all of which are habitually swallowed by spider monkeys (Figure 6.3, Table 6.3). These plants include *Garcinia macrophylla*, *Guarea kunthiana*, *Matisia cordata*, *Otoba parvifolia*, *Persea* sp., *Prunus debilis*, *Socratea exorrhiza*, *Spondias mombin*, *Trichilia* sp., and two species of *Virola*.

I examined this seed spitting difference by performing a regression comparing the mean seed sizes of multiple fruit taxa eaten by the two monkeys with the numbers of those seeds swallowed. The results were clear. Woolly monkeys showed a very significant relationship: the larger the seed, the less likely woolly monkeys are to swallow them ($r^2 = 53.9\%$, $p = 0.002$). Spider monkeys in contrast showed no significant seed size preferences ($r^2 = 24.7\%$, $p = 0.06$).

Seed predation

Although woolly monkeys, like spider monkeys, disperse the seeds of nearly all ingested fruit species, woolly monkeys prey on the seeds of more plants than spider monkeys, regularly crushing and digesting the seeds of at least

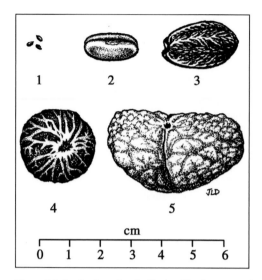

Figure 6.3 A selection of seeds dispersed by the two monkeys. Only a small proportion (16%) of monkey dung samples contained small seeds, such as *Cecropia* seeds (#1) whereas at least 96% of samples contained large seeds. Both monkey species dispersed large numbers of *Inga* seeds (#2). The seeds of *Spondias mombin* (#3) were the largest seeds passed by woolly monkeys, but spider monkeys dispersed seeds from *Ireartea deltoidea* (#4) and *Iryanthera* sp. (#5) as large as 27 mm in diameter.

nine plant species (Figure 6.3). Woolly monkeys also eat the unripe seeds of at least six fleshy-fruited species, including *Allophylus* sp. (Sapindaceae), *Cissus* sp. (Vitaceae), *Inga* sp. (Fabaceae), and two palms, *Socratea exorrhiza* and *Ireartea deltoidea* (Arecaceae). Approximately 80% of one *I. deltoidea* tree's entire crop of more than 600 fruits was destroyed by woolly monkeys within one 42-minute group feeding period.

A peak in seed-eating occurred in April, when both woollies and spiders fed on the abundant unripe seeds of two species of palms, *Ireartea deltoidea* and *Socratea exhorriza* (Arecaceae). Overall, however, seed predation by woollies and spiders was lower at Yasuní than at any other site, averaging only 5% of woollies' feeding records and 0.07% of spider monkey feeding records.

Unlike woolly monkeys, spider monkeys were never seen feeding on nuts or wind-dispersed fruits. Spider monkeys regularly prey upon the seeds of only one plant species, the palm *Ireartea deltoidea*, which is one of the most common plants in the monkeys' diets. Single incidents of spider monkey seed predation were observed for two additional plant species, *Astrocaryum chambira* (Arecaceae) and an unidentified Sapotaceous fruit. In both cases spider monkeys preyed upon seeds by removing them from unripe fleshy fruits.

Table 6.3 *Medium and large seeded (>5 mm) plant species dispersed by the two monkey species*

Plant family	Genus and species	Monkey species	Mean seed diameter (mm)
Anacardiaceae	*Spondias mombin*	*A. belzebuth, L. lagotricha*	16
Annonaceae	*Guatteria* sp.	*A. belzebuth, L. lagotricha*	5
Bombacaceae	*Matisia obliquifolia*	*A. belzebuth, L. lagotricha*	8
Bombacaceae	*Quararibea* sp.	*A. belzebuth, L. lagotricha*	12
Cecropiaceae	*Pourouma* sp.	*A. belzebuth, L. lagotricha*	10
Cecropiaceae	*Pourouma bicolor*	*A. belzebuth, L. lagotricha*	9
Clusiaceae	*Clusia* sp.	*A. belzebuth, L. lagotricha*	5
Clusiaceae	*Garcinia* sp.	*A. belzebuth, L. lagotricha*	15
Fabaceae	*Brownea* sp.	*A. belzebuth, L. lagotricha*	13
Fabaceae	*Inga* sp.	*A. belzebuth, L. lagotricha*	9
Fabaceae	*Parkia* sp.	*A. belzebuth, L. lagotricha*	9
Loganiaceae	*Strychnos* sp.	*A. belzebuth, L. lagotricha*	8
Meliaceae	*Paullinia* sp.	*A. belzebuth, L. lagotricha*	10
Moraceae	*Clarisia racemosa*	*A. belzebuth, L. lagotricha*	15
Myristicaceae	*Virola sp. #1*	*A. belzebuth, L. lagotricha*	7
Rosaceae	*Prunus debilis*	*A. belzebuth, L. lagotricha*	12
Sapindaceae	*Allophylus* sp.	*A. belzebuth, L. lagotricha*	7
Violaceae	*Leonia glycicarpa*	*A. belzebuth, L. lagotricha*	6
Vitaceae	*Cissus* sp.	*A. belzebuth, L. lagotricha*	5
Arecaceae	***Oenocarpus bataua***	*A. belzebuth*	21
Arecaceae	*Ireartea deltoidea*	*A. belzebuth*	25
Bombacaceae	*Matisia cordata*	*A. belzebuth*	14
Lauraceae	*Ocotea* sp.	*A. belzebuth*	15.3
Meliaceae	***Guarea kunthiana***	*A. belzebuth*	17.5
Myristicaceae	*Virola sp. #2, #3, #4*	*A. belzebuth*	2–25
Myristicaceae	***Iryanthera jurvensis***	*A. belzebuth*	27
Sapotaceae	*Pouteria* sp.	*A. belzebuth*	15

While spider monkeys habitually swallowed and dispersed the seeds of all of these plants, seeds from the final eight taxa listed were never found in woolly monkey dung. Woolly monkeys typically ignored plants shown in bold type. The remaining five taxa were often eaten by woolly monkeys, but were never witnessed being dispersed by them. Unlike spider monkeys, woollies often appear to act as "fruit thieves" for these plants, at least for the larger-seeded members of these genera.

Seed ingestion and dispersal

Unlike many African and Asian monkeys, primates in the Neotropics have no cheek pouches, and the two species discussed in this chapter rarely move away from feeding sources in order to masticate foods and spit seeds. Instead, they usually swallow them. Spider monkeys ingested and dispersed large seeds from at least 17 plant families and 25 genera. Nearly all of the dung samples produced by spider monkeys contain intact seeds (99%, $n = 176$). Seeds dispersed by spider monkeys were significantly larger than those of woolly monkeys (mean volumes 2303 mm^3 versus 756 mm^3; Mann–Whitney $z = -11.296, p < 0.0001$). The largest seeds dispersed by woolly monkeys were 3.6 cm^3 in volume, while the largest seeds dispersed by spider monkeys had volumes of more than 14.8 cm^3. The two taxa disperse similar proportions of small seeds less than 5 mm in length, such as those from *Ficus* spp. and *Cecropia* spp. (Moraceae). Woolly monkeys, however, disperse significantly more medium-sized seeds (5–10 mm) and significantly fewer large seeds (>10 mm) per deposit than do spider monkeys (Figures 6.4 and 6.5; $\chi^2 = 60.6$, df $= 1, p < 0.0001$; $\chi^2 = 13.17$, df $= 1, p < 0.0003$).

Spider monkeys habitually swallow the fruits in their diet whole, with a minimum of mastication. Woolly monkeys typically do the same, but they are more selective of the seeds they will swallow. Spider monkeys swallow and disperse the seeds of all of the plants listed in Table 6.3. Woolly monkeys fed on all but two of these plants, but acted as "fruit thieves" for at least seven of them, stripping the large seeds of flesh and dropping the vast majority, if not all of them, unswallowed. Underneath parent crowns these dropped seeds are at high risk of predation by weevils, fungal infection, or other density-dependent factors (Howe, 1980, 1990). The seeds of some large-seeded plants such as *Ocotea* sp. (Lauraceae) and several *Virola* spp. (Myristicaceae) were common in spider monkey dung but never found in woolly monkey dung despite often high proportions of these fruits in the woolly monkey diet. Woolly monkeys did not swallow seeds larger than 17 mm in diameter, but spider monkeys swallowed and dispersed seeds as large as 27 mm in diameter. Plants at Yasuní with seed diameters larger than 17 mm may therefore depend on spider monkeys exclusively among primates for endozoochorous dispersal.

Three very large-seeded plant species in the study were eaten and dispersed by spider monkeys alone. Woolly monkeys were never observed feeding on the lipid-rich fruits of *Oenocarpus bataua* (Arecaceae) or *Iryanthera jurvensis* (Myristicaceae) and they only rarely sampled those of *Guarea kunthiana* (Meliaceae). Sympatric white-fronted capuchin monkeys, *Cebus albifrons*, fed on two of these plants, but these much smaller-bodied monkeys (*c.* 2.8 kg) never swallowed these seeds. Woolly monkeys typically passed through fruit-laden crowns of these trees without stopping. The spider monkey therefore appears to be the only primate disperser of these plants at Yasuní, and possibly the only endozoochorous disperser, because no other vertebrate species was seen to swallow these seeds.

(a)

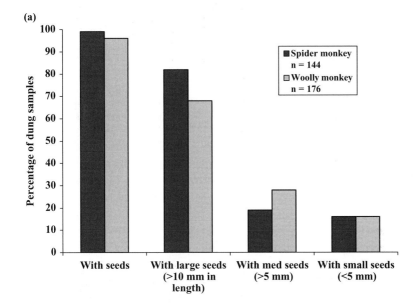

(b)

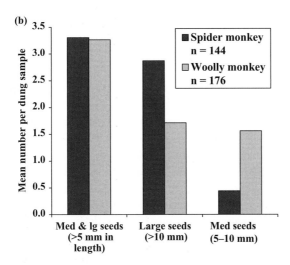

Figure 6.4 Comparison of size and number of dispersed seeds. (a) Sizes of dispersed seeds and (b) number of seeds dispersed per dung sample. Spider monkey dung samples are more likely to contain large seeds (>10 mm), and spider monkeys disperse significantly higher numbers of large seeds per deposit than do woolly monkeys ($\chi^2 = 13.17$, df $= 1$, $p < 0.0003$).

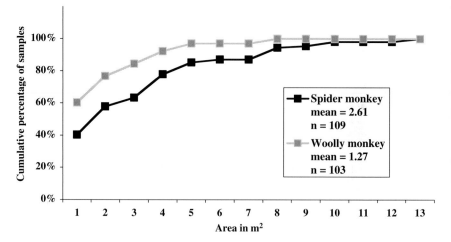

Figure 6.5 Deposition site spread. A typical spider monkey dung deposit is spread over an area more than twice as large as a typical woolly monkey deposit. The largest woolly monkey deposits covered 9 m^2, while the largest spider monkey deposits covered 13 m^2.

Deposition site spread

The seeds dispersed by these two primates, unlike those of howlers (*Alouatta* spp.) or chimpanzees (*Pan* spp.), are not usually deposited in fecal clumps, but are spread over several meters of forest floor (Andresen, 1999; Chapman, 1989; Estrada and Coates-Estrada, 1986; Howe, 1980, 1989; Julliot, 1994, 1996; Lambert, 1997). Figure 6.5 demonstrates that the spread of seeds in a typical spider monkey deposit covers an area more than double that of a typical woolly monkey deposit. The largest woolly monkey deposition sites covered 9 m^2, while the largest spider monkey sites covered as much as 13 m^2. This difference in spread may be explained partly because spider monkeys are typically two or three meters higher in the canopy than woolly monkeys (17 m versus 14 m), so their dung hits the ground with more momentum and hits more branches along the way (see Dew and Wright, 1998). Tightly clustered seeds may be more vulnerable to density-dependent mortality, just like seeds clustered under a parent plant (Howe, 1980, 1989, 1990; Howe and Smallwood, 1982; Janzen, 1982; Chapman, 1989; Notman *et al.*, 1996; Lambert, 1997; Andresen, 1999; Forget *et al.*, 2000). Seeds in a spider monkey deposit may thus be more likely to survive than those in other monkey deposits, because the seeds land further from each other on the ground.

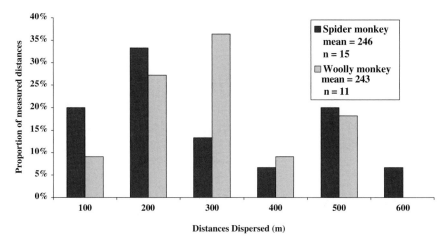

Figure 6.6 A comparison of spider monkey and woolly monkey seed dispersal distances.

Dispersal distances

The seed dispersal distances measured for the two primate species are illustrated in Figure 6.6. The distances dispersed averaged around 245 m, and ranged between 50 and 500 m. Of all Neotropical primates studied, only howlers (*Alouatta* spp.), and capuchins (*Cebus* spp.) are capable of dispersing seeds comparable distances (see Stevenson, 2000 for review).

A second measure of potential dispersal distance, straight-line distance between sleep sites, is provided in Figure 6.7. Animals' convoluted day paths can limit the correlation between day path length and dispersal distance (Stevenson, 2000). Since these primates often defecate at dawn, and since many swallowed seeds are not deposited until the morning following ingestion, distances between sleeping sites may be good indicators of dispersal distance. The two species have mean inter-sleep-site distances between 400 and 450 m and similar ranges from approximately 75 m to 600 or 700 m, except for one spider monkey outlier at 1200 m. The outlier is this study's only measured sleeping site distance measurement for male spider monkeys. Male spider monkeys range over multiple female home ranges and therefore move much farther than females. The home range of a female spider monkey is around 80 ha, but a male's range is larger than the 350-ha area of the entire Proyecto Primates trail system. If the seed dispersal distances of male spider monkeys were accurately measured, spider monkey mean seed dispersal distances could be significantly longer (see Link and Di Fiore, 2006).

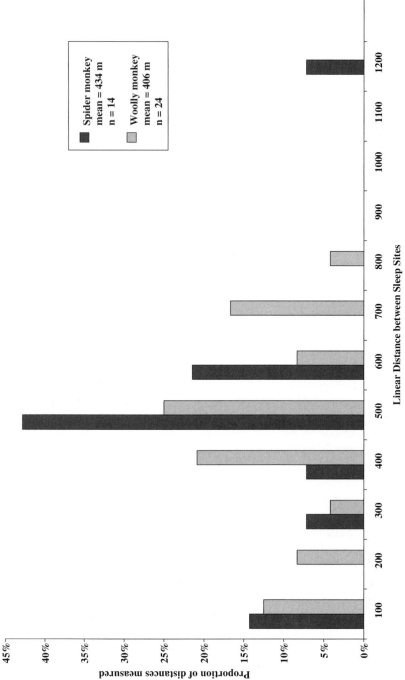

Figure 6.7 Distance between sleeping sites. Distances between sleep sites may affect seed dispersal distance because many seeds are deposited at dawn under sleep sites. These distances appear to be similar for the two monkeys, but only one of these measurements, the longest, is from male spider monkeys. The very large ranges of male spider monkeys make them capable of dispersing seeds over several square kilometers (Link and Di Fiore, 2006).

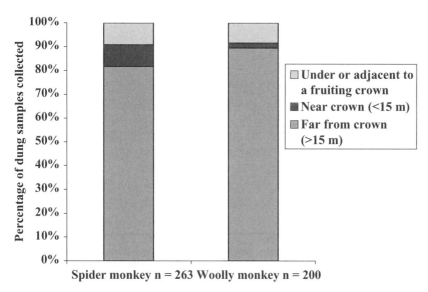

Figure 6.8 Locations of dung samples. The great majority of seeds dispersed by both woollies and spiders (80%–90%) end up far from fruiting crowns. Spider monkeys, however, deposit a significantly higher proportion (9% versus 2%) of seeds in the "hot zone" of high density-dependent mortality within 15 m of fruiting crowns ($\chi^2 = 10.31$, df $= 1, p < 0.0002$).

Directed dispersal

When the locations of seed deposits are compared (Figure 6.8), the great majority of seeds dispersed by these monkeys (roughly 90%) end up tens of meters from parent crowns. The same proportion of seeds (9%) are gut-dispersed by the two monkeys to locations underneath fruiting crowns. Spider monkeys, however, deposit a significantly higher proportion (9% versus 2%) of seeds in the zone of high density-dependent mortality within 15 m of fruiting crowns ($\chi^2 = 10.31$; df $= 1, p < 0.002$; Howe, 1990, 1993; Howe and Smallwood, 1982). This is because spider monkeys have a habit of moving rapidly between food sources, feeding heavily then resting in nearby trees, sometimes for several hours, before returning to feed again. Spider monkeys also appear to have a greater tendency to choose the same individual trees repeatedly as sleeping sites, and this behavior could further bias the distribution of seeds they disperse (Russo *et al.*, 2005).

Numbers dispersed

The estimated numbers of seeds dispersed per animal per day are similar for the two species (Figure 6.9a). There may be a trend for a typical spider monkey to

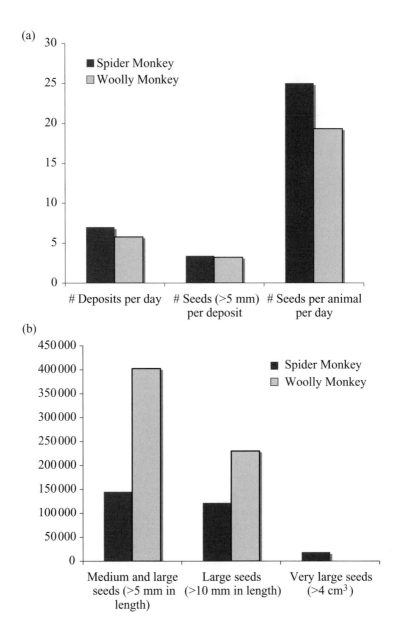

Figure 6.9 Numbers of seeds dispersed per animal per day (a), and per primate species per year (b). Spider monkeys disperse more seeds per animal per day than do woolly monkeys. Woolly monkeys overall, however, disperse far more medium and large seeds per unit of time at this site. This is because woolly monkeys live at a density nearly three times that of spider monkeys at our site. Only spider monkeys, however, disperse seeds larger than 4 cm^3. Some large-seeded plants therefore appear to have no primate seed dispersers other than spider monkeys. The same cannot be said for any known plants dispersed by woolly monkeys.

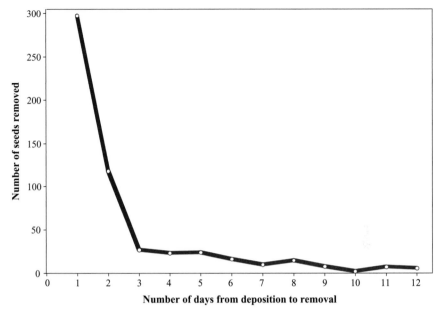

Figure 6.10 Timing of primate-dispersed seed removal events: number of days from seed deposition to destruction or removal. The most dangerous time for primate-dispersed seeds was the first three days following deposition, when 41% of all removal occurred. Average half-life for all primate-dispersed seeds and seedlings was 21 days.

disperse more seeds, but with these sample sizes (woolly monkey $n = 210$ defecation events, spider monkey $n = 264$ defecation events) it is not a statistically significant difference. A clear quantitative difference comes to light, however, when the number of seeds dispersed across the study area by the two species overall is estimated (Figure 6.9b). Comparing the numbers of medium and large seeds dispersed per species per unit of time in the 234-ha study area shows that woolly monkeys disperse far more. This is because woolly monkeys live at a density roughly three times that of spider monkeys at Yasuní. They therefore disperse more seeds, roughly 3193 versus 2378 medium and large seeds/ha/yr.

Seed removal

Sixty-two percent of all primate-dispersed seeds were destroyed or removed by secondary dispersers or seed predators before germination. Seed removal was highest during the first three days after deposition, when 41% of all removal occurred (Figure 6.10). It was impossible to determine what proportions of removed seeds were destroyed or successfully secondarily dispersed, but 6% of all seeds were clearly destroyed at deposition sites.

Mean half-life for primate-dispersed seeds was 21 days. Mortality quickly declined after the first few days and became asymptotic as it approached 15%. If seed removal by secondary dispersers equals mortality, then more than 15% of primate-dispersed seeds can be expected to survive to the germination stage. Since some secondarily dispersed seeds presumably survive, the overall survivorship of primate-dispersed seeds should be somewhat higher.

Using nonparametric Kaplan–Meier survivorship analysis I found that survivorship curves of different plant species were highly statistically significantly different (Trend Rank Test, Logrank (Mantel–Cox) $\chi^2 = 13.7$; df = 18; $p < 0.0002$; Breslow–Gehan Wilcoxon $\chi^2 = 28.23$, df = 18, $p < 0.0001$). Removal is correlated with seed taxon, but not with seed size (Notman *et al.*, 1996). Removal agents do not appear to detect larger seeds more easily or remove them preferentially within any particular taxon (General Linear Model, GLM, $F_{1,329} < 0.005$; $p = 0.972$). Instead they appear to prefer particular large-seeded plant taxa (GLM, $F_{1,7} = 5.83$; $p < 0.0005$). Plants such as *Ocotea* sp. (Lauraceae), *Inga* sp. (Fabaceae), and *Pouteria* sp. (Sapotaceae), have average half-lives of only a single day, while *Spondias mombin* (Anacardiaceae) has a half-life of 122 days (Table 6.3). Within plant taxa there were no significant differences in removal or survivorship of primate-dispersed seeds.

Germination and predation of seedlings

Germination success, seedling mortality and fates of primate-dispersed seeds are listed in Table 6.4. A total of 238 (23%) of 1056 primate-dispersed seeds germinated, and of these 111, or 11% of all primate-dispersed seeds, successfully established as seedlings and survived until the study's end. These seedlings grew to heights of up to 68 cm over the course of the study. An additional 164 seeds, or 16%, remained in place and ungerminated. Of 47 identified seedling mortality agents, 94% appeared to be vertebrates. Twenty-six (55%) of these were unidentified vertebrate herbivores, 18 (39%) were rodents (13 [28%] caviomorph rodents and 5 [11%] mouse-sized rodents) and only three (6%) were invertebrate herbivores.

More than 37% of all the seeds dispersed in the guts of these monkeys remained in place and viable long after the initial period of high seed removal. These are considerably higher seed survival rates than those observed in three previous studies of seed dispersal by ateline primates (Chapman, 1989; Estrada and Coates-Estrada, 1991; Andresen, 1999). I propose two possible reasons

Table 6.4 *Mortality and germination of seeds dispersed by the two monkey species*

	Woolly monkey		Spider monkey	
Fate	#	%	#	%
Destroyed or removed pre-germination	243	58	412	65
Germinated	97	23	144	23
Seed intact and viable at study's end	84	20	80	13
Unknown	5	1	3	0.5
Total	424	100	636	100

Germination, removal and mortality rates of seeds dispersed by the two monkeys were similar.

for this difference in survivorship. The first is the fact that unlike seeds dispersed by the howlers (*Alouatta* spp.) in the previous studies, these seeds were not deposited in dense fecal clumps, but were instead scattered across several meters of forest floor and thus better protected from density-dependent mortality (Price and Jenkins, 1986; Notman *et al.*, 1996; but see Forget *et al.*, 2000). A second possible reason for these seeds' higher survivorship is the ever-wet Amazonian climate in Yasuní, which might facilitate germination and reduce seed mortality.

Discussion

This study has demonstrated that spider monkeys rank among the best known seed dispersers on several levels: they prey upon very few seeds, they swallow large quantities of seeds in a wide range of sizes, and they deposit most of these seeds in areas quite far from parent trees (see Schupp, 1993; Schupp and Fuentes, 1995; Hulme, 1998). The seedling trials indicate that many seeds dispersed by spider monkeys are not destroyed or secondarily dispersed. Instead, high proportions of them successfully sprout *in situ* and survive the establishment stage.

Past studies of seed dispersal erred in assuming the seed shadows produced by different primate species to be equivalent (Howe, 1989, 1990; Levey *et al.*, 1994). This study confirms that this is not the case, even among sympatric ripe fruit specialists (see Mittermeier and van Roosmalen, 1981; Garber, 1986; Rowell and Mitchell, 1991; Gautier-Hion *et al.*, 1993; Zhang and Wang, 1995; Lambert, 1997, 1999; Dew and Wright, 1998; Kaplin and Moermond, 1998; Overdorff and Strait, 1998; Andresen, 1999; Knogge, 1999). Because frugivorous primates differ in seed processing, digestive physiology, population density, ranging patterns, habitat selection and defecation patterns, they may produce quite different seed shadows. These differences may affect the subsequent movement, survival and recruitment of dispersed plant embryos. Although the species of fleshy-fruited plants eaten by these two primate species overlapped almost entirely (Dew, 2001, 2005), the monkeys differed in their roles as seed dispersers.

One is unlikely to find any Neotropical mammal dispersing higher quantities of large canopy plant seeds per kilogram of biomass, or one with longer mean seed dispersal distances than *Ateles* (see Knogge, 1999; and see review in Stevenson, 2000). Amazonian spider monkeys disperse the large seeds of multiple plants which are poorly dispersed by or even ignored by their closest relatives and ecological competitors, the woolly monkeys. Spider monkeys also scatter seeds further from each other upon deposition. Spider monkeys can move more than three kilometers in an hour – farther than woolly monkeys travel in a typical day. They therefore have the potential to disperse seeds much greater distances than woolly monkeys. When male spider monkeys are accurately tracked, I expect that they will be found to disperse some seeds as far as several kilometers away from a parent plant.

Spider monkeys are not, however, the most effective primate dispersers of all the fleshy-fruited plants in their diet. Sympatric woolly monkeys, with their higher population densities, disperse larger numbers of seeds from a variety of plant species with seeds that fall within woolly monkeys' preferred size range. Spider monkeys' effectiveness may be further limited by their habit of resting and defecating in crowns adjacent to feeding trees. Such deposition sites along food source "traplines" may be very near or very far from a parent plant, but they are likely to be near the "hot zones" of simultaneously fruiting conspecific plants. Finally, spider monkeys are often solitary, and this independence from group movement may allow them more opportunity to regularly revisit the same sleeping sites. Since defecation often occurs at dawn at sleeping sites, this behavior offers an additional possibility for "contagious" or clumped seed dispersal by spider monkeys. Thus, behavioral differences between dispersers may produce more dramatic contrasts in seed shadows than differences in passage time, dispersal distance or population density.

Frugivore population density is clearly one of the most important determinants of seed removal. Yasuní's dense populations of woolly monkeys thus remove and disperse many more seeds on average than its spider monkeys. There is a second, possibly very important, relationship between population density and seed dispersal, however. If a frugivore's population density is high in a particular area, that species is also likely to live in relatively smaller home ranges than in areas where its density is lower. Yasuní's woolly monkeys, for example, live at their highest known densities, and have their smallest and least overlapping home ranges (Peres, 1996, 1997; Di Fiore, 1997; P. Stevenson, personal communication).

The effects of frugivore social systems on seed dispersal may be very important and merit additional study. The seed dispersal capabilities of male spider monkeys are much larger than those of females, because male spider monkeys travel widely across multiple females' home ranges. The maximum seed dispersal distances of this "fission–fusion" species is therefore potentially much greater than that of group-living monkeys. A solitary spider monkey, however, may remove a day's entire offering of ripe fruits from a crown, then disperse the seeds in only three or four defecation sites. An individual woolly monkey would disperse these seeds over roughly the same number of droppings, but since woolly monkeys live in troops of 20–25 animals, more individual woolly monkeys will feed in that crown. They will then disperse the seeds from many more individual locations across the hundreds of square meters of forest occupied by the troop as it moves.

Plants with seeds dispersed by *both* monkey species may benefit from the combination of two distinct and complementary seed shadows. Spider monkeys appear to disperse seeds to areas both near and very far from parent trees, while woolly monkeys fill in the regions in between. This combination of seed shadows should diffuse the "contagious" or directed dispersal patterns of both primates, resulting in more even dispersal overall (Fleming *et al.*, 1993; Schupp, 1993; Schupp and Milleron, 2000). Plants may therefore be selected to produce fruits that are attractive to both primate species, because the larger the coterie of complementary seed dispersers, the more even and widespread the seed dispersal curve will be.

The results of this study indicate that despite the taxonomic, morphological and dietary similarities of these two monkeys, they are not functionally equivalent seed dispersers. Spider monkeys have a greater tolerance for swallowing large-seeded and lipid-rich fruits. This facilitates the dispersal of several species of plants that are dispersed rarely, if at all, by other animals.

If woolly monkeys were removed from the system, there appear to be no plant species that would not be dispersed by spider monkeys. The reverse is not true. Without spider monkeys, there are multiple Amazonian plants that would lose their only known dispersal vector. A host of animals may rely on these plants for food, but only spider monkeys have been found to disperse them. Since spider monkeys range across the Neotropics, while woolly monkeys are confined to Amazonia, the seed dispersal importance of spider monkeys outside of Amazonia is presumably even more dominant. Pound for pound, the spider monkey is likely to be the most important disperser of large fleshy-fruited seeds wherever it is found. With this key ecological role and its sensitivity to hunting pressure, this threatened primate can serve as a Neotropical forest indicator species whose removal signifies a disturbed ecosystem.

Acknowledgements

Financial support for this research came from an NSF Doctoral Dissertation Improvement Grant, a generous grant from the Douroucouli Foundation, and a grant from Primate Conservation Incorporated. Many people helped this project. My deepest thanks go to Peter Rodman, Scott Suarez, Tony Di Fiore and Kristin Phillips. I would also like to thank the Huaorani, the Quichua, and the Ecuadorean people whose reserve at Yasuní is a gem.

References

Ahumada, J.A. (1989). Behavior and social structure of free ranging spider monkeys (*Ateles belzebuth*) in La Macarena. *Field Studies of New World Monkeys, La Macarena, Colombia*, **2**, 7–31.

Altmann, J. (1974). Observational study of behaviour: sampling methods. *Behaviour*, **49**, 227–267.

Andresen, E. (1999). Seed dispersal by monkeys and the fate of dispersed seeds in a Peruvian rain forest. *Biotropica* **31**, 145–158.

Bernstein, I. S., Balcaen, P., Dresdale, L., *et al.* (1976) Differential effects of forest degradation on primate populations. *Primates*, **17**, 401–411.

Birkinshaw, C. R. (1999). The importance of black lemur (*Eulemur macaco*) for seed disersal in Lokobe Forest, Nosy Be. In *New Directions in Lemur Studies*, ed. B. Rakotosamimanana, New York: Plenum Press, pp. 189–199.

Bourliere, F. (1985). Primate communities: their structure and role in tropical ecosystems. *Int. J. Primatol.*, **6**, 1–26.

Cant, J. G. H. (1977). Ecology, locomotion and social organization of spider monkeys (*Ateles geoffroyi*). Unpublished Ph.D. thesis, University of California, Davis.

Cant, J. G. H. (1990). Feeding ecology of spider monkeys (*Ateles geoffroyi*) at Tikal, Guatemala. *Hum. Evol.*, **5**, 269–281.

Castellanos, H. G. and Chanin, P. (1996). Seasonal differences in food choice and patch preference of long-haired spider monkeys (*Ateles belzebuth*). In *Adaptive Radiations of Neotropical Primates*, ed. M. A. Norconk, A. L. Rosenberger and P. A. Garber, New York: Plenum Press, pp. 451–466.

Chapman, C. (1987). Flexibility in diets of three species of Costa Rican primates. *Folia Primatol.*, **49**, 90–105.

Chapman, C. A. (1989). Primate seed dispersal: the fate of dispersed seeds. *Biotropica*, **21**, 148–154.

Chapman, C. A. (1995). Primate seed dispersal: coevolution and conservation implications. *Evol. Anthropol.*, **4**, 74–82.

Chapman, C. A. and Onderdonk, D. A. (1998). Forests without primates: primate/plant codependency. *Am. J. Primatol.*, **45**, 127–142.

Cox, P. A., Elmqvist, T., Pierson, E. D. and Rainey, W. E. (1991). Flying foxes as strong interactors in South Pacific island ecosystems: a conservation hypothesis. *Conserv. Biol.*, **5**, 448–454.

Defler, T. R. and Defler, S. B. (1996). Diet of a group of *Lagothrix lagotricha lagotricha* in southeastern Colombia. *Int. J. Primatol.*, **17**, 161–190.

Dew, J. L. (2001). Synecology and seed dispersal in woolly monkeys (*Lagothrix lagotricha poeppigii*) and spider monkeys (*Ateles belzebuth belzebuth*) in Parque Nacional Yasuní, Ecuador. Unpublished Ph.D. thesis, University of California, Davis.

Dew, J. L. (2005). Foraging, food choice and food processing by sympatric ripe-fruit specialists: *Lagothrox lagotricha poepigii* and *Ateles belzebuth belzebuth*. *Int. J. Primatol.*, **26**, 1107–1135.

Dew, J. L. and Wright, P. C. (1998). Frugivory and seed dispersal by four species of primates in Madagascar's eastern rainforest. *Biotropica*, **30**, 425–437.

Di Fiore, A. F. (1997). Ecology and behavior of lowland woolly monkeys (*Lagothrix lagotricha poeppigii*, Atelinae) in eastern Ecuador. Unpublished Ph.D. thesis, University of California, Davis.

Di Fiore, A. and Campbell, C. J. (2007). The atelines: variation in ecology, behavior, and social organization. In *Primates in Perspective*, ed. C. J. Campbell, A. Fuentes, K. C. MacKinnon, M. Panger and S. K. Bearder, New York: Oxford University Press, pp. 155–185.

Emmons, L. H. (1990). *Neotropical Rainforest Mammals: A Field Guide*. Chicago: University of Chicago Press.

Estrada, A. and Coates-Estrada, R. (1986). Frugivory in howling monkeys (*Alouatta palliata*) at Los Tuxtlas, Mexico: dispersal and the fate of seeds. In *Frugivores and Seed Dispersal*, ed. A. Estrada and T. H. Fleming, Dordrecht: Dr W. Junk, pp. 93–104.

Estrada, A. and Coates-Estrada, R. (1991). Howler monkeys (*Alouatta palliata*), dung beetles (Scarabaeidae) and seed dispersal: ecological interactions in the tropical rain forest of Los Tuxtlas, Mexico. *J. Trop. Ecol.*, **7**, 459–474.

Fleming, T. H., Venable, D. L. and Herrera, L. G. (1993). Opportunism vs. specialization: the evolution of dispersal strategies in fleshy-fruited plants. *Vegetatio*, **107–108**, 107–120.

Forget, P. M., Milleron, T., Feer, F., Henry, O. and Dubost, G. (2000). Effects of dispersal pattern and mammalian herbivores on seedling recruitment for *Virola michelii* (Myristicaceae) in French Guiana. *Biotropica*, **32**, 452–462.

Garber, P. A. (1986). The ecology of seed dispersal in two species of callitrichid primates (*Saguinus mystax* and *Saguinus fuscicollis*). *Am. J. Primatol.*, **10**, 155–170.

Gautier-Hion, A. and Michaloud, G. (1989). Are figs always keystone resources for tropical frugiorous vertebrates? A test in Gabon. *Ecology*, **70**, 1826–1833.

Gautier-Hion, A., Gautier, J. P. and Maisels, F. (1993). Seed dispersal versus seed predation: an inter-site comparison of two related African monkeys. In *Frugivory and Seed Dispersal: Ecological and Evolutionary Aspects*, ed. T. H. Fleming and A. Estrada, Dordrecht: Kluwer, pp. 237–244.

Gentry, A. H. (1983). Dispersal ecology and diversity in neotropical forest communities. In *Dispersal and Distribution*, ed. K. Kubitzki, Berlin: Verlag Paul Parey, pp. 303–314.

Gentry, A. H. (1988). Tree species richness of upper Amazonian forests. *Proc. Natl. Acad. Sci. USA*, **85**, 156–159.

Gilbert, L. E. (1980). Food web organization and conservation of neotropical diversity. In *Conservation Biology: An Evolutionary Perspective*, ed. M. E. Soule and B. Wilcox, Sunderland, MA: Sinauer Associates, pp. 11–34.

Given, D. R. (1995). Biological diversity and the maintenance of mutualisms. In *Islands, Biological Diversity and Ecosystem Function*, ed. L. L. Loope and H. Adersen, Berlin: Springer-Verlag, pp. 149–162.

Gonzalez-Kirchner, J. P. (1999). Habitat use, population density and subgrouping pattern of the Yucatan spider monkey (*Ateles geoffroyi yucatanensis*) in Quintana Roo, Mexico. *Folia Primatologica*, **70**(1), 55–60.

Howe, H. F. (1977). Bird activity and seed dispersal of a tropical wet forest tree. *Ecology*, **58**, 539–550.

Howe, H. F. (1980). Monkey dispersal and waste of a neotropical fruit. *Ecology*, **6**, 944–959.

Howe, H. F. (1984). Implications of seed dispersal by animals for tropical reserve management. *Biol. Conserv.*, **30**, 261–281.

Howe, H. F. (1989). Scatter versus clump dispersal and seedling demography: hypotheses and implications. *Oikos*, **79**, 417–426.

Howe, H. F. (1990). Seed dispersal by birds and mammals: implications for seedling demography. In *Reproductive Ecology of Tropical Forest Plants*, ed. K. S. Bawa and M. Hadley, Park Ridge, NJ: Parthenon Publishing, pp. 191–218.

Howe, H. F. (1993). Aspects of variation in a neotropical seed dispersal system. In *Frugivory and Seed Dispersal: Ecological and Evolutionary Aspects*, ed. T. H. Fleming and A. Estrada, Dordrecht: Kluwer Academic Publishers, pp. 149–162.

Howe, H. F. and Smallwood, J. (1982). Ecology of seed dispersal. *Ann. Rev. Ecol. Syst.*, **13**, 201–228.

Hulme, P. E. (1998). Post-dispersal seed predation: consequences for plant demography and evolution. *Persp. Plant Ecol. Evol. Syst.*, **1**, 32–46.

Janzen, D. H. (1982). Removal of seeds from horse dung by tropical rodents: influence of habitat and amount of dung. *Ecology*, **63**, 1887–1990.

Janzen, D. H. (1985). *Spondias mombin* is culturally deprived in megafauna-free forest. *J. Trop. Ecol.*, **1**, 131–155.

Julliot, C. (1994). Frugivory and seed dispersal by red howler monkeys: evolutionary aspects. *Revue d'Ecologie (Terre et Vie)*, **49**, 331–341.

Julliot, C. (1996). Seed dispersal by the red howler monkey (*Alouatta seniculus*) in the tropical rain forest of French Guiana. *Int. J. Primatol.*, **17**, 239–258.

Kaplin, B. A. and Moermond, T. C. (1998). Variation in seed handling by two species of forest monkey in Rwanda. *Am. J. Primatol.*, **45**, 83–101.

Klein, L. L. and Klein, D. J. (1976). Neotropical primates, aspects of habitat usage, population density, and regional distribution in La Macarena, Colombia. In

Neotropical Primates: Field Studies and Conservation, ed. R. W. Thorington and P. G. Heltne, Washington DC: National Academy of Sciences, pp. 70–78.

Klein, L. L. and Klein, D. J. (1977). Feeding behavior of the Colombian spider monkey. In *Primate Ecology*, ed. T. H. Clutton-Brock, London: Academic Press, pp. 504–539.

Knogge, C. (1999). Seed dispersal by two sympatric tamarin species *Saguinus mystax* and *Saguinus fuscicollis. Neotrop. Primates*, **7**, 91–93.

Lambert, J. E. (1997). Digestive strategies, fruit processing, and seed dispersal in the chimpanzees (*Pan troglodytes*) and redtail monkeys (*Cercopithecus ascanius*) of Kibale National Park, Uganda. Unpublished Ph.D. thesis, University of Illinois at Urbana-Champaign, Urbana, IL.

Lambert, J. E. (1999). Seed handling in chimpanzees (*Pan troglodytes*) and redtail monkeys (*Cercopithecus ascanius*): implications for understanding hominoid and cercopithecine fruit-processing strategies and seed dispersal. *Am. J. Phys. Anthropol.*, **109**, 365–386.

Lambert, J. E. and Garber, P. A. (1998). Evolutionary and ecological implications of primate seed dispersal. *Am. J. Primatol.*, **45**, 9–28.

Levey, D. J., Moermond, T. C. and Denslow, J. S. (1994). Frugivory: an overview. In *La Selva: Ecology and Natural History of a Neotropical Rain Forest*, ed. L. A. McDade, H. A. Bawa, K. S. Hespenheide and G. S. Hartshorn, Chicago: University of Chicago Press, pp. 282–294.

Lieberman, M. and Lieberman, D. (1986). An experimental study of seed ingestion and germination in a plant-animal assemblage in Ghana. *J. Trop. Ecol.*, **2**, 113–126.

Link, A. and Di Fiore, A. (2005). Seed dispersal by spider monkeys and its importance in the maintenance of neotropical rain-forest diversity. *J. Trop. Ecol.*, **22**, 235–246.

Link, A. and Di Fiore, A. (2006). Seed dispersal by spider monkeys and its importance in the maintenance of neotropical rain-forest diversity. *J. Trop. Ecol.*, **22**, 335–346.

Martin, P. and Bateson, P. (1993). *Measuring Behaviour: An Introductory Guide.* 2nd edn. Cambridge: Cambridge University Press.

Milton, K. (1981). Food choice and digestive strategies of two sympatric primate species. *Am. Nat.*, **117**, 496–505.

Mittermeier, R. A. (1987). Effects of hunting on rain forest primates. In *Monographs in Primatology.* Vol. 9: *Primate Conservation in the Tropical Rain Forest*, ed. C. W. Marsh and R. A. Mittermeier, New York: Alan R. Liss, Inc., pp. 109–146.

Mittermeier, R. A. and van Roosmalen, M. G. M. (1981). Preliminary observations of habitat utilization and diet in eight Suriname monkeys. *Folia Primatol.*, **36**, 1–39.

Mittermeier, R. A., Kinzey, W. G. and Mast, R. B. (1989). Neotropical primate conservation. *J. Hum. Evol.*, **18**, 587–610.

Muckenhirn, N. A., Mortensen, B. K., Vessey, S., Fraser, C. E. O. and Singh, B. (1975). Report on a primate survey in Guyana. Unpublished report to the Pan American Health Organization.

Muskin, A. and Fischgrund, A. J. (1981). Seed dispersal of *Stemmadenia* (Apocynaceae) and sexually dimorphic feeding strategies by *Ateles* in Tikal, Guatemala. *J. Reprod. Bot.*, **13**, 78–80.

Myers, N., Mittermeier, R. A., Mittermeier, C. G., da Fonseca, G. A. B. and Kent, J. (2000). Biodiversity hotspots for conservation priorities. *Nature*, **403**, 853–858.

Notman, E., Gorchov, D. L. and Cornejo, F. (1996). Effect of distance, aggregation, and habitat on levels of seed predation for two mammal-dispersed neotropical rain forest tree species. *Oecologia*, **106**, 221–227.

Nunes, A. (1998). Diet and feeding ecology of *Ateles belzebuth belzebuth* at Maracá Ecological Station, Roraima, Brazil. *Folia Primatol.*, **69**, 61–76.

Overdorff, D. J. and Strait, S. G. (1998). Seed handling by three prosimian primates in southeastern Madagascar: implications for seed dispersal. *Am. J. Primatol.*, **4**, 69–82.

Peres, C. A. (1990). Effects of hunting on western Amazonian primate communities. *Biol. Conserv.*, **54**, 47–59.

Peres, C. A. (1991). Humboldt's woolly monkeys decimated by hunting in Amazonia. *Oryx*, **25**, 89–95.

Peres, C. A. (1994). Primate responses to phenological changes in an Amazonian terra firme forest. *Biotropica*, **26**, 98–112.

Peres, C. A. (1996). Use of space, spatial group structure, and foraging group size of gray woolly monkeys (*Lagothrix lagotricha cana*) at Urucu, Brazil: a review of the Atelinae. In *Adaptive Radiations of Neotropical Primates*, ed. M. A. Norconk, P. A. Garber and A. L. Rosenberger, New York: Plenum Press, pp. 467–488.

Peres, C. A. (1997). Effects of subsistence hunting and forest types on the structure of Amazonian primate communities. In *Primate Communities*, ed. J. G. Fleagle, C. H. Janson and K. E. Reed, Cambridge: Cambridge University Press, pp. 268–283.

Peres, C. A. (2000a). Effects of subsistence hunting on vertebrate community structure in Amazonian forests. *Conserv. Biol.*, **14**, 240–253.

Peres, C. A. (2000b). Evaluating the impact and sustainability of subsistence hunting at multiple Amazonian forest sites. In *Hunting for Sustainability in Tropical Forests*, ed. J. G. Robinson and E. L. Bennett, New York: Columbia University Press, pp. 83–115.

Peres, C. A. and Dolman, P. M. (2000). Density compensation in neotropical primate communities: evidence from 56 hunted and nonhunted Amazonian forests of varying productivity. *Oecologia*, **122**, 175–189.

Pijl, L., van der (1957). The dispersal of plants by bats (Cheiropterochory). *Acta Bot. Nederlandica*, **6**, 291–315.

Pitman, N. C. A., Terborg, J., Silman, M. R. and Núñez, P. V. (1999). Tree species distributions in an upper Amazonian forest. *Ecology*, **80**, 2651–2661.

Price, M. V. and Jenkins, B. (1986). Rodents as seed consumers and dispersers. In *Seed Dispersal*, ed. D. R. Murray, Sydney: Academic Press, pp. 191–235.

Richard, A. (1970). A comparative study of the activity patterns and behavior of *Alouatta villosa* and *Ateles geoffroyi*. *Folia Primatol.*, **12**, 241–263.

Redford, K. H. (1992). The empty forest. *BioScience*, **42**, 412–422.

Rowell, T. E. and Mitchell, B. J. (1991). Comparison of seed dispersal by guenons in Kenya and capuchins in Panama. *J. Trop. Ecol.*, **7**, 269–274.

Russo, S. E. (2003). Responses of dispersal agents to tree and fruit traits in *Virola calophylla* (Myristicaceae): implications for selection. *Oecologia*, **136**, 80–87.

Russo, S. E., Campbell, C. J., Dew, J., Stevenson, P. and Suarez, S. (2005). A multi-forest comparison of dietary preferences and seed dispersal by *Ateles* spp. *Int. J. Primatol.*, **26**, 1017–1037.

Schupp, E. W. (1993). Quantity, quality and the effectiveness of seed dispersal by animals. In *Frugivory and Seed Dispersal: Ecological and Evolutionary Aspects*, ed. T. H. Fleming and A. Estrada, Dordrecht: Kluwer Academic Publishers, pp. 15–29.

Schupp, E. W. and Fuentes, M. (1995). Spatial patterns of seed dispersal and the unification of plant population ecology. *Ecoscience*, **2**, 267–275.

Schupp, E. W. and Milleron, T. (2000). Dispersal limitation in tropical forests: consequences for species richness. Third International Symposium-Workshop on Frugivores and Seed Dispersal: Biodiversity and Conservation Perspectives, held in Brazil.

Stevenson, P. R. (2000). Seed dispersal by woolly monkeys (*Lagothrix lagothricha*) at Tinigua National Park, Colombia: dispersal distance, germination rates, and dispersal quantity. *Am. J. Primatol.*, **50**, 275–289.

Stevenson, P. R., Castellanos, M. C., Pizarro, J. C. and Garavito, M. (2002). Effects of seed dispersal by three ateline monkey species on seed germination at Tinigua National Park, Colombia. *Int. J. Primatol.*, **23**, 1187–1204.

Stevenson, P. R., Link, A. and Ramírez, B. H. (2005). Frugivory and seed fate in *Bursera inversa* (Burseraceae) at Tinigua Park, Colombia: implications for primate conservation. *Biotropica*, **37**, 431–438.

Symington, M. M. (1988). Food competition and foraging party size in the black spider monkey (*Ateles paniscus chamek*). *Behaviour*, **105**, 117–132.

Terborg, J. (1983). *Five New World Primates: A Study in Comparative Ecology.* Princeton, NJ: Princeton University Press.

Terborgh, J. (1986). Keystone plant resources in the tropical forest. In *Conservation Biology: The Science of Scarcity and Diversity*, ed. M. Soule. Sunderland, MA: Sinauer Associates, pp. 330–334.

van Roosmalen, M. G. M. (1985a). Habitat preferences, diet, feeding strategy and social organization of the black spider monkey (*Ateles paniscus paniscus* Linnaeus 1758) in Surinam. *Acta Amazonica*, **15**, 1–238.

van Roosmalen, M. G. M. (1985b). *Fruits of the Guianan Flora.* Utrecht: Institute for Systematic Botany, Utrecht University.

van Roosmalen, M. G. M. and Klein, L. L. (1988). The spider monkeys, genus *Ateles*. In *Ecology and Behavior of Neotropical Primates*, ed. R. A. Mittermeier, A. B. Rylands, A. F. Coimbra-Filho and G. A. B. da Fonseca, Washington, DC: World Wildlife Fund, pp. 455–537.

White, F. (1986). Census and preliminary observations on the ecology of the black-faced black spider monkey (*Ateles paniscus chamek*) in Manu National Park, Peru. *Am. J. Primatol.*, **11**, 125–132.

Willson, M. F. and Traveset, A. (2000). The ecology of seed dispersal. In *Seeds: The Ecology of Regeneration in Plant Communities* (2nd edn.), ed. M. Fenner, Wallingford, UK: CAB International, pp. 85–110.

Yost, J. and Kelley, P. (1983). Shotguns, blowguns, and spears: the analysis of technological efficiency. In *Adaptive Responses of Native Amazonians*, ed. R. B. Hames and W. T. Vickers, New York: Academic Press, pp. 189–224.

Zhang, S. Y. and Wang, L. X. (1995). Fruit consumption and seed dispersal of *Ziziphus cinnamomum* (Rhamnaceae) by two sympatric primates (*Cebus apella* and *Ateles paniscus*) in French Guiana. *Biotropica*, **27**, 397–401.

Part III

Behavior and reproduction

7 Locomotion and positional behavior of spider monkeys

DIONISIOS YOULATOS

Introduction

Living atelines form a unique group that occupies an important place in the adaptive radiation of New World monkeys. The four (or five) genera, *Alouatta*, *Ateles*, *Lagothrix* (and *Oreonax*) and *Brachyteles*, that compose the group are distinguished from other platyrrhines by their relatively large size (*c*. 5–11 kg), strong inclination to herbivory (fruits and leaves), tail-assisted forelimb-dominated suspensory positional behavior with associated postcranial morphology, and a prehensile tail with a naked ventral grasping surface (Rosenberger and Strier, 1989; Strier, 1992). Although consistent in its phylogenetic unity, atelines are morphologically and behaviorally heterogeneous.

More precisely, howler monkeys, *Alouatta* spp., appear to have opted for an energy minimizing foraging strategy, feeding on high proportions of leaves through sitting and tail–hindlimb assisted postures, and exploiting small home ranges traveling short daily distances mainly by quadrupedally walking and clambering above branches (Rosenberger and Strier, 1989; Strier, 1992). On the other hand, the rest of the genera, grouped as atelins, appear to share several derived features related to more agile and suspensory locomotor and postural behaviors as well as a more energy maximizing foraging strategy (Rosenberger and Strier, 1989; Strier, 1992). Phylogenetic relationships between these genera have not been established yet, especially under the light of new data deriving from molecular studies (Harada *et al.*, 1995; Schneider *et al.*, 1996; Cavanez *et al.*, 1999; von Dornum and Ruvolo, 1999; Meireles *et al.*, 1999; Collins, 2004). Based on ecomorphological and behavioral data, woolly monkeys *Lagothrix*, appear to be the more primitive atelin, whereas spider monkeys *Ateles*, and woolly spider monkeys *Brachyteles* are considered as a highly derived group, sharing postcranial adaptations to agile tail-assisted forelimb suspensory locomotion, similar energy expenditure foraging strategies, and fission–fusion societies (Rosenberger and Strier, 1989; Strier, 1992). In contrast, molecular data, deriving from the study of many different loci, have produced

Spider Monkeys: Behavior, Ecology and Evolution of the Genus Ateles, ed. Christina J. Campbell. Published by Cambridge University Press. © Cambridge University Press 2008.

identical results grouping *Lagothrix* and *Brachyteles* together and leaving *Ateles* as a sister group to the former (Harada *et al.*, 1995; Schneider *et al.*, 1996; Cavanez *et al.*, 1999; von Dornum and Ruvolo, 1999; Meireles *et al.*, 1999) or even unresolved trichotomy (Collins, 2004; see also Rosenberger *et al.*, this volume).

These differences, although contradictory, provide fertile ground for reconsidering the values and weights of different sets of data (Hartwig, 2005). These arguments are further supplemented by the rich fossil material (complete skeletons of *Caipora* and *Protopithecus*) that provides new insights into the evolution of morphological traits in atelines, within and/or outside assumed ecomorphological complexes. Among such complexes, positional behavior (i.e. locomotion, postures, manipulation), the way body weight is transferred across substrates, the active maintenance of immobility, and the way food items are collected and processed to mouth, may play a significant role in inferring and interpreting the interrelationships between historical and functional factors that shape morphology, as well as the adaptive significance of behavioral ecology. In effect, positional behavior is a signal of the constant and versatile interaction between the architecture of the postcranium and that of the surrounding environment. When quantified, well described and analyzed, positional behavior can provide a strong basis for interpreting homoplasies and homologies within evolutionary scenarios (Lockwood, 1999).

Among platyrrhines, atelines have enjoyed a relatively large number of studies of positional behavior. This is mainly because the postcranial morphology of the group is repeatedly compared to that of early anthropoids (Fleagle and Simons, 1982) and Miocene hominoids (Rose, 1996). More particularly, the positional behavior of *Ateles* especially has received much attention both in quantitative analysis of positional behavior in the wild (Mittermeier, 1978; Fleagle and Mittermeier, 1980; Cant, 1986; Fontaine, 1990; Bergeson, 1998; Cant *et al.*, 2001, 2003; Youlatos, 2002; Campbell *et al.*, 2005), as well as kinematic and kinetic analyses of selected frequent and critical positional modes (see previous papers and Turnquist *et al.*, 1999; Hirasaki *et al.*, 2000; Isler, 2004; Schmitt *et al.*, 2005). This chapter aims to review the available data and to provide stimuli for further and more focused research on the positional behavior of *Ateles* in the field.

Locomotor and postural modes performed by *Ateles*

In order to study the positional behavior of *Ateles* and its evolution in association with postcranial morphofunctional correlates, there is a need to know exactly how extant ateline species locomote in the wild. Different modes have been defined in different ways across observers and this requires a clarification of this system by detailed description, kinematic (and kinetic) analysis, and tests of performance of the different positional categories. This will aid in two ways. A detailed analysis of definition of modes will help in the comprehension of eventual differences between *Ateles* species or field sites. Second, each species appears to execute each mode in a different manner. Therefore lumping categories together may mask subtle differences that can be crucial to the understanding of adaptive variability and evolutionary trends. Although there has been an attempt to standardize applied definitions of modes used in studies of positional behavior (Hunt *et al.*, 1996), most researchers tend to use their own system mainly for the sake of comparability. For these reasons, I shall initially compare the terminology applied to describe the modes that compose positional categories and provide a kinematic analysis where available, based on both field and laboratory research.

Leap and drop

Leaping and dropping involve an airborne phase and are the main means of crossing gaps across the canopy. Leaping is used for crossing gaps that are longer than an animal can navigate with other, more secure, means. In leaping, the horizontal component of the airborne phase is usually longer than the vertical one, and functions as a rapid way to cross tree crowns or within-tree sites. Leaping requires the action of rapid and simultaneous extension of the hindlimbs in order to provide the necessary thrust for propelling the body. Initiation may be from a moving or a stationary position, above or below one or multiple supports. Most takeoffs are initiated from above a support by adopting a quadrupedal fast or slow progression or an analogous posture. When below the support, forelimb-assisted progression is usually the predominant takeoff mode. During the airborne phase, the animal usually assumes the spread-eagled position, but other postures may be used (Mittermeier, 1978). Direction of the body displacement can be upward, horizontal or downward. However, in all cases the horizontal component is more important than the vertical displacement of the body. Most leaps cover distances between 2 and 5 m, rarely longer (Mittermeier, 1978; Fontaine, 1990; Cant *et al.*, 2001). Landing usually takes place above the support, by adopting a quadrupedal stance, where the forelimbs land first. Below support landing is also common and usually terminates with a tail–forelimb-assisted suspensory posture or progression (Fontaine, 1990; Cant *et al.*, 2001).

In dropping, the vertical component of the airborne phase is longer than the horizontal one. Thus, dropping serves for rapid downward vertical displacement within the canopy. Dropping usually lacks propulsive effort from the hindlimbs and usually initiates from a position above or below the take-off support. When above the support, the animal assumes a quadrupedal or bipedal posture or a cautious quadrupedal or bipedal progression. Suspensory takeoffs are more frequent (Fontaine, 1990; Cant *et al.*, 2001) and usually involve a combination of forelimbs, hindlimbs, and the tail. Tail-assisted forelimb and tail-only suspension appear to be the most common takeoff modes (Fontaine, 1990) and are passive as the animal simply leaves the grasps. During the airborne phase the animal assumes the spread-eagled position, and it can be either head- or tail-first. Dropping usually covers relatively short vertical distances between 1 and 3 m (Fontaine, 1990; Cant *et al.*, 2001). In most cases, landing occurs above the support, with the hindlimbs touching the landing support first, waiting for the stabilization of the flexible supports to continue progression.

Bipedalism

Bipedal activities mainly involve bipedal walk and assisted bidepal walk, where forelimbs and the prehensile tail help support the orthograde stance of the body. During this infrequent mode, which occurs mainly on medium-sized horizontal or subhorizontal supports (Mittermeier, 1978; Cant *et al.*, 2001) and on the ground (Campbell *et al.*, 2005), progression is relatively slow. There is significant flexion and abduction of the thigh, and when it is forelimb assisted, the knee is importantly flexed.

Quadrupedalism

Quadrupedalism or quadrupedal activities are variably defined in different studies. Mittermeier (1978), in his study of *A. geoffroyi* and *A. paniscus*, limited quadrupedal activities to quadrupedal walk and run involving diagonal or lateral gaits on relatively stable horizontal or inclined supports. Fontaine (1990), in his study of *A. geoffroyi*, included quadrupedal walk, bound, slow bound, gallop, and run as well as tripedal walk and bound in his category of quadrupedal locomotion, and all these modes were described similarly to Mittermeier's. However it is not clear as to whether these modes occur only on single or multiple parallel-sided supports and not on multiple intertwined ones. Similarly, in the study of *A. belzebuth belzebuth* in Ecuador, the quadrupedal locomotion described by Cant *et al.* (2001) involved quadrupedal walk along single or multiple more or less parallel horizontal supports utilizing all four limbs, in a slightly or highly flexed posture, in a diagonal sequence. Following the same lines, in my study of *A. paniscus* in French Guiana (Youlatos, 2002), I also separated quadrupedal walk and run from other locomotor categories. In contrast, the quadrupedalism of Cant (1986) stands out as it appears to incorporate quadrupedal walk and run on single horizontal supports, as well as all climbing activities on vertical single supports, and thus included all quadrupedal activities irrespective of support orientation.

In all cases, quadrupedal walk along single horizontal or subhorizontal supports (Figure 7.1a) involves a diagonal sequence in *Ateles* (Youlatos, 1994). The gait appears to provide locomotor stability on relatively unstable supports, by ensuring that the hind foot grasps in a protracted position on a known substrate at the moment when the contralateral forefoot touches down on a new substrate (Cartmill *et al.*, 2002). Hindlimbs are usually the first to leave and touch down on the support (Vilensky, 1989; Cartmill *et al.*, 2002). The hip joint is maintained adducted and progressively extends throughout the support phase, aided by the constant firing of the adductor magnus (Fleagle *et al.*, 1981). Abduction is minimal and reaches a maximum at the mid-swing phase. The knee does not show complete extension prior to touchdown. In the hind foot the toes touch the substrate first and are subsequently followed by the heel (Schmitt and Larson, 1995). The foot either grasps or poses itself laterally upon the substrate, depending on its size. Excursion in the shoulder joint is ample, reaching a minimum in the propulsive phase and a maximum at touchdown, when the forelimb is also slightly abducted during touchdown (Stern *et al.*, 1980a, 1980b; Schmitt, 1994; Larson *et al.*, 2000). Propulsion is provided by the recruitment of several extrinsic scapular muscles, such as the trapezius, serratus and pectoralis major, that pull the body forward relative to the supported forelimb (Stern *et al.*, 1976, 1980a, 1980b; Fleagle *et al.*, 1981; Konstandt *et al.*, 1982). Arm abduction is especially pronounced during the mid-swing and mid-support phases (Schmitt, 1994). The elbow joint is usually kept semi-extended during the support phase, but reaches particularly extended positions prior to touchdown when a new farther reach is attained (Schmitt, 1994; Larson, 1998a, 1998b; Larson *et al.*, 2000), at which point the body is primarily supported by the active firing of the triceps muscles (Fleagle *et al.*, 1981). The forearm is variably positioned during the support phase. In most cases, the forearm is pronated and the hand is posed in a palm down position and laterally deviated on the substrate depending on its size (Lemelin and Schmitt, 1998). In contrast, in phases when the forearm is maintained supinated, a hook-like grasp is used around the underside of the support. In all cases, the wrist is passively hyperextended during the mid-support phase (Schmitt, 1994). On small flexible substrates (e.g. lianas or terminal branches), a crouched form of quadrupedal walk is assumed. This is referred to as "compliant walking" and protraction, flexion and abduction of the forelimb and hindlimb joints are more pronounced, providing a longer and more stable reach, improving balance on unstable substrates, and reducing peak

(a)

(b)

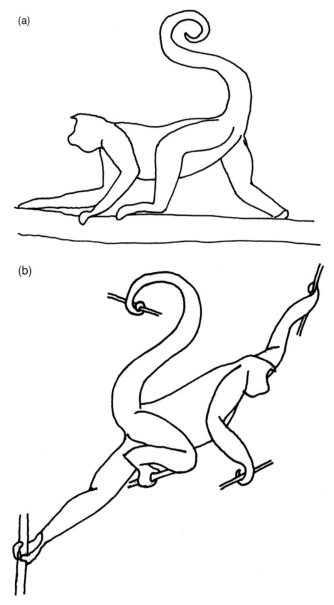

Figure 7.1 Locomotor modes in *Ateles belzebuth* and *A. paniscus*: (a) quadrupedal walk, (b) clamber, (c) brachiation, and (d) bridging.

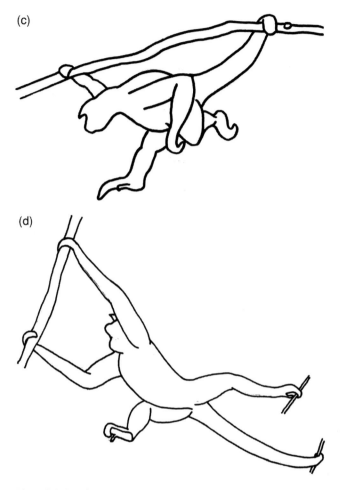

Figure 7.1 (*cont.*)

reaction stresses (Schmitt, 1999; Schmitt and Hanna, 2004). In such instances, both diagonal and lateral gaits can be used, the latter mainly as transitional (Youlatos, 1994).

Vertical ascent and descent

When substrates deviate from the horizontal or subhorizontal plane, most authors tend to recognize a separate locomotor mode – vertical climb (contra Cant, 1986). Mittermeier (1978) and Fontaine (1990) included what they termed "quadrupedal climbing" in their climbing category. I recognized a similar mode, defined as vertical climb (Youlatos, 2002). In contrast, Cant *et al.* (2001) recognized two separate modes depending on the inclination of the used support and termed them as "vertical and oblique ascent" respectively, but included them in a separate ascend/descend locomotor category.

Vertical climbing is characterized by diagonal sequence/diagonal couplets (Isler, 2004), where the foot leaves the support first, followed by the contralateral forelimb. Three limbs usually provide support (Isler, 2004) and the tail is rarely used (Youlatos, 2002). Substrate characteristics appear to influence footfall patterns, and vertical climbing on large diameter supports is mainly performed by lateral sequences (Isler, 2004). Cycle duration and relative stride lengths are long, with that of the forelimb being longer (Hirasaki *et al.*, 1993; Isler, 2004). In effect, the arm, with the aid of middle serratus and middle deltoid, is raised well above the shoulder to reach a new and higher hold (Stern *et al.*, 1980a, 1980b; Fleagle *et al.*, 1981) and the elbow may fully extend prior to touchdown. During support, the elbow is flexed on thin substrates, while fully extended on larger ones (Isler, 2004) and shows a momentary peak of flexion mainly in the early support phase (Hirasaki *et al.*, 2000). The elbow joint appears to act in flexion throughout the support phase, while it extends passively in the late support phase (Fleagle *et al.*, 1981; Hirasaki *et al.*, 2000). The wrist shows an abduction peak in the late support phase (Hirasaki *et al.*, 2000), while the hand uses a hook-like prehension around the substrate. In general, kinematic and electromyographic analyses of muscle recruitment, during the support phase, provide evidence that the major role of the forelimb is to keep the body close to the substrate (Stern *et al.*, 1976, 1980a, 1980b; Hirasaki *et al.*, 2000). In contrast, the hindlimb plays a preponderant role in pushing the animal off and providing thrust for propulsion (Fleagle *et al.*, 1981; Hirasaki *et al.*, 2000; Isler, 2004). In general, hindlimb joints exhibit greater excursions than forelimb ones, probably in order to reduce joint moments (Hirasaki *et al.*, 1993). In addition, the hip and knee joints act mainly in extension and may be abducted (Hirasaki *et al.*, 2000). The ankle acts primarily in plantar flexion (Hirasaki *et al.*, 2000) and the foot is characterized by strong prehension while the heel never touches the support. Lastly, it appears that total muscle power in the hindlimb peaks in the late support phase, most likely related to intense muscle activity that is important for the push-off phase (Hirasaki *et al.*, 2000). In fact, the climbing style of *Ateles* resembles that of hominoids, and more particularly the bonobo, *Pan paniscus*, and appears to be quite efficient requiring less energy than other quadrupedal monkeys (Hirasaki *et al.*, 2000; Isler, 2004).

Clamber

When spider monkeys move in various directions along and across multiple variably oriented supports, Cant (1986) defines it as clambering, but transitional, which is a mode between clambering and the initiation of tail-assisted forelimb suspension. The former mode is termed as "horizontal climbing" by Mittermeier (1978), and involves climbing on a roughly horizontal plane upon flexible supports, combining quadrupedal walk and suspensory posturing. On the other hand, Fontaine (1990) most likely includes it within the quadrupedal walk category. Cant *et al.* (2001) recognize several clambering modes related to the direction of body displacement along the network of the multiple flexible supports, as well as the adopted posture (orthograde or pronograde). Lastly, I recognize clamber as a separate locomotor mode but have included all directions of body displacement and postures within the same category (Youlatos, 2002).

By definition, clambering in various directions occurs on multiple supports that are intertwined in variable orientations (Figure 7.1b). During progression, there is no particular gait as limbs grasp hold of the available supports. The body can be directed upwards, horizontally, or downwards. The elbow and knee joints are frequently under complete extension and abduction, especially during the stance phase. The forelimb is usually placed above the shoulder level during the swing phase depending on support availability. The hindlimb grasps the support via the toes and the heel does not contact the support. Hand grasps are hook-like. On fewer occasions, the body is held orthograde and both forelimbs and tail support the posture during progression. The arms are usually raised above the shoulder showing significant extension and abduction. The hindlimbs appear to support the body weight and may be under complete knee extension, especially prior to touchdown. Abduction and flexion at the thigh can also be significant. In all directions and postures, the tail is used often (Cant, 1986; Youlatos, 2002) and is placed primarily above the animal, most certainly supporting part of the body weight. No detailed analysis exists of tail use during clambering bouts but comparable data of tail use in tail–arm brachiation (Schmitt *et al.*, 2005) imply a significant and active role.

Bridge

Mittermeier (1978) and Cant (1986) distinguish bridging as a mode closely related to horizontal climbing (or clambering) that is used during crossings between tree peripheries. Bridging begins by securing the initial supports and grasping the terminal supports, followed by pulling them over, and finally climbing across the terminal supports (Figure 7.1d). Grasping of terminal supports is mainly achieved by an incomplete leap, but bridging never involves an airborne phase. I also define bridging in a similar manner without distinguishing any finer categories that involve body direction or initiation method (Youlatos, 2002). Fontaine (1990) identifies two different bridging modes based on direction of body crossing: descending bridging which involves a tail-hindlimb suspensory posture between primary and terminal supports, and pronograde bridging. In contrast to previous authors, the study of Ecuadorian *A. belzebuth belzebuth* identifies many bridging modes that are grouped under bridge/hoist locomotor category (Cant *et al.*, 2001). All these modes are defined the same way as previous authors, but are distinguished by the direction of the movement, as well as the relative placement of the body. Bridging represents a significant proportion of the traveling locomotion of *A. b. belzebuth*, but the dominant modes are pronograde horizontal and downward oblique bridging. Furthermore, bridging is the dominant mode for crossing gaps in the canopy. In all studies, the frequency of bridging behavior, which incipiently involves traversing of many tree crowns, is significantly reduced during feeding locomotion.

Irrespective of the direction of the movement during bridging, the mode always involves the use of the prehensile tail that grasps firmly on a support above the animal and bears initially part, but finally the whole of the body weight prior to the grasping of the terminal supports. In all cases, the forelimbs leave the initial supports first, are usually characterized by frequent abduction and extension at the elbow, and are raised well above the shoulder. The forearm appears to be under considerable pronosupinatory rotation and the hands anchor the terminal supports by a hook-like grasp. In many cases, bridging is initiated by a lunging movement, but never involves an airborne phase. More precisely, there are at least two limbs secured at any time, the tail excluded. The hindlimbs leave the initial support last and are usually under adduction and complete extension at the thigh and knee, and plantar flexion at the ankle joint.

Suspensory locomotion

Ateles engages in a variety of suspensory modes of locomotion. The most common ones are tail-assisted forelimb swing and tail-assisted brachiation (Figure 7.1c), with tail swing, inverted quadrupedal walk and inverted clamber being used less frequently.

Tail-assisted forelimb swing is described as suspensory progression using alternating forelimbs that may be held in an extended or partly flexed position at the elbow joint during the support phase. Trunk rotation is absent or particularly limited and the tail is frequently, but not always, used; however, when in use, it may grasp behind the trailing hand, or between hands. In this mode it is possible that only the initially trailing forelimb swings forward to take a grip, whereas the initially leading forelimb remains in place. In this case, Cant *et al.* (2003) identify a separate mode, the half-stride forelimb swing.

Tail-assisted brachiation is a more dynamic and usually faster form of fore-limb swing that involves a trunk rotation of over 180 degrees between hand contacts (Jenkins *et al.*, 1978; Jenkins, 1981). Half-stride brachiation, when only the initially trailing forelimb swings forward to take a grip, while the initially leading forelimb remains in place, can also occur, but trunk rotation also remains close to 180 degrees. Half-stride suspensory locomotion is rela-tively uncommon in *Ateles* (Cant *et al.*, 2003). *Ateles* also engages in ricochetal brachiation that usually involves variable periods of noncontact between hand-holds. This suspensory mode is less frequently used, but can be much faster, as horizontal velocity is supposed to remain constant during the ballistic aerial phase, resulting in an increased forward speed, providing access to farther hand-holds by means of an unusual pendulum that has substantially more kinetic than potential energy (Preuschoft and Demes, 1984; Bertram, 2004).

Tail-assisted brachiation, forelimb swing, and inverted quadrupedalism have been identified as separate modes in most studies of positional behavior of *Ateles*, although the latter mode has been given different names (Mittermeier, 1978; Youlatos, 2002; Cant *et al.*, 2003). Only Cant (1986) included tail-assisted brachiation and forelimb swing under his tail–arm-suspension category, and identified a separate one, "transitional," that can be placed somewhere between tail-assisted forelimb suspension and clamber.

The way tail-assisted brachiation is performed by *Ateles* has been well doc-umented in both the wild and under laboratory settings. Detailed analysis of videos shot in the wild in both French Guiana and Ecuador show that brachiation is usually employed below and along one or sometimes across multiple sup-ports. The body is usually held relatively orthograde, but body position largely depends on tail use and the placement of the tail hold (Turnquist *et al.*, 1999). It appears that *Ateles* keeps the body relatively pronograde mainly due to the consistent and active use of the prehensile tail (Turnquist *et al.*, 1999). How-ever, the more horizontal body postures in *Ateles* appear to be associated with the addition of potential energy within the brachiating system, in a different way from that observed in brachiating hylobatids (Turnquist *et al.*, 1999). In addition, lateral body sway is lowest when the tail grasps the substrate (Schmitt *et al.*, 2005). The tail is capable of retarding and lifting the body (Jenkins *et al.*, 1978) and its liftoff precedes that of the leading hand (Turnquist *et al.*, 1999). In effect, the tail grasps behind the trailing hand or between forelimbs, and is used with every other handhold, resulting in a more flowing progression

(Turnquist *et al.*, 1999). The active recruitment of the tail in tail-assisted brachiation kinematics in *Ateles* appears to be related to its capacity of hyperextension that is functionally associated with certain osteomuscular features such as its length, longer proximal caudal region, more acute sacrocaudal angle, and wider transverse processes (Lemelin, 1995; Schmitt *et al.*, 2005). This results in significant differences in forelimb joint kinematics between tail use and nonuse. The forelimbs are raised well above the shoulder, particularly when the tail is simultaneously engaged, whereas shorter excursions are observed when not using the tail (Turnquist *et al.*, 1999). The morphology of the shoulder girdle allows the efficiency of the swing, by engaging in scapular rotation, and caudal and medial translation (Jenkins *et al.*, 1978). This morphology, in combination with complex muscle recruitment (Stern *et al.*, 1976, 1980a, 1980b; Fleagle *et al.*, 1981; Konstandt *et al.*, 1982), permits a caudal and medial shoulder movement towards the median plane and appears to facilitate the dynamics of pendular swinging (Jenkins *et al.*, 1978; Schmitt *et al.*, 2005). In addition, the elbows are only partially extended when the tail grasps, whereas higher values of elbow extension are encountered when the tail is not used (Turnquist *et al.*, 1999). The forearm engages in ample pronatory and supinatory movements that appear to be facilitated by midcarpal rotation during the support phase (Jenkins, 1981; Jungers and Stern, 1981). In fact, in *Ateles* these distal rotatory movements appear to be responsible for rotation of the trunk (Jenkins, 1981). During tail-assisted brachiation, the trunk always displays a great rotation between hand contacts (Jenkins *et al.*, 1978). In addition, strides are relatively high in length and duration, with long support and swing phases, as well as a relatively short free flight phase per stride duration (Turnquist *et al.*, 1999). In higher speeds, both swing and support phase duration decrease, resulting in an insignificant reduction of the duty factor. This pattern is supposed to accommodate longer periods of contact and thus to reduce peaks of substrate reaction forces, as in primate quadrupeds (Turnquist *et al.*, 1999). Initiation of a brachiation bout can be from above, below or the same level as the substrate and termination can be in a suspended position below the support, in a supported position above the support, or on top of a lower support. The fluidity of these movements as analyzed under controlled conditions and as observed under real naturalistic conditions in the canopy demonstrates that *Ateles* can perform, readily and proficiently, long bouts of tail-assisted brachiation with variable periods of free flight. This mode may be energetically efficient and may provide shorter travel routes within the canopy (Parsons and Taylor, 1977; Cant, 1986, 1992). On the other hand, *Ateles* appear to use a different pendular system of energy storage and use, but are kinematically closer to hominoids than the other atelin brachiators (Turnquist *et al.*, 1999; Bertram, 2004).

(a)

Figure 7.2 Suspended feeding postures in *Ateles belzebuth*: (a) tail–two hindlimbs hang, (b) tail–forelimb–hindlimb hang, (c) tail–forelimb hang, (d) tail-only hang, (e) tripod, and (f) sit.

Supported postures

Among the anthropoids the seated posture is the most frequently used (Rose, 1974). While sitting, the weight of the animal is usually supported by either the ischia or the hindlimbs (Figure 7.2f). Most studies fail to distinguish between these two categories, grouping together all sitting observations into one major mode, termed as sit (Mittermeier, 1978; Cant, 1986; Fontaine, 1990). Youlatos (2002) and others (D. Youlatos, J. G. H. Cant and D. M. Rose, unpublished data) distinguish two modes: (a) sit or ischial sit where the majority of the animal's weight is borne by the rump, and the hindlimbs may be either flexed or extended; (b) squat or nonischial sit, where the animal sits at right angles to the

(b)

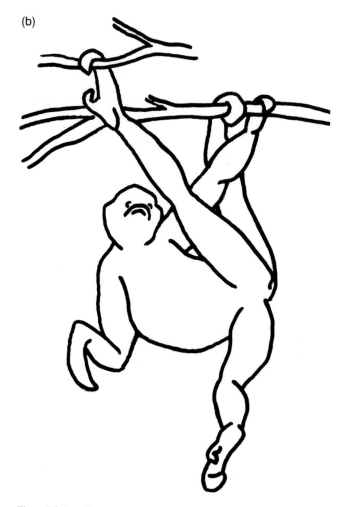

Figure 7.2 (*cont.*)

length of the substrate, while the extremely flexed hindlimbs firmly grasp the substrate and no weight is borne by the rump. In both cases, extra stabilization may be provided by any combination of forelimbs and tail.

In both quadrupedal and tripedal stand the body is held pronograde and mainly parallel to the substrate. Contact with the substrate is provided by firm grasps of both hands and feet. When one forelimb is not in contact with the supportive substrate(s) the posture is defined as tripedal. This hand may be either free, manipulating a food item, or providing additional support upon another substrate, which is not in parallel with the supportive substrate(s).

(c)

Figure 7.2 (*cont.*)

Most studies classify both postures under stand (Mittermeier, 1978; Bergeson, 1998; Youlatos, 2002) or distinguish between tripedalism and quadrupedalism (Fontaine, 1990). D. Youlatos, J. G. H. Cant and D. M. Rose (unpublished data) further discern these postures on the basis of limb flexion, which implies the action of applied forces upon the supportive limbs. When both forelimbs and hindlimbs are substantially flexed in elbows and knees, the posture (either quadrupedal or tripedal) is identified as crouch. This posture allows the animal to stand closer to the substrate and very likely provides an equilibrium advantage. The posture represents variable proportions of the above substrate postural repertoire of spider monkeys but appears to be the second more frequent supported posture.

When only the hindlimbs support the body weight, the standing posture is defined as bipedal stand and is seldom used by *Ateles* (Mittermeier, 1978; Cant, 1986; Fontaine, 1990; Bergeson, 1998, Youlatos, 2002). The feet may be either posed on the substrate or engaged in a firm grasp. The body may be held

(d)

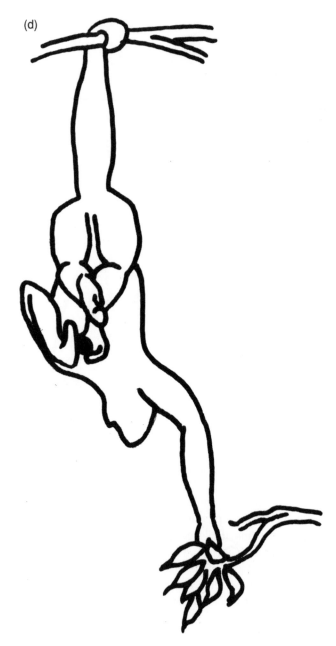

Figure 7.2 (*cont.*)

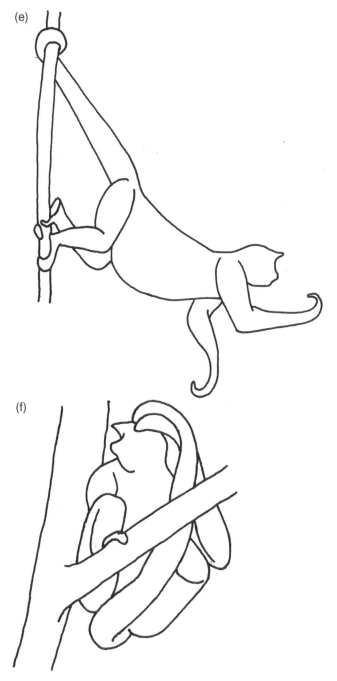

Figure 7.2 (*cont.*)

either orthograde or semipronograde. Additional support may be provided by a combination of forelimbs and tail. In this case, the forelimbs may be actively involved, holding the substrate under tension, or they may be simply placed upon a substrate at the level of the body, stabilizing the posture of the animal. When the knees are substantially flexed, D. Youlatos, J. G. H. Cant and D. M. Rose (unpublished data) distinguish a different mode referred to as bipedal crouch.

Vertical cling is a posture that occurs on vertical and subvertical substrates, and is characterized by the attachment of the animal by all four limbs. The body is held orthograde and parallel to the substrate and the tail may or may not provide additional support, grasping either above or below the animal on the same substrate or one that is nearby. All limbs are highly flexed and act to maintain the body close to the substrate. Clinging postures are identified in most studies (Mittermeier, 1978; Fontaine, 1990; Bergeson, 1998; Youlatos, 2002; D. Youlatos, J. G. H. Cant and D. M. Rose, unpublished data) but contribute insignificantly to the overall postural repertoire of *Ateles*.

Recline, lie and sprawl incorporate a variety of supported postures where the body, either pronograde or supinograde, is supported by the substrate. These postures are seldom used during feeding activities (Mittermeier, 1978; Youlatos, 2002; D. Youlatos, J. G. H. Cant and D. M. Rose, unpublished data) but are extensively adopted during short or long resting sessions (Fontaine, 1990). This variety of postures is usually grouped under recline (Mittermeier, 1978) or lie (Youlatos, 2002), but Fontaine (1990) and D. Youlatos, J. G. H. Cant and D. M. Rose (unpublished data) distinguish detailed categories, depending on the side of the body that is in contact with the substrate (ventral, dorsal, flank, etc.) and the position of the limbs.

Suspended postures

Ateles exhibits a remarkable variety of suspended postures that involve several combinations of forelimbs, hindlimbs and the tail. The latter is always used during suspensory postural behavior and depending on the posture may or may not play a primary role.

Tail-only hang is commonly used by *Ateles*. In this posture, the tail exclusively supports the entire weight of the body (Figure 7.2d). The body may be held either vertically downward, with the hindlimbs particularly flexed at the hips, or at an angle, with the hindlimbs held somewhat behind the suspended animal. The forelimbs may be hanging free or manipulating food items or holding twigs that are found at a level which is lower to the center of mass of the animal. Most authors recognize this posture as a separate mode (Cant, 1986; Fontaine, 1990; Youlatos, 2002; D. Youlatos, J. G. H. Cant and D. M. Rose, unpublished data).

Tail–hindlimb hang involves the action of the tail and one or both hindlimbs (Figure 7.2a). This posture appears to be used extensively among atelines (Schön Ybarra, 1984; Cant, 1986; Gebo, 1992; Bergeson, 1998; Youlatos, 1998; D. Youlatos, J. G. H. Cant and D. M. Rose, unpublished data). The hindlimbs are under tension, and the foot is particularly plantar flexed. The hip and knee are extended and can be variably abducted. Most authors distinguish a tail–hindlimb hang posture, and Fontaine (1990) goes even further to distinguish tail–one-hindlimb-hang and tail–two-hindlimb hang.

Horizontal tripod is a posture that employs the tail and the two hindlimbs, the former being under tension, while the latter are pressed against the substrates and are under compression (Figure 7.2e). In this posture, the body is usually held more or less pronograde or at an acute angle to the horizontal. The tail grasps above the animal but the hindlimbs are usually positioned below the level of the rump. The hindlimbs are usually extended and adducted at the hip joint, but can be either flexed or semiextended at the knee. Cant (1986) initially identified the posture as tripod; Youlatos (2002) and others (D. Youlatos, J. G. H. Cant and D. M. Rose, unpublished data) use the same term, whereas Fontaine (1990) and Bergeson (1998) use the term "inverted bipedal" to describe it.

Tail–forelimb hang is a relatively common feeding posture in *Ateles*. The tail grasps above the animal and one forelimb grasps in full extension below the same or a nearby substrate (Figure 7.2c). Depending on the relative position of the two grasping points, the body is usually held pronograde and only rarely orthograde. Conversely, tail–two-forelimb hang is a common posture during short-term or long-term resting. During this posture, the tail grasps beside the grasping forelimbs that are both extended and may be either abducted or adducted at the shoulder. In this case, the trunk is held more or less pronograde as in tail–forelimb hang.

Tail–forelimb–hindlimb hang is similar to tail–forelimb hang, with the exception that one foot grasps on the side, so that the trunk of the animal faces partly to that side (Figure 7.2b). The forelimb is usually abducted at the shoulder, and extended at the elbow. The hindlimb is particularly abducted and extended at the hip, while extended at the knee. A strongly plantar flexed and distally inverted foot secures hindlimb grasp; however, this posture is not very common in *Ateles* (D. Youlatos, J. G. H. Cant and D. M. Rose, unpublished data).

Lastly, a quite uncommon posture is quadrupedal hang, when all forelimbs and hindlimbs are involved in supporting the suspended supinograde trunk. The limbs may be suspended below the same or different substrates, and the tail may or may not be grasping above the animal. Mittermeier (1978) indirectly describes the posture when he considers the use of all five limbs, in his analysis

Table 7.1 *Summary of percentages of gross locomotor categories in positional studies of three species of* Ateles

	A. geoffroyi			A. paniscus		A. belzebuth
	1	2	3	1	4	5
Site	Panama	Guatemala	Panama	Suriname	Fr. Guiana	Ecuador
Rain forest	wet	dry	wet	wet	wet	wet
Method	bout	10-sec	30-sec	bout	20-sec	20-sec
Quadrupedal	22.0%	52.0%	51.07%	25.4%	20.1%	20.8%
Climb/clamber	24.2%	19.0%	6.08%	17.0%	28.1%	37.9%
Bridge	4.6%	4.0%	1.63%	8.6%	8.8%	12.5%
Suspended	25.7%	25.0%	21.89%	38.6%	34.2%	23.3%
Airborne	10.9%	1.0%	10.88%	4.2%	2.6%	2.8%

1: Mittermeier (1978); 2: Cant (1986); 3: Fontaine (1990); 4: Youlatos (2002); 5: Cant et al. (2001).

of limb use in suspensory postures, and it is also described by Fontaine (1990) and Bergeson (1998).

Positional behavior and profile of *Ateles* in the wild

Thus far, the positional behavior of *Ateles* has been the subject of several quantitative studies. These studies, although they have used different methodologies of data collection and mode definition, and were carried out in different sites during different periods, have provided quantitative data for three different species (Tables 7.1 and 7.2). The Central American spider monkey (*A. geoffroyi*) has been studied in both wet tropical forest in Barro Colorado Island (BCI), Panama (Mittermeier, 1978; Fontaine, 1990) and Costa Rica (Bergeson, 1998), and dry forest in Tikal, Guatemala (Cant, 1986). The black spider monkey (*A. paniscus*) has been observed in two similar high wet terra-firme forests from the Guianan plateau, in Raleighvalen-Voltzberg, Suriname (Mittermeier, 1978) and Nouragues, French Guiana (Youlatos, 2002). Finally, the white-bellied spider monkey (*A. belzebuth belzebuth*) has been studied in a wet terra firme forest in Yasuní National Park, Ecuador (Cant et al., 2001, 2003; D. Youlatos, J. G. H. Cant and D. M. Rose, unpublished data). Below I compare the quantitative aspects of the locomotion and postures of the different species across the different habitats, aiming to set a background for further discussions concerning phylogenetic or habitat structural aspects.

All species of *Ateles* appear to employ relatively high rates of quadrupedal activities. Quadrupedal walk and run varies from around 21%–25% for

Table 7.2 *Summary of percentages of postural categories in positional studies of species of* Ateles

	A. geoffroyi				A. paniscus		A. belzebuth
	1	2	3	4	1	5	6
Site	Panama	Guatemala	Panama	Costa Rica	Suriname	Fr. Guiana	Ecuador
Rain forest	wet	dry	wet	wet	wet	wet	wet
Sampling post	bout	10-sec	30-sec	bout	bout	bout	20-sec
Sit	33.8%	45.0%	55.10%	25.7%	43.5%	34.9%	26.1%
Stand	5.5%	3.0%	5.62%	12.6%	13.1%	4.1%	16.3%
Suspended	53.7%	52.0%	30.71%	33.3%	41.2%	55.4%	55.5%
Tail-only hang	(3.5%)	20.0%	7.44%	13.6%	(4.6%)	18.8%	17.4%
Tail–hindlimbs			2.59%	11.3%		16.0%	13.5%
Tripod		21.0%	1.58%	13.1%		7.9%	8.9%
Tail–forelimbs		11.0%	20.11%	8.4%		8.2%	13.9%

1: Mittermeier (1978); 2: Cant (1986); 3: Fontaine (1990); 4: Bergeson (1998); 5: Youlatos (2002); 6: D. Youlatos, J. G. H. Cant and D. M. Rose, unpublished data.
Percentages in brackets correspond to Mittermeier's (1978) "one-limb hang."

A. geoffroyi in Panama, *A. paniscus* in both French Guiana and Suriname, and *A. belzebuth* in Ecuador, up to 50% for *A. geoffroyi* in Guatemala, where vertical quadrupedal climb is included in the quadrupedal category (Cant, 1986). Proportions of vertical climbing for *A. geoffroyi* in Panama range between 9% and 16% (Mittermeier, 1978; Fontaine, 1990). Comparable rates are observed in *A. belzebuth* (12.5%; Cant *et al.*, 2001) and *A. paniscus* in Suriname (9%; Mittermeier, 1978), but yet much lower for the latter species in French Guiana (2.3%; Youlatos, 2002).

Pronograde, ascending and descending clamber also varies significantly across species and forests. *Ateles belzebuth* in Ecuador and *A. paniscus* in French Guiana engage in quite high proportions (28.2% and 25.8% respectively; Cant *et al.*, 2001; Youlatos, 2002), and a slightly lower rate (20%) is encountered for *A. geoffroyi* in Guatemala, if the transitional mode is included in this category (Cant, 1986). However, proportions stay as low as 8–9% in both Panama and Suriname (Mittermeier, 1978; Fontaine, 1990). Bipedalism is quite uncommon, scoring particularly low in Panama, Suriname, and Ecuador (0.6–0.8%; Mittermeier, 1978; Cant *et al.*, 2001) to 1.7% in French Guiana (Youlatos, 2002) and 2.23% in Panama (Fontaine, 1990).

Leap and drop are used commonly by *A. geoffroyi* in Panama (Table 7.1; Mittermeier, 1978; Fontaine, 1990), while very rarely in Guatemala (Cant, 1986). Both *A. paniscus* and *A. belzebuth* show consistently lower rates (2.3–4.2%) in all forests (Table 7.1; Mittermeier, 1978; Cant *et al.*, 2001; Youlatos,

Table 7.3 *Tail–limb(s) hang/tail-only hang ratio (TL/T), tail–arm brachiation/forelimb swing ratio (TAB/FSW), and leap/drop ratio (L/D) across studies of positional behavior of* Ateles

Ratios	TL/T	TAB/FSW	L/D
Ateles geoffroyi			
Panama[1]	14.15	1.44	–
Guatemala[2]	1.60		
Panama[3]	3.34	16.67	4.68
Costa Rica[4]	1.44		
Ateles paniscus			
Suriname[1]	7.93	1.49	–
French Guiana[5]	1.94	10.00	6.5
Ateles belzebuth			
Ecuador[6,7,8]	2.19	4.17	6.8

1: Mittermeier (1978); 2: Cant (1986); 3: Fontaine (1990); 4: Bergeson (1998); 5: Youlatos (2002); 6: Cant *et al.* (2001); 7: Cant *et al.* (2003); 8: D. Youlatos, J. G. H. Cant and D. M. Rose, unpublished data.

2002). These proportions, however, mask the use of active (leap and jump) versus passive (drop) airborne locomotion, and leaping dominates and is used over five times more than dropping in all studies where detailed data are available (Table 7.3; Fontaine, 1990; Cant *et al.*, 2001; Youlatos, 2002).

Suspensory locomotion is employed in similar proportions in *A. geoffroyi* and *A. belzebuth* (23–26%; Mittermeier, 1978; Cant, 1986; Fontaine, 1990; Cant *et al.*, 2001, 2003). In contrast, *A. paniscus* appears to be the most suspensory species with rates up to 38.6% in Suriname and 35.2% in French Guiana. In addition, significant differences exist in the ratio of use between tail–arm brachiation and forelimb swinging across species and sites. Brachiation is used over 10 times more than forelimb swing in Panama (Fontaine, 1990) and French Guiana (Youlatos, 2002; Table 7.3) or can be more equally shared in Panama and Suriname (Mittermeier, 1978). In *A. belzebuth*, the same ratio is intermediate (Table 7.3), and bouts of both modes are mainly complete or full stride (Cant *et al.*, 2003). *Ateles belzebuth* bridges gaps extensively (12.5%), while lower proportions are encountered in *A. paniscus*, and even lower in *A. geoffroyi* (Table 7.1).

In terms of postural behavior, only feeding postures that are active and necessary for forage procurement are considered here, Fontaine's (1990) study being an exception by not including behavioral contexts. In *A. geoffroyi*, sitting is by far the most dominant posture, and it may or may not involve the ischia, and rates of use may be as moderate as 25.7% in Costa Rica (Berge-son, 1998) to as high as 55.1% in Panama (Fontaine, 1990). *Ateles belzebuth* also show moderate rates (26.1%) in Ecuador (D. Youlatos, J. G. H. Cant and D. M. Rose, unpublished data) whereas sitting was particularly frequent for *A. paniscus* in French Guiana and especially in Suriname (Table 7.2; Mittermeier, 1978; Youlatos, 2002). Stand also shows variable rates across species and sites, scoring high in Ecuador, Costa Rica and Suriname (16.3%, 12.6%, and 13.1% respectively; Mittermeier, 1978; Bergeson, 1998; D. Youlatos, J. G. H. Cant and D. M. Rose, unpublished data) and particularly low in Panama, French Guiana and Guatemala (Table 7.2).

On the other hand, rates of suspended postures are variable across sites for the different species. Thus, *A. geoffroyi* show high rates in Panama (30.7%; Fontaine, 1990) and Costa Rica (33.3%; Bergeson, 1998), and especially high ones in Panama (53.7%; Mittermeier, 1978) and Guatemala (52%; Cant, 1986). Similarly, particularly high rates are also exhibited by *A. paniscus* in both French Guiana (55.4%; Youlatos, 2002) and Suriname (41.2%; Mittermeier, 1978), and *A. belzebuth* in Ecuador (55.5%; D. Youlatos, J. G. H. Cant and D. M. Rose, unpublished data).

In a finer analysis of suspensory postural behavior, tail-only suspension (7.44%; Fontaine, 1990) or one-limb hanging (6.6%; Mittermeier, 1978) accounts for a small proportion of the suspensory profile of *A. geoffroyi* in Panama (Table 7.2). Low rates are also encountered in *A. paniscus* in Suriname (Mittermeier, 1978). In contrast, tail-only hang represents 20% of total postural behavior of *A. geoffroyi* in Guatemala (Cant, 1986), and reaches a total of 41% of the suspensory feeding behavior in Costa Rica (Bergeson, 1998). Similarly, *A. paniscus* in French Guiana (18.8% of all postures), and *A. belzebuth* in Ecuador (31.3% of suspensory subsample) exhibit comparably high rates of tail-only suspension (Table 7.2; Youlatos, 2002; D. Youlatos, J. G. H. Cant and D. M. Rose, unpublished data).

In *A. geoffroyi*, tail–hindlimbs-assisted suspension appears to be very common in Costa Rica (13.1%) and Guatemala (21%), defined as tripod and inverted bipedal respectively (Cant, 1986; Bergeson, 1998). In a similar manner, tripod scores significant rates (16.1%) in *A. belzebuth* in Ecuador, and when added to the particularly high rates of tail–hindlimbs-assisted hanging (24.4%) the total use of the hindlimbs and the tail is predominant among suspensory postural behavior. On the other hand, in French Guiana, rates of tripod are low (7.9%) but proportions of tail–hindlimbs-assisted suspension can reach high rates, should the former be added to the tail–hindlimb(s) hang category (Table 7.2; Youlatos, 2002).

The involvement of forelimbs under tension in suspensory postural behavior is relatively consistent. In *A. geoffroyi*, rates range from 11% in Guatemala to

16% in Panama, while data from Panama and Costa Rica (Mittermeier, 1978; Bergeson, 1998) are hard to decipher and it is therefore difficult to gauge fore-limb use under suspension. *Ateles paniscus*, in French Guiana, show compara-ble rates when both tail–forelimb and tail–forelimb–hindlimb hang are included (total of 13%; Youlatos, 2002) that are similar to *A. belzebuth* from Ecuador (21.5% of suspensory subsample; D. Youlatos, J. G. H. Cant and D. M. Rose, unpublished data). Comparing the ratios of tail-only to tail–limb hanging across species, it is evident that most studies show that the latter is used from 1.5 to 2.2 times more than the former (Table 7.3). This shows a more or less equal use of these suspensory modes and provides evidence for the important role of the prehensile tail supporting whole body weight during feeding suspensory behavior. Particularly high ratios are found in Mittermeier's (1978) study, but this may be due to his definitional differences (Table 7.3).

These quantitative data on the positional behavior of the three species of *Ateles* can only be considered as indicative for each species and the genus in general. The use of different methods of data collection during different study seasons and periods in addition to different definitional systems of positional modes may render intra- and interspecific analyses and comparisons difficult. Moreover, variability in rates of use of different locomotor modes may also be related to the rapid and fluid locomotion of *Ateles*, occurring within and across trees, making them difficult to observe in detail. On the other hand, the relative consistency in rates of supported versus suspended postural behavior may be related to the fact that these postures occur for longer periods within a single tree and on a specific fruit patch. They are therefore easier to observe and sample than some other behaviors. In all cases, a gross profile of the locomotion and feeding postural behavior of the genus can be outlined.

Spider monkeys extensively employ suspensory forelimb dominated locomo-tion that mainly involves tail–arm brachiation and forelimb swing in variable rates. Pronograde quadrupedal activities, involving either walk/run on single substrates or clamber across multiple ones, also represent a major component of their locomotor profile. Vertical climb is used variably but in relatively notable proportions. Lastly, active and passive airborne locomotion appears to play a minor role in the overall repertoire of *Ateles* spp. During feeding, suspensory and nonsuspensory postures appear to be shared relatively equally. Sitting is the dominant supported posture. The rates of suspensory feeding are particularly high and outrank all other nonhominoid anthropoids. In contrast to hominoid suspensory feeding, however, the use of the tail by *Ateles* appears to be pri-mordial during suspended feeding. This is depicted by the notably high rates of tail-only hanging in most species, but also in the constant use of the tail in all below-substrate postures.

These observations support the fact that *Ateles* is by far the most suspensory of all atelines and of all platyrrhines. Among atelines, the positional behavior of *Lagothrix* (woolly monkeys) appears to approximate that of *Ateles*, containing a diversity of forelimb and hindlimb suspensory modes in relatively high proportions but is still dominated by pronograde above support behavior (Defler, 1999; Cant *et al.*, 2001; 2003). Howler monkeys (*Alouatta* spp.) depend mainly on pronograde modes during locomotion, albeit showing respectable rates of tail–hindlimb suspensory feeding (Fleagle and Mittermeier, 1980; Schön Ybarra, 1984; Cant, 1986; Schön Ybarra and Schön, 1987; Bicca Marques and Calegaro Marques, 1995; Bergeson, 1998; Youlatos, 1998). The morphology of the postcranial skeleton of *Brachyteles*, coupled with limited postural data, advocate behavioral similarities with *Ateles* that remain to be examined in the field (Erikson, 1963; Nishimura *et al.*, 1988; Jones, 2004).

Although the above studies provide a general positional profile for *Ateles* that differentiates it from the rest of the atelines, a finer analysis suggests several intra- and interspecific differences. As noted above, these differences may result from diverse methodological approaches adopted during the different studies. However, they may be also related to the utilization of specific structural features of the different habitats due to intrinsic forest architecture or seasonal exploitation of different food sources, or to phylogenetic constraints.

In terms of phylogeny, it appears that *A. paniscus* is the most suspensory species employing high rates of both suspensory locomotion and bridging, as well as suspensory feeding, with *A. belzebuth* being intermediate, with relatively intermediate rates of suspensory locomotion but high rates of suspensory feeding. The same studies indicate that *A. geoffroyi* may be the least suspensory species, showing the lowest rates of suspensory locomotion coupled with variable rates of suspensory feeding. It also appears that the latter species may be the most terrestrial one, as it exhibits the higher rates of terrestriality, although multiple reasons may account for this (Campbell *et al.*, 2005).

Locomotion, postures, and canopy use in *Ateles*

Locomotion and postures are developed through a constant interaction with the physical structure of the surrounding environment of the tropical forest canopy. Locomotion inflicts constraints as animals are called to cope with continuously emerging problems within time and space as they travel within the canopy. The problems may be multiple and require a diversity of solutions to help the animal get through the canopy and attain food sources and mates (Cant, 1992). On the other hand a different set of somewhat more "static" problems emerge during feeding. Although these problems may be more demanding to solve, there is likely a larger diversity of alternatives of appropriate behaviors available to the animal. The above studies of the positional behavior of *Ateles* have revealed a diversity of both locomotor and postural behavior. As discussed above some of the differences among the different species of *Ateles* may reside in phylogenetic constraints, but the study of the same species in different habitats may provide evidence for a differential use of structural differences in relation to anatomical design.

The main habitat that *Ateles* exploit is the higher levels of the canopy of tropical forests (Mittermeier and Roosmalen, 1981; Roosmalen, 1985; Youlatos, 2002). These parts of the forest are characterized by intertwining tree crowns that are mainly composed of small, flexible branches. This more or less continuous structure is however interrupted by the presence of vertically and horizontally distributed gaps of variable sizes between adjacent trees. In addition, emergent crowns that are disconnected from, and communicate with, the rest of the canopy via large vertical or subvertical substrates and wide gaps are also a component of this part of the forest. *Ateles* need to deal with these features of the canopy in order to find patches of ripe fruit trees and to gain access to fruit- or flush leaf-bearing substrates in the tree crown peripheries (Grand, 1972, 1984; Cant, 1992). In effect, the diversity of locomotion and postures that they exhibit are the behavioral means that enable them to deal with all aspects of habitat structure that may facilitate or impede their access to valuable food sources.

Initially, monkeys need to obtain access to these fruit trees that are randomly dispersed in time and space in the canopy of the tropical forest. Spider monkeys are ripe fruit specialists, obliged to look for trees that bear fruit at the right moment (Roosmalen, 1985; Symington, 1988b; Cant, 1990; Strier, 1992; Guillotin *et al.*, 1994; Simmen and Sabatier, 1996; Nunes, 1998; Iwanaga and Ferrari, 2001; Pozo, 2001; Dew, 2005; Wallace, 2005; Di Fiore and Campbell, 2007). Recent research has demonstrated that they appear to use a Levy walk pattern of traveling within their home ranges (Ramos-Fernández *et al.*, 2004). In this way, they show spatial scale invariance in the length of constituent steps and temporal invariance in the duration of intervals between steps. This appears to contribute more to the exploitation of these random scarce food sources that are far from uniform in their distribution and show a temporal and spatial variation. Levy walking aids in visiting more new sites and revisiting less previously visited sites. This pattern is strongly related to fruiting tree distribution that appears to follow a more or less fractal manner. Long infrequent steps could be those between known patches and shorter more frequent steps would be given while foraging within a patch. Avoiding revisiting previously visited fruit patches and/or revisiting them after appropriately long periods of time enhances the possibility of finding them bearing more ripe fruit (Ramos-Fernández *et al.*, 2004).

This strategy would imply fast traveling within the canopy, the shortening of pathway distances, and a variety of locomotor modes that could provide safe crossing of canopy gaps (Cant, 1992). Spider monkeys do appear to travel fast in the canopy in a similar but probably less acrobatic way than gibbons do (Cant, 1992). I have reported that travel speeds for brachiation bouts were significantly higher than those for quadrupedal walk (1.72 m/s vs. 0.84 m/s) in French Guiana (Youlatos, 1994). In addition, Cant (1986) found that spider monkeys traveled faster than howlers in all contexts in Guatemala. Brachiation is a form of locomotion that is energetically maximizing but may compensate body displacement in the forest canopy by enabling its performers to travel faster and, perhaps, to shorten pathways within and across tree crowns (Parsons and Taylor, 1977). It is thus very likely, that the consistently high proportions of tail-arm brachiation of spider monkeys do contribute to faster traveling within and across tree crowns. Moreover, suspensory *sensu lato* locomotor modes (i.e. tail-arm brachiation, forelimb swing, clambering, bridging) would help relatively large-bodied animals to move quite safely on slender flexible supports (Grand, 1984). Within a single crown, these modes would allow for rapid movements across supports avoiding zigzags that follow the more stable main branch forks in the central part of the crown (Grand, 1984). This helps reduce certain pathways within the same tree or across trees within intertwined crowns that bear no gaps.

On the other hand, crossing gaps between trees appears to be a very important determinant of shortening pathways, and can be extremely important for primates with long day ranges such as spider monkeys (Cant, 1992). Although smaller arboreal mammals tend to leap in order to traverse gaps, larger ones use more cautious and deliberate ways such as bridging (Youlatos, 1993), tree sways or vine crossings (Cant, 1992). This may account for the relatively low rates of leaping behavior generally observed in *Ateles*. However, studies have shown that, at least for *A. geoffroyi*, leaping is quite common in Panama and relatively infrequent in Guatemala, further accompanied by low rates of quadrupedalism in the former forest versus high ones in the latter (Mittermeier, 1978; Cant, 1986; Fontaine, 1990). Such observations may imply differences in habitat structure between the two forests. The dry forests of Guatemala may be more discontinuous with larger canopy gaps and a higher frequency of relatively "isolated" trees. This would eventually compel spider monkeys to use less leaping, as gaps would be larger and almost impossible to cross with such means, and impose a higher use of quadrupedal or suspensory modes to travel within single trees in order to obtain access to inter-crown passages that are negotiable for the animals. Thus, spider monkeys appear to heavily employ bridging and suspensory modes such as tail-arm brachiation in order to traverse gaps (Bergeson, 1996; Cant *et al.*, 2001). However, it is not clear as to how and under what conditions certain modes are opted or avoided. Could this actually provide path shortening or fast crossing within the canopy? This is hard to answer, but it may very likely help reduce pathways within and across trees in order to give spider monkeys a selective advantage for reaching more fruit trees within their large home ranges. To date, no study has specifically explored succession and shifting of locomotor modes in relation to habitat structure availability, in order to assess alternative solutions that appear to allow the animals to avoid time- and energy-consuming detours.

In effect, there are no detailed data on the different modes that spider monkeys employ during traveling within and across canopy trees in order to gain access to the desired fruit sources. However, it would be interesting to determine whether there are differences in both the frequency as well as sequence of locomotor modes employed during these differential approaches of known and unknown fruit patches. This would demand a large and short scale analysis of the sequence of modes so that a pattern could arise of the different solutions that spider monkeys use in respect with the structure of the different trees. This approach would help in the investigation of locomotor behavior in a wider adaptive context. The sequence of modes could elucidate the range of eventual selections that enable spider monkeys to cope with the structural diversity of their environment while also acknowledging the adaptive significance of foraging patterns from the locomotion-postcranium complex point of view (Cant, 1992). Such a study should cover both selection at the broader scale (i.e. locomotion across trees and overall traveling patterns) as well as at the finer scale (i.e. locomotion within a single tree), thus allowing a test of the proximate causes of mode selection. Would spider monkeys select for speed and quicker displacement across the canopy? Would they prefer or avoid large vertical structures that provide direct and rapid changes of vertical location in the canopy? Could all these provide shortening of pathway distances? How are gaps traversed, and is there any relationship between the 3-dimensional structure of the gap and locomotor options selected? These problems, although thoroughly described and partially tackled by Cant (1992) are far from having a tangible answer. Thus far, most studies focus on anatomical correlates, rather than providing answers to the questions I have outlined.

In addition, it is well established that spider monkey males have larger home ranges and travel further per day than females (van Roosmalen, 1985; Symington, 1988a; Strier, 1992; Nunes, 1995; Shimooka, 2005). It would thus be interesting to test whether there are differences in the frequencies and sequences of locomotor modes between males and females. Because spider monkeys are monomorphic in body size (Di Fiore and Campbell, 2007), eventual sex differences could be related to differential canopy use rather than body size differences. A similar comparative study could shed light on the modes that are actually related to path shortening or faster traveling within the forest canopy.

Another major problem that arboreal dwellers face is the way that food is actually acquired, manipulated, and consumed (Cant, 1992). As ripe fruit specialists and flush leaf consumers, spider monkeys are compelled to negotiate the terminal branches of tree crown peripheries. Such a microhabitat involves small, slender and fragile substrates, whereupon relatively large-bodied arboreal animals face difficulty in maintaining equilibrium and face a high risk of toppling over and falling. In this context, below-branch suspensory patterns or above-branch distribution of body weight through cautious body displacement or dynamic maintenance may provide substantial solutions to overcome this problem (Cartmill, 1985; Cant, 1987; Bergeson, 1998; Dunbar and Badam, 2000).

Such modes appear to be used at quite high rates during feeding activities in most primate species and may provide quicker and safer access to food-bearing substrates, enabling them to explore, detect, locate, approach, acquire and ingest the desired food items (Grand, 1972; Cant, 1992). In spider monkeys this is further achieved by the increased use of the prehensile tail in both suspended and supported locomotor and postural modes offering substantial support during both foraging movements and food acquisition. Sarmiento (1995) asserted that the deliberate climbing (= clambering) heritage of atelines (and hylobatids) evolved in more suspensory habits mainly to solve fruit-eating problems. This may hold true for siamangs in Malaysia (Fleagle, 1976) but data for the atelines do not support similar assumptions (Bergeson, 1998; Youlatos, 2002). The use of high rates of suspensed positional modes during both fruit and flush leaf acquisition in *Ateles* most likely supports the idea that similar behaviors are associated with the characteristic energy-maximizing strategy that the proto-atelines must have adopted (Jones, 2004). However, postural data for *A. paniscus* in similar forests reveal strong differences in rates of sit, stand, and suspensory postural behavior (Mittermeier, 1978; Youlatos, 2002). This may be related to differences in food sources acquired during different seasons. A similar pattern has been observed in *Alouatta* and has been related to differences in microhabitat and temporal and spatial distribution of food sources (Youlatos, 1998). These observations denote the importance of structural differences of habitat features on the postural behavior of *Ateles*, providing evidence for further research focusing on the interaction between postural behavior and microhabitat structure, as well as the options of positional modes selected within such contexts.

Acknowledgements

I am particularly grateful to Christina Campbell who invited me to participate in and contribute to this volume. However, without the help and financial support of the following persons and institutions none of these thoughts could have been typed. Thus, field research and acquaintance with these amazing animals was made feasible through financial support by "Action Spécifique Guyane" of the Muséum National d'Histoire Naturelle in Paris, France and the C.N.R.S.-U.M.R. 8570 Laboratoire d'Anatomie Comparée (French Guiana) and NSF SBR 9222526 and the School of Medicine, University of Puerto Rico (Ecuador). For these reasons, I am greatly indebted to P. Charles-Dominique, J.-P. Gasc, B. Simmen, B. de Thoisy (French Guiana), and L. Albuja V., L. Arcos Terén, J. G. H. Cant, I. Cornejo, A. Di Fiore, J. L. Dew, W. E. Pozo R., P. S. Rodman, M. D. Rose (Ecuador).

References

Bergeson, D. J. (1996). The positional behavior and prehensile tail use of *Alouatta palliata*, *Ateles geoffroyi, and Cebus capucinus*. Unpublished Ph.D. thesis, Washington University, St. Louis, MO.

Bergeson, D. J. (1998). Patterns of suspensory feeding in *Alouatta palliata, Ateles geoffroyi*, and *Cebus capucinus*. In *Primate Locomotion. Recent Advances*, ed. E. Strasser, J. G. Fleagle, A. Rosenberger and H. McHenry, New York: Plenum Press, pp. 45–60.

Bertram, J. E. A. (2004). New perspectives on brachiation mechanics. *Yrbk. Phys. Anthropol.*, **47**, 100–117.

Bicca-Marques, J. C. and Calegaro-Marques, C. (1995). Locomotion of black howlers in a habitat with discontinuous canopy. *Folia Primatol.*, **64**, 55–61.

Campbell, C. J., Aureli, F., Chapman, C. A., *et al.* (2005). Terrestrial behavior of *Ateles* spp. *Int. J. Primatol.*, **26**, 1039–1051.

Cant, J. G. H. (1986). Locomotion and feeding postures of spider and howling monkeys: field study and evolutionary interpretation. *Folia Primatol.*, **46**, 1–14.

Cant, J. G. H. (1987). Positional behavior of female Bornean orangutans (*Pongo pygmaeus*). *Am. J. Primatol.*, **12**, 71–90.

Cant, J. G. H. (1990). Feeding ecology of spider monkeys (*Ateles geoffroyi*) at Tikal, Guatemala. *Hum. Evol.*, **5**, 269–281.

Cant, J. G. H. (1992). Positional behavior and body size of arboreal primates: a theoretical framework for field studies and an illustration of its application. *Am. J. Phys. Anthropol.*, **88**, 273–283.

Cant, J. G. H., Youlatos, D. and Rose, M. D. (2001). Locomotor behavior of *Lagothrix lagothricha* and *Ateles belzebuth* in Yasuni National Park, Ecuador: general patterns and nonsuspensory modes. *J. Hum. Evol.*, **41**, 141–166.

Cant, J. G. H., Youlatos, D. and Rose, M. D. (2003). Suspensory locomotion of *Lagothrix lagothricha* and *Ateles belzebuth* in Yasuni National Park, Ecuador. *J. Hum. Evol.*, **44**, 685–699.

Cartmill, M. (1985). Climbing. In *Functional Vertebrate Morphology*, ed. M. Hildebrand, D. M. Bramble, K. F. Liem and D. B. Wake, Cambridge: Belknap Press, pp. 73–88.

Cartmill, M., Lemelin, P. and Schmitt, D. (2002). Support polygons and symmetrical gaits in mammals. *Zool. J. Linn. Soc.*, **136**, 401–420.

Cavanez, F. C., Moreira, M. A. M., Ladasky, J. J., *et al.* (1999). Molecular phylogeny of New World primates (Platyrrhini) based on β2-microglobulin DNA sequences. *Mol. Phyl. Evol.*, **12**, 74–82.

Collins, A. C. (2004). Atelinae phylogenetic relationships: the trichotomy revived? *Am. J. Phys. Anthropol.*, **124**, 284–296.

Defler, T. R. (1999). Locomotion and posture in *Lagothrix lagothricha. Folia Primatol.*, **70**, 313–327.

Dew, J. L. (2005). Foraging, food choice, and food processing by sympatric ripe-fruit specialists: *Lagothrix lagotricha poeppigii* and *Ateles belzebuth belzebuth. Int. J. Primatol.*, **26**, 1107–1135.

Di Fiore, A. and Campbell, C. J. (2007). The Atelines: variation in ecology, behavior and social organization. In *Primates in Perspective*, ed. C. J. Campbell, A. F. Fuentes, K. C. MacKinnon, M. Panger and S. Bearder, Oxford: Oxford University Press, pp. 155–185.

Dunbar, D. C. and Badam, G. L. (2000). Locomotion and posture during terminal branch feeding. *Int. J. Primatol.*, **21**, 649–669.

Erikson, G. E. (1963). Brachiation in New World primates and in anthropoid apes. *Symp. Zool. Soc. Lond.*, **10**, 135–163.

Fleagle, J. G. (1976). Locomotion and posture of the Malayan siamang and implications for hominoid evolution. *Folia Primatol.*, **26**, 245–269.

Fleagle, J. G. and Mittermeier, R. A. (1980). Locomotor behavior, body size, and comparative ecology of seven Suriname monkeys. *Am. J. Phys. Anthropol.*, **52**, 301–314.

Fleagle, J. G. and Simons, E. L. (1982). The humerus of *Aegyptopithecus zeuxis*: a primitive anthropoid. *Am. J. Phys. Anthropol.*, **59**, 175–193.

Fleagle, J. G., Stern, J. T., Jr., Jungers, W. L., Susman, R. L., Vangor, A. K. and Wells, J. P. (1981). Climbing: a biomechanical link with brachiation and with bipedalism. *Symp. Zool. Soc. Lond.*, **48**, 359–375.

Fontaine, R. (1990). Positional behavior in *Saimiri boliviensis* and *Ateles geoffroyi. Am. J. Phys. Anthropol.*, **82**, 485–508.

Gebo, D. L. (1992). Locomotor and postural behavior of *Alouatta palliata* and *Cebus capucinus. Am. J. Primatol.*, **26**, 277–290.

Grand, T. I. (1972). A mechanical interpretation of terminal branch feeding. *J. Mammal.*, **53**, 198–201.

Grand, T. I. (1984). Motion economy within the canopy: four strategies for mobility. In *Adaptations for Foraging in Nonhuman Primates*, ed. P. S. Rodman and J. G. H. Cant, New York: Columbia University Press, pp. 54–72.

Guillotin, M., Dubost, G. and Sabatier, D. (1994). Food choice and food competition among the three major primate species of French Guiana. *J. Zool. Lond.*, **233**, 551–579.

Harada, M. L., Schneider, H., Schneider, M. P. C., *et al.* (1995). DNA evidence on the phylogenetic systematics of New World monkeys: support for the sister grouping of *Cebus* and *Saimiri* from two unlinked nuclear genes. *Mol. Phyl. Evol.*, **4**, 331–349.

Hartwig, W. (2005). Implications of molecular and morphological data for understanding ateline phylogeny. *Int. J. Primatol.*, **26**, 999–1015.

Hirasaki, E., Kumakura, H. and Matano, S. (1993). Kinesiological characteristics of vertical climbing in *Ateles geoffroyi* and *Macaca fuscata*. *Folia Primatol.*, **61**, 148–156.

Hirasaki, E., Kumakura, H. and Matano, S. (2000). Biomechanical analysis of vertical climbing in the spider monkey and the Japanese macaque. *Am. J. Phys. Anthropol.*, **113**, 455–472.

Hunt, K. D., Cant, J. G. H., Gebo, D. L., *et al.* (1996). Standardized descriptions of primate locomotor and postural modes. *Primates*, **37**, 363–387.

Isler, K. (2004). Footfall patterns, stride length and speed of vertical climbing in spider monkeys (*Ateles fusciceps robustus*) and woolly monkeys (*Lagothrix lagotricha*). *Folia Primatol.*, **75**, 133–149.

Iwanaga, S. and Ferrari, S. F. (2001). Party size and diet of syntopic atelids (*Ateles chamek* and *Lagothrix cana*) in southwestern Brazilian Amazonia. *Folia Primatol.*, **72**, 217–227.

Jenkins, F. A., Jr. (1981). Wrist rotation in primates: a critical adaptation for brachiators. *Symp. Zool. Soc. Lond.*, **48**, 429–451.

Jenkins, F. A., Jr., Dombrowski, P. J. and Gordon, E. P. (1978). Analysis of the shoulder in brachiating spider monkeys. *Am. J. Phys. Anthropol.*, **48**, 65–76.

Jones, A. L. (2004). The evolution of brachiation in atelines: a phylogenetic comparative study. Unpublished Ph.D. thesis, University of California, Davis.

Jungers, W. L. and Stern, J. T., Jr. (1981). Preliminary electromyographical analysis of brachiation in gibbon and spider monkey. *Int. J. Primatol.*, **2**, 19–33.

Konstandt, W., Stern, J. T., Jr., Fleagle, J. G. and Jungers, W. L. (1982). Function of the subclavius muscle in a nonhuman primate, the spider monkey (*Ateles*). *Folia Primatol.*, **38**, 170–182.

Larson, S. G. (1998a). Parallel evolution in the hominoid trunk and forelimb. *Evol. Anthropol.*, **6**, 87–99.

Larson, S. G. (1998b). Unique aspects of quadrupedal locomotion in nonhuman primates. In *Primate Locomotion. Recent Advances*, ed. E. Strasser, J. G. Fleagle, A. Rosenberger and H. McHenry, New York: Plenum Press, pp. 157–173.

Larson, S. G., Schmitt, D., Lemelin, P. and Hamrick, M. (2000). Uniqueness of primate forelimb posture during quadrupedal locomotion. *Am. J. Phys. Anthropol.*, **112**, 87–101.

Lemelin, P. (1995). Comparative and functional myology of the prehensile tail in New World monkeys. *J. Morphol.*, **224**, 1–18.

Lemelin, P. and Schmitt, D. (1998). The relation between hand morphology and quadrupedalism in primates. *Am. J. Phys. Anthropol.*, **105**, 185–197.

Lockwood, C. A. (1999). Homoplasy and adaptation in the atelid postcranium. *Am. J. Phys. Anthropol.*, **108**, 459–482.

Meireles, C. M., Czelusniak, J., Schneider, M. P. C., *et al.* (1999). Molecular phylogeny of ateline New World monkeys (Platyrrhini, Atelinae) based on γ-globin gene sequences: evidence that *Brachyteles* is the sister group of *Lagothrix*. *Mol. Phyl. Evol.*, **12**, 10–30.

Mittermeier, R. A. (1978). Locomotion and posture in *Ateles geoffroyi* and *Ateles paniscus*. *Folia Primatol.*, **30**, 161–193.

Mittermeier, R. A., van Roosmalen, M .G. M. (1981). Preliminary observations on habitat utilization and diet in eight Surinam monkeys. *Folia Primatol.*, **36**, 1–39.

Nishimura, A., Fonseca, G. A. B., Mittermeier, R. A., *et al.* (1988). The muriqui, genus *Brachyteles*. In *Ecology and Behavior of Neotropical Primates*, Vol. 2., ed. R. A. Mittermeier, A. B. Rylands, A. F. Coimbra-Filho and G. A. B. Fonseca, Washington, DC: World Wildlife Fund, pp. 577–610.

Nunes, A. (1995). Foraging and ranging patterns in white-bellied spider monkeys. *Folia Primatol.*, **65**, 85–99.

Nunes, A. (1998). Diet and feeding ecology of *Ateles belzebuth belzebuth* at Maraca Ecological Station, Roraima, Brazil. *Folia Primatol.*, **69**, 61–76.

Parsons, P. E. and Taylor, C. R. (1977). Energetics of brachiation versus walking: a comparison of a suspended and an inverted pendulum mechanism. *Physiol. Zool.*, **50**, 182–188.

Pozo R., W. E. (2001). Composición social y costumbres alimenticias del "mono araña oriental" (*Ateles belzebuth belzebuth*) en el Parque Nacional Yasuní, Ecuador. Unpublished Ph.D. thesis, Universidad Central del Ecuador, Quito, Ecuador.

Preuschoft, H. and Demes, B. (1984). Biomechanics of brachiation. In *The Lesser Apes*, ed. H. Preuschoft, D. J. Chivers, W. Y. Brockelman and N. Creel, Edinburgh: Edinburgh University Press, pp. 96–118.

Ramos-Fernández, G., Mateos, J. L., Miramontes, O., *et al.* (2004). Levy walk patterns in the foraging movements of spider monkeys (*Ateles geoffroyi*). *Behav. Ecol. Sociobiol.*, **55**, 223–230.

Rose, M. D. (1974). Postural adaptations in New and Old World monkeys. In *Primate Locomotion*, ed. F. Jenkins, New York: Academic Press, pp. 201–222.

Rose, M. D. (1996). Functional morphological similarities in the locomotor skeleton of Miocene catarrhines and platyrrhine monkeys. *Folia Primatol.*, **66**, 7–14.

Rosenberger, A. L. and Strier, K. B. (1989). Adaptive radiation of the ateline primates. *J. Hum. Evol.*, **18**, 717–750.

Sarmiento, E. E. (1995). Cautious climbing and folivory: a model of hominoid differentiation. *Hum. Evol.*, **10**, 289–321.

Schmitt, D. (1994). Forelimb mechanics as a function of substrate type during quadrupedalism in two anthropoid primates. *J. Hum. Evol.*, **26**, 441–457.

Schmitt, D. (1999). Compliant walking in primates. *J. Zool. Lond.*, **248**, 149–160.

Schmitt, D. and Hanna, J. B. (2004). Substrate alters forelimb and hindlimb peak force ratios in primates. *J. Hum. Evol.*, **46**, 237–253.

Schmitt, D. and Larson, S. G. (1995). Heel contact as a function of substrate type and speed in primates. *Am. J. Phys. Anthropol.*, **96**, 39–50.

Schmitt, D., Rose, M. D., Turnquist, J. E. and Lemelin, P. (2005). The role of the prehensile tail during ateline locomotion: experimental and osteological evidence. *Am. J. Phys. Anthropol.*, **126**, 435–446.

Schneider, H., Sampaio, I., Harada, M. L., *et al.* (1996). Molecular phylogeny of the New World monkeys (Platyrrhini, Primates) based on two unlinked nuclear genes: IRPB Intron 1 and ε-globin sequences. *Am. J. Phys. Anthropol.*, **100**, 153–180.

Schön Ybarra, M. A. (1984). Locomotion and postures of red howlers in a deciduous forest–savanna interface. *Am. J. Phys. Anthropol.*, **63**(1), 65–76.

Schön Ybarra, M. A. and Schön, M. A. (1987). Positional behavior and limb bone adaptations in red howling monkeys (*Alouatta seniculus*). *Folia Primatol.*, **49**, 70–89.

Shimooka, Y. (2005). Sexual differences in ranging of *Ateles belzebuth belzebuth* at La Macarena, Colombia. *Int. J. Primatol.*, **26**, 385–406.

Simmen, B. and Sabatier, D. (1996). Diets of some French Guianan primates: food composition and food choices. *Int. J. Primatol.*, **17**, 661–693.

Stern, J. T., Jr., Wells, J. P., Jungers, W. L., Vangor, A. K. and Fleagle, J. G. (1980a). An electromyographic study of serratus anterior in atelines and *Alouatta*: implications for hominoid evolution. *Am. J. Phys. Anthropol.*, **52**, 323–334.

Stern, J. T., Jr., Wells, J. P., Jungers, W. L., Vangor, A. K. and Fleagle, J. G. (1980b). An electromyographic study of the pectoralis major in atelines and *Hylobates*, with special reference to the evolution of a pars cranialis. *Am. J. Phys. Anthropol.*, **52**, 13–25.

Stern, J. T., Jr., Wells, J. P., Vangor, A. K. and Fleagle, J. G. (1976). Electromyography of some muscles of the upper limb in *Ateles* and *Lagothrix*. *Yrbk. Phys. Anthropol.*, **20**, 498–507.

Strier, K. B. (1992). Atelinae adaptations: behavioral strategies and ecological constraints. *Am. J. Phys. Anthropol.*, **88**, 515–524.

Symington, M. M. (1988a). Demography, ranging patterns, and activity budgets of black spider monkeys (*Ateles paniscus chamek*) in the Manu National Park, Peru. *Am. J. Primatol.*, **15**, 45–67.

Symington, M. M. (1988b). Food composition and foraging party size in the black spider monkey (*Ateles paniscus chamek*). *Behaviour*, **105**, 117–134.

Turnquist, J. E., Schmitt, D., Rose, M. D. and Cant, J. G. H. (1999). Pendular motion in the brachiation of captive *Lagothrix* and *Ateles*. *Am. J. Primatol.*, **48**, 263–281.

van Roosmalen, M. G. M. (1985). Habitat preferences, diet, feeding strategy and social organization of the black spider monkey (*Ateles paniscus paniscus* Linnaeus 1758) in Surinam. *Acta Amazonica*, **15**, 1–238.

Vilensky, J. A. (1989). Primate quadrupedalism: how and why does it differ from that of typical quadrupeds? *Brain Behav. Evol.*, **34**, 357–364.

von Dornum, M. and Ruvolo, M. (1999). Phylogenetic relationships of the New World monkeys (Primates, Platyrrhini) based on nuclear G6PD DNA sequences. *Mol. Phyl. Evol.*, **11**, 459–476.

Wallace, R. B. (2005). Seasonal variations in diet and foraging behavior of *Ateles chamek* in a southern Amazonian tropical forest. *Int. J. Primatol.*, **26**, 1053–1075.

Youlatos, D. (1993). Passages within a discontinuous canopy: bridging in the red howler monkey (*Alouatta seniculus*). *Folia Primatol.*, **61**, 144–147.

Youlatos, D. (1994). Maîtrise de l'espace et accès aux ressources chez le singe hurleur roux (*Alouatta seniculus*) de la Guyane Française. *Etude morpho-fonctionnelle.* Unpublished Ph.D. thesis, Muséum National d'Histoire Naturelle, Paris, France.

Youlatos, D. (1998). Seasonal variation in the positional behavior of red howling monkeys (*Alouatta seniculus*). *Primates*, **39**(4), 449–457.

Youlatos, D. (2002). Positional behavior of black spider monkeys (*Ateles paniscus*) in French Guiana. *Int. J. Primatol.*, **23**, 1071–1093.

8 Communication in spider monkeys: the function and mechanisms underlying the use of the whinny

GABRIEL RAMOS-FERNÁNDEZ

Introduction

Several aspects of the biology of spider monkeys (*Ateles* spp.) make their communication system particularly interesting. First, as canopy-dwelling, frugivorous primates, spider monkeys must forage in a complex environment consisting of fruit patches of variable size that are spatially distributed over large areas (Klein, 1972; van Roosmalen and Klein, 1987). Such an environment places a high demand on those mechanisms by which individuals can efficiently coordinate their movements, exploit their food resources and avoid danger (Milton, 2000). Additionally, because of their fluid grouping and association patterns (Aureli and Schaffner, this volume), individual spider monkeys may spend long periods of time away from others in the group (Symington, 1990; Ramos-Fernández, 2005). This implies that at times, long-distance vocalizations, which can overcome spatial separation, or olfactory marks, which can overcome temporal separation, may be the only means by which spider monkeys can locate group members and, in general, maintain their social relationships (Ramos-Fernández, 2005; Aureli *et al.*, in press).

This chapter reviews the studies carried out so far on spider monkey communication, placing more emphasis on the vocal mode, as it has received the most attention. The first section reviews the early, more descriptive studies of communication in spider monkeys and provides a set of definitions of the different vocal types used by different species. Among the different vocalizations used by spider monkeys, only the whinny has been studied in any detail. Therefore, in the second section of this chapter I review the studies that have tested specific hypotheses about the proximate mechanisms underlying production, transmission and perception of whinnies in spider monkeys. In the third section I review the studies that have attempted to determine the functional significance of vocal signals. In the final section I place the evidence reviewed into

Spider Monkeys: Behavior, Ecology and Evolution of the Genus Ateles, ed. Christina J. Campbell. Published by Cambridge University Press. © Cambridge University Press 2008.

the current framework of communication networks, emphasizing the role that communication may have in the social system of spider monkeys.

Background information

After observing a nonhabituated group of *Ateles geoffroyi* in the Coto region in western Panama for one and a half months, Carpenter (1935) published the first account of spider monkey communication. He noted that the most common vocalization was the "terrier-like bark," which the monkeys gave when approached by an observer on the ground and upon taking defensive action or flight. He proposed that this vocalization functioned as a warning signal to others in the group. The next most common vocalization was the "growl," which was only heard from "greatly aroused males" or from subgroups that were "contending with each other" (Carpenter, 1935, p. 180). Finally, the "whinny" was reported as a higher pitched, lesser amplitude version of a horse's whinny, emitted when subgroups or individuals became separated. Carpenter (1935) interpreted that this vocalization functioned to coordinate movements of subgroups and assumed it was equivalent in function to the capuchin's "caws."

The first published study to present sonograms of spider monkey vocalization recordings was Klein's (1972) report of his long-term study of *Ateles belzebuth* in La Macarena, Colombia. Klein used vocalizations to locate spider monkey subgroups and described the use of six vocalization types: "whoops" and "wails," given in bouts and assumed to be used to locate group members, often followed by a merger of subgroups; "barks," which were given to predators and conspecific disturbances; "squaks," given by juveniles that had become separated from the rest of the subgroup; "tschooks," which were exchanged by spider monkeys after a separation and before joining subgroups; and "whinnys," emitted by members of approaching and approached subgroups before and during a subgroup merger. Klein (1972) provides a particularly interesting description of subgroup dispersals (fissions) and mergers (fusions) that is very suggestive of the potential role of vocal communication during these important events in the fission–fusion system of *Ateles*. Briefly, during fissions, "tschook" vocalizations from lone juveniles could be followed by "whinny" vocalizations from adult individuals that apparently were waiting for the juveniles to follow them. During fusions, "whinny" vocalizations were almost always heard and "appeared to be initiated when at least one of the members of two different subgroups were first able to see one another directly" (Klein, 1972, p. 179). The exchanges of "whinny" vocalizations during fusions were often reciprocal between individuals in the joined and the joining subgroup. He also noted that the same vocalizations also occurred in contexts that were not related to subgroup fusions.

Table 8.1 *Vocal repertoire of spider monkeys with some suggested functions*

Call type	Suggested function	Reference
Long call (harsh)	Position indicator and identifier	Eisenberg 1976
Long call (clear)	Position indicator and potential identifier; may promote assembly	Eisenberg 1976
Bark	Position indicator; promotes assembly; warning signal	Eisenberg 1976; Symington 1987
Chitter	Warns conspecifics about disturbances in the environment	Eisenberg 1976
Tee-tee	Position indicator; friendly greetings	Eisenberg 1976
Whinny	Position indicator; may indicate availability of food; informs others about individual identity	Eisenberg 1976; Chapman and Lefebvre 1988; Teixidor and Byrne 1997, 1999; Ramos-Fernández 2005
Ook-ook-ak-ak	Invitation to play	Eisenberg 1976
Gutural whinny	Friendly approach	Eisenberg 1976
Growl	Hostile approach	Eisenberg 1976
Whoop	Male position indicator	van Roosmalen and Klein 1987

In 1976, J. F. Eisenberg published the first and only systematic study of the complete communication system of any *Ateles* species. Many of the following studies have focused only on a single communication mode or a set of vocalizations. This study organized the vocal signals of captive *A. fusciceps* (now classified as *A. geoffroyi fusciceps*) into different types based on acoustic analyses. The study was complemented by a field study of *A. geoffroyi*. Eisenberg (1976) distinguished 13 different vocal types depending on their loudness, their tonal quality, the presence of frequency modulation and their length. This classification is based on extensive acoustic analyses and a thorough description of the possible combinations of elements between call types. It also includes the context (in captivity) in which each vocalization type occurs. It is this classification that will be used throughout the rest of this chapter (Table 8.1).

Adding to Eisenberg's classification, van Roosmalen and Klein (1987) described loud "whoops," long-distance calls performed exclusively by adult males that, although sounding the same, can be given in different contexts: when a male has been recently separated from a subgroup, when monkeys hear a sudden, unfamiliar movement or during agonistic intergroup encounters.

Both Klein's (1972) study and van Roosmalen and Klein's (1987) review include a description of the olfactory communication system in *Ateles*. They suggest that the primary function of the elongated clitoris in spider monkeys is to leave urine marks on branches. Based on the fact that it is mostly males who smelled these marks, these authors inferred that females are announcing their reproductive status in this manner (see Campbell, 2006; Campbell and Gibson, this volume for additional functions of the clitoris length). Another potential signal may consist of the secretions from the sternal gland, which monkeys rub with saliva (Klein, 1972) or an aromatic leaf–saliva mix (Campbell, 2000; Laska *et al.*, 2007) using their hand. The gland is smelled mutually when two individuals, regularly adults, meet after a separation and engage in a series of embraces, licking and sniffing their sternal gland areas and grunting. The function of this social interaction, which involves several communicative signals, was proposed to be the strengthening of the social bonds (van Roosmalen and Klein, 1987; see also Aureli and Schaffner, this volume).

In describing the communication signals of *Ateles* spp., the studies cited above laid the framework for hypothesis-driven research. Subsequent researchers have asked questions regarding the mechanisms underlying the production, transmission and perception of communication signals as well as the possible functions that these signals may play in the social and ecological environment of spider monkeys. Although vocal signals have received far more attention than olfactory signals, most studies examining vocalizations have focused only on the most frequently heard call type in wild spider monkeys, the whinny. Accordingly, in the next two sections I review what is known about the production and function of this call specifically.

Most studies have made extensive use of field recordings and spectrographic analyses (Chapman and Weary, 1990; Teixidor and Byrne, 1997; Ramos-Fernández, 2000). Briefly, spectrograms are graphic depictions of compound signals indicating variation in their frequency and amplitude with respect to time. They allow researchers to distinguish and measure fine features of vocalizations, quantifying distinctive parameters and studying how they vary among contexts or callers. Another important tool that primatologists use to uncover which aspects of the acoustic variation in natural calls are meaningful to primates are playback experiments (e.g. Waser, 1977; Seyfarth *et al.*, 1980). Using hidden speakers, calls recorded from known subjects in known contexts are played in the field, and the the subjects' behavior before and after the playback is observed and measured. These experiments have been performed only recently on habituated groups of spider monkeys (Teixidor and Byrne, 1997, 1999; Ramos-Fernández, 2005).

The whinny – proximate mechanisms

Questions pertaining to the proximate mechanisms underlying vocal communication in *Ateles* deal with those properties of the vocal organs that produce signals, how the signals transmit through the environment and finally, how they are perceived by the auditory organs of recipients. Although sonograms of

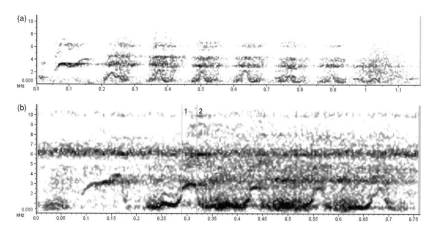

Figure 8.1 Spectrograms of whinny exemplars, recorded in the Yucatán peninsula by the author (see methods in Ramos-Fernández, 2005). (a) A typical whinny composed of several elements, in this case two distinct elements at the beginning and then a series of seven "arches" (see text). (b) A whinny composed of two distinct elements up to 0.2 seconds, and then a series of four "arches" and "interarch elements." Label 1 shows a point of period doubling and label 2 shows the return to the original fundamental frequency at the end of the arch (see text for description). The continuous line at about 6 kHz corresponds to insect noise. Ordinate in kHz, abscissa in seconds.

several of the spider monkey's calls are presented in many studies, the mechanisms of production and perception of these calls have received very little attention. This is unfortunate, given the fact that the study of these mechanisms may provide important insights into the sources of variation in calls and, therefore, the potential information that recipients may be able to extract from this variation as well as the call's potential function (Hauser, 1996). This is exemplified by the following analysis of the acoustic structure and the transmission properties of whinnies (M. Y. Liberman, personal communication, 1996).

Whinnies consist of a series of 2–12 rapid rises and falls in pitch, about 100 ms from peak to peak, with a fundamental frequency (F_0) moving from about 1000 Hz to about 2300 Hz and back (Figure 8.1). These "arches" have a fundamental component that is much stronger than the higher harmonics. In between the arches, a typical whinny has other sounds, here called "interarch elements." These sound (and look, on a spectrogram) quite different. Their pitch is a rise or fall–rise, with a short fall (if any) and a longer rise, all at a much lower fundamental frequency, with a minimum of about 400 Hz and a maximum of about 1000 Hz. The first (and sometimes second) harmonics are just about as strong as the fundamental. The rapid alternation of arches and interarch elements gives whinnies the "grunt-like" quality as well as the similarity to other, higher-pitched calls like trills and tweeters. Because the arches and interarch elements sound so different, and because they sometimes come and go separately, yielding for instance, calls with no interarch elements (like trills and tweeters), or with intermittent ones, one could think that the two kinds of sounds are produced by different articulatory mechanisms. Alternatively, interarch elements could arise naturally by period-doubling in the lower-frequency regions of a call in which the fundamental frequency of laryngeal oscillation is being rapidly modulated up and down, a phenomenon well known in human phonetics (Herzel, 1993). This is suggested by the fact that the first harmonic of the interarch elements can be seen in spectrograms to be continuous with the fundamental frequency of the arches. That is, near the bottom of the falling limb of the arch, a new fundamental suddenly appears an octave lower, but the former fundamental remains as a strong component of the signal. This period doubling occurs as the fundamental is falling, at or near the bottom of its cycle. As the pitch rises again, the new double-period component vanishes, but this change occurs typically at a much higher frequency than the one at which the double-period component earlier appeared. This hysteresis (Herzel *et al.*, 1995) also seems consistent with the view that the interarch elements are due to a phase transition in laryngeal oscillation. Such phenomena have been observed in human babies' cries and in normal voice characterized as hoarse or creaky and in subharmonic vocalization or "vocal fry" (Herzel, 1993).

Some insight into the functional significance of this alternation of arches and interarch elements in whinnies can be obtained by studying the transmission loss that results from the specific environment in which they occur. Simultaneous observations of two subgroups have determined that spider monkeys can hear whinnies and respond to them when at a distance of up to 300 m from the caller (Ramos-Fernández, 2005). Wahlberg *et al.* (2002) studied the transmission loss exerted by the canopy environment on whinnies using playbacks and recordings of whinnies at different distances from the speaker. These authors found a strong excess attenuation due to friction in the canopy, up to 20 dB per 100 m for the 2000 Hz frequency component of whinnies. This means that a recipient of a whinny located 300 m from the caller can be assumed to obtain much less information in the frequency domain than would a recipient close to the caller. However, if the call contains abrupt changes in fundamental frequency such as those described above, the temporal information will remain despite the loss of energy due to the forest canopy in some frequency bands.

No studies have been carried out on the acoustic perception mechanisms in spider monkeys. However, studies in other nonhuman primate species, like Japanese macaques (*Macaca fuscata*), have shown that their auditory organs are highly sensitive to fundamental frequency modulations such as those found in several of the calls of *Ateles* (Hauser, 1996). Moreover, macaques show categorical perception of calls that grade temporally in the peak frequency of a modulated call, in a manner akin to humans' speech perception (May *et al.*, 1989). With regard to the localization of long-distance vocalizations in forest environments, the classic playback experiments by Waser (1977) on grey-cheeked mangabeys (*Cercocebus albigena*) showed that these primates were able to localize a single instance of a "whoop-gobble" call at several hundred meters of tropical forest, with a median error of only 6 degrees. Whether spider monkeys can also do this is not known, but anecdotal evidence suggests that they are able to perceive and locate other subgroups in their home range based on their long-distance calls (Ramos-Fernández, unpublished data). Another open question is whether spider monkeys, like other animals, are able to estimate the distance from the source of a call from the acoustic structure as it degrades through their environment (reviewed in Naguib and Wiley, 2001).

Closely related to the production and transmission mechanisms is the search for consistent patterns of variability among signals that could provide recipients with information about the identity of the signaler or other contextual features. Masataka (1986) studied trill vocalizations in a group of captive spider monkeys (*Ateles geoffroyi*) and found that these calls contain information about the caller's identity in the frequencies of the strongest energy band, the second strongest energy band and the lowest energy band. Moreover, trills that had been responded to by particular individuals could be discerned on the basis of their acoustic features. The individual that had originally responded to a trill with another trill of their own could be predicted on the basis of the trill's peak frequency, the amplitude differences between peaks, the frequency modulation of amplitude peaks between 3–5 and 5–8 kHz and the frequency of the strongest energy band (Masataka, 1986). A recorded trill was then played back to those individuals that had originally vocalized after the trill, as well as those that had originally not vocalized. Subjects that originally gave a trill vocalization after the playback call were significantly more likely to vocalize after its played-back version than other individuals. Masataka (1986) interpreted this result as implying that each group member is able to distinguish trills to which they originally responded, so that these vocalizations may indeed consist of "rudimentary" representations for the identity of each group member. If replicated in field conditions, this result could imply that these calls actually consist of labels for one another and that they address each other in this manner.

Chapman and Weary (1990) studied individual differences in the acoustic properties of what seems to be the same vocalization type as Masataka's (1986; based on the sonogram presented), although in this case the authors call it a whinny. Chapman and Weary (1990) found that the duration of the first, middle and ending arches of these vocalizations vary consistently among individuals and that the calls of mothers and their offspring resembled each other more than all other pairwise combinations of calls from different individuals.

Two additional field studies investigating the actual call known by most as the whinny report similar results on the important role of individual identity as a source of consistent variation between calls. Teixidor and Byrne (1999) found significant differences in nine different acoustic parameters among different individuals' calls and then used discriminant function analysis to identify the parameters that best predicted individual identity. These were the maximum peak in the fundamental frequency, the number of frequency modulations and the frequency range of the call. These parameters alone could assign between 50% and 94% of the calls to the correct caller (Teixidor and Byrne, 1999). Similarly, I have found that there are consistent differences between individuals in the call duration, the maximum and minimum frequencies, the frequency range and the duration of the ending element (Ramos-Fernández, 2000). Furthermore, these acoustic parameters could assign 30–70% of the calls to their correct caller using the same discriminant function analysis as the study by Teixidor and Byrne (1999). These results are qualitatively confirmed by the fact that experienced observers of spider monkeys after some time are able to identify some study animals by their whinnies alone (Ramos-Fernández, unpublished data; C. Campbell, personal communication).

Based on the results of these studies, it can be concluded that, among the many vocalizations used by *Ateles*, at least trills and whinnies contain sufficient information to allow identification of the caller. Some of the sources of consistent variation lie in the frequency domain (maximum and minimum frequencies of elements, frequencies of strongest energy bands) and others in the temporal domain (number and duration of elements). The results of the transmission loss experiments cited above, as well as those of perceptual abilities in other primate species, suggest that the individual identity information contained in some of the long-distance calls of spider monkeys would be transmitted successfully even through a closed canopy environment.

The whinny: functional significance

A different question posed by animal communication studies has to do with the function of a communication act; that is, the way in which a signal benefits both signaler and recipients (in the case of cooperative communication acts) or only the signaler (in the case of communication as a manipulation of recipients by the signaler; Dawkins and Krebs, 1978). In primate societies, any communication signal will often have multiple functions, for there are often multiple recipients, each with a different social relationship with the signaler. Thus, in order to understand the function of a communication act, it is important to consider the different social strategies that individuals use to deal with conflict and cooperation within their group (Walters and Seyfarth, 1987; Aureli and Schaffner, this volume).

The first field study to test an explicit hypothesis about the functional significance of a spider monkey call was performed by Chapman and Lefebvre (1990) on *Ateles geoffroyi* in Costa Rica. Assuming that there is a benefit for spider monkeys to forage with others, mainly by decreasing the risk of predation, these authors postulated that whinnies are used to attract group members to fruiting trees. They predicted that high-ranking individuals should call more than lower-ranking ones and males should call more than females (based on the assumption that feeding competition has a stronger effect on low-rank compared with high-rank individuals and on females compared with males). Additionally monkeys should call more when food is abundant than in other times of the year. Indeed, high-ranking individuals gave more whinnies than low-ranking ones, and there was more calling per subgroup when there was more food in the environment or in the tree they were foraging. However, there was no evidence that subgroups containing only females gave fewer whinnies than mixed-sex subgroups. In addition, other monkeys joined vocalizing subgroups more often (17%) than nonvocalizing subgroups (12%). In those occasions when a subgroup was joined, the more calls there were, the more individuals joined. Subsequently, the whinny of spider monkeys has been cited as an example of a food call used to recruit others to a newly found food source (e.g. Hauser, 1996, p. 449; Heinrich and Marzluff, 1991). However, the fact that there is a very low proportion of subgroups that are actually joined after calling should warn us against concluding that the sole purpose of whinnies would be to recruit others (Cheney *et al.*, 1996, p. 507).

Noting that spider monkeys are considered to be territorial (Klein, 1972; van Roosmalen and Klein, 1987; Symington, 1987), Texidor and Byrne (1997) proposed that spider monkeys may use whinnies to distinguish among group members and strange individuals from other groups. In order to test this hypothesis, they performed a series of experiments in which whinnies from known and unknown animals were played back to subgroups of *A. geoffroyi*. While the subjects of these experiments responded vocally by giving another whinny in 11/13 trials with stranger whinnies and only in 3/10 trials with familiar whinnies, the frequency of calling per individual did not differ between both trial types. Only in two of the 13 trials with stranger whinnies was there any aggressive response, while none of the 10 trials with familiar whinnies elicited any aggression from the subjects. While these results suggest that spider monkeys are able to distinguish between those in their group and those that are not, this result does not demonstrate conclusively that it is because they are territorial that spider monkeys recognize the stranger whinnies as different from the familiar ones (Teixidor and Byrne, 1997). This is because there was no strong, overt response toward the stranger whinnies, which would be expected if spider monkeys were in fact detecting an intrusion into their territory (Teixidor and Byrne, 1997).

In their second study, Teixidor and Byrne (1999) searched for the function of *A. geoffroyi* whinnies using observational data as well as acoustic analyses and playback experiments. They found that whinnies emitted when traveling provoked more active responses, such as calling back or approaching the caller, than whinnies emitted when resting or feeding, while whinnies emitted at the presence of an observer did not elicit any active response. This result held true when analyzing the number of responses by any individual to a particular vocalization or the number of responses by a given individual to any vocalization. Next, through acoustic analysis, Teixidor and Byrne (1999) searched for significant sources of variation in the whinny that could inform recipients about what the caller was doing when calling (traveling or feeding). While whinnies emitted when traveling tended to be longer and contained a larger number of elements than those emitted when feeding or resting, there was no consistent difference in the context from which whinnies from different individuals were emitted. In the same study, Teixidor and Byrne (1999) demonstrated that whinnies contain significant acoustic information about the caller's identity (see previous section), but the same was not true for the caller's context. Finally, a series of playback experiments tested the response of subjects to 10 whinnies by 4 different individuals who were moving when vocalizing and to 10 whinnies by 3 different individuals who were feeding when vocalizing. Of all the behaviors observed, scanning was the only behavior to show a difference between the two playback sets. Subjects scanned more when they had heard whinnies emitted while feeding than when they had heard whinnies emitted when traveling.

Recently, I have hypothesized that one of the main functions of whinnies was to maintain contact between individuals in different subgroups (Ramos-Fernández, 2005). To test this I employed simultaneous observations of two different subgroups of *Ateles geoffroyi* Punta Laguna, Mexico. Subgroups that were within 300 m of each other tended to maintain the same distance or come together more often, as well as exchange more whinnies, than subgroups that were farther apart. As reviewed in the previous section, 300 m is close to what can be considered as the active space of the whinny. Also, to test the related hypothesis that whinnies were used to maintain contact between close associates, I included a series of paired playback experiments in which the same whinny was played at the same distance to a close associate and to another, nonclose associate of the caller. While both types of subjects were equally likely to respond to the playback with another whinny, only the close associates ever approached the speaker. This suggested to me that whinnies play a role in maintaining vocal contact between close associates in a loose aggregation system where even monkeys in the same subgroup may lose visual contact with one another.

Another study that touched upon the functional significance of an *Ateles* call was the field study of *A. belzebuth belzebuth* at Cocha Cashu National Park, Peru, done by Margaret Symington (1987). She used the aerial alarm call, Eisenberg's (1976) ook-bark call, to test the hypothesis that these calls are emitted in the presence of aerial predators in order to warn collateral kin about the presence of an aerial predator. Thus, Symington (1987) predicted that males, who are the philopatric sex, would give more calls than females, who disperse upon reaching adulthood (van Roosmalen and Klein, 1987). In order to simulate the presence of an aerial predator, she played back the call of a harpy eagle (*Harpia harpyja*), a common primate predator in Cocha Cashu. Contrary to her prediction, females gave more calls than males and neither the number of dependent offspring nor the presence of very young affected females' calling rates. While these alarm calls may not be used to warn particular individuals about the presence of a predator in order to provoke an escape reaction, they could be used to mount a mobbing response. Mobbing congregations have been observed in several studies (van Roosmalen and Klein, 1987; Symington, 1987; Ramos-Fernández, personal observation). In support of this, the distance at which these calls can be heard in the forest is up to 300 m (Symington, 1987), although her observational data did not allow her to determine conclusively whether subgroups are actually joined by others in the group after long bouts of alarm calling.

Synthesis

Traditionally, animal communication acts are considered to occur among dyads of one signaler and one recipient (Smith, 1977; Dawkins and Krebs, 1978; Guilford and Dawkins, 1991; Endler, 1993). However, more recent studies have stressed that most communication acts occur between a signaler and a multiplicity of receivers, each of them holding a different relationship to the caller (McGregor, 1993; McGregor and Dabelsteen, 1996). In this communication network, potential recipients of a signal include all individuals who extract information from a signal and that happen to be within its active space. This includes those recipients who "eavesdrop" on the communication interactions of pairs of others, extracting information about their social relationship (Cheney and Seyfarth, 2004), or even members of other species who detect the presence of the signaler through its communication signal (e.g. predators detecting spider monkeys through their whinnies). It is clear that both the proximate mechanisms and the functional significance of a signal will be determined by the presence of a network of several potential recipients (McGregor and Peake, 2000).

Take the best-studied signal in the spider monkeys' repertoire, for example. In a typical spider monkey habitat, the monkeys can hear whinnies for up to 300 m (Ramos-Fernández, 2005). The average distance between different individuals in the same subgroup is less than 30 m, while the distance between different subgroups is usually between 170 and 370 m (Ramos-Fernández, 2005). This means that most whinnies will reach all individuals in the caller's subgroup and, most of the time, several individuals in other subgroups as well. All the different functions that have been attributed to this call (contact maintenance: Eisenberg, 1976; Ramos-Fernández, 2005; recruitment to food patches: Chapman and Lefebvre, 1990; defense of territory: Teixidor and Byrne, 1997; coordination of group movements: Teixidor and Byrne, 1999; Ramos-Fernández, 2005) are possible if one considers the fact that there are multiple recipients within the communication network of a group of spider monkeys. How these multiple functions have shaped the acoustic structure of whinnies and their perception by recipients is still an open question, but they all may be equally important.

The change in emphasis of communication studies from dyads to networks implies that signalers will only provide information when there is an appropriate audience, or will attempt to direct information at suitable recipients. The former situation, known as the "audience effect" (Evans and Marler, 1984) has not been demonstrated in spider monkeys, but the fact that more whinnies are heard in subgroups that have another one within the active space of the whinny (Ramos-Fernández, 2005) is suggestive. As for directing signals at particular individuals, Masataka's (1986) results suggest that whinnies may actually contain information about the intended recipient, which would certainly make communication more specific within the network, permitting monkeys to stay in touch with certain individuals. While suggestive, these experiments need to be replicated in natural conditions, where the social and spatial relationships among different individuals are precisely those in which this individual label would be required.

Another shift in emphasis in communication studies concerns the different perspectives of signalers and recipients, particularly with regard to the information they are able to extract from a signal (Seyfarth and Cheney, 2003). While the information contained in an alarm call, for example, is relatively easy to extract by an observer from the stimulus that elicits the call (i.e. the presence of a predator or a source of danger) and from the responses of recipients (e.g. run for cover), there are other animal signals for which this is not so simple. So-called "contact" and "food" calls do not seem to have such a clear external referent as alarm calls do (Cheney *et al.*, 1996). Recipients of these calls may simply keep doing what they were before hearing the call. In such instances, if there has been no overt response, how can we know what recipients have been informed about? In dealing with this issue, it becomes crucially important to distinguish between the cognitive mechanisms that underlie the production and perception of a call. An individual that gives a contact call upon being separated from the group can be said to call regarding its own position with respect to the group. Recipients of this signal, however, are not separated from the group; therefore the information content of the signal cannot be about the separation from the group, unless they are able to place themselves into the signaler's perspective. The available evidence, however, suggests that primates cannot do this (Seyfarth and Cheney, 2003).

What, then, is the information content of whinnies? The only conclusion that can be drawn from the available evidence is that they contain information about the individual identity of the caller. The playback experiments performed by Teixidor and Byrne (1999) demonstrate that recipients use this information and those by Ramos-Fernández (2005) that they respond according to the social relationship they have with the caller. It is possible that, communication being a dynamic phenomenon, with the identities of signaler and recipient changing continuously within a network of related individuals, the information content of a signal may not be as "fixed" and may not reside so much in the signal as in the relationship between two or more communicating individuals (Aureli and Schaffner, this volume).

The social system of spider monkeys (and other so-called "fission–fusion" species) has been cited as complex, and as a selective pressure on the evolution of complex cognitive abilities (Milton, 2000; Barrett *et al.*, 2003). However, where this complexity lies is not clear. Societies with "fission–fusion" properties have been found in so many different animal groups, from elephants (*Loxodonta africana*) to dolphins (*Tursiops truncatus*), from hyenas (*Crocuta crocuta*) to bats (*Myotis bechsteinii*), from chimpanzees (*Pan* spp.) to galagos (Galagonidae; F. Aureli *et al.*, in press) that it is hard to define a common evolutionary path or a set of cognitive abilities that are necessary for fission–fusion to function properly. In a recent study by Ramos-Fernández *et al.* (2006), a set of simple agents searching for food in a realistic environment are seen to form subgroups and associate among them in ways that resemble the spider monkeys' and other fission–fusion species' grouping and association patterns, without any rules by which they interact among them. The authors suggest that it in fission–fusion societies, it is the complexity of the environment that causes variation in the size and composition of subgroups and not the social relationships themselves (Ramos-Fernández *et al.*, 2006). Long-distance vocal and olfactory communication may be used to maintain contact with animals that become separated as they search for food. If they can keep track of their social relationships, the simplest hypothesis to consider is that the signal simply informs them about their own identities.

References

Aureli, F., Schaffner, C. M., Boesch, C., *et al.* (in press). Fission–fusion dynamics: new research frameworks. *Current Anthropology.*

Barrett, L., Henzi, P. and Dunbar, R. (2003). Primate cognition: from 'what now?' to 'what if?' *Trends Cogn. Sci.*, **7**, 494–497.

Campbell, C. J. (2000). Fur rubbing behavior in free-ranging black-handed spider monkeys (*Ateles geoffroyi*) in Panama. *Am. J. Primatol.*, **51**, 205–208.

Campbell, C. J. (2006). Copulation in free-ranging black-handed spider monkeys (*Ateles geoffroyi*). *Am. J. Primatol.*, **68**, 507–511.

Carpenter, C. R. (1935). Behavior of red spider monkeys in Panama. *J. Mammal.*, **16**, 171–180.

Chapman, C. A and Lefebvre, L. (1990). Manipulating foraging group size: spider monkey food calls at fruiting trees. *Anim. Behav.*, **39**, 891–896.

Chapman, C. A. and Weary, D. M. (1990). Variability in spider monkey's vocalizations may provide basis for individual recognition. *Am. J. Primatol.*, **22**, 279–284.

Cheney, D. L. and Seyfarth, R. M. (2004). Social complexity and the information acquired during eavesdropping by primates and other animals. In *Animal Communication Networks*, ed. P. K. McGregor, Cambridge: Cambridge University Press, pp. 583–603.

Cheney, D. L., Seyfarth, R. M. and Palombit, R. (1996). The function and mechanisms underlying baboon "contact" barks. *Anim. Behav.*, **52**, 507–518.

Dawkins, R. and Krebs, J. R. (1978). Animal signals: information or manipulation? In *Behavioural Ecology: An Evolutionary Approach*, ed. J. R. Krebs and N. B. Davies, Oxford: Blackwell, pp. 282–309.

Eisenberg, J. F. (1976). Communication mechanisms and social integration in the black spider monkey, *Ateles fusciceps robustus*, and related species. *Smithson. Contrib. Zool.*, **213**.

Enlder, J. A. (1993). Some general comments on the evolution and design of animal communication systems. *Philos. Trans. Roy. Soc. Lond. B*, **340**, 215–225.

Evans, C. S. and Marler, P. (1984). Food calling and audience effects in male chickens, *Gallus gallus*: their relationships to food availability, courtship and social facilitation. *Anim. Behav.*, **47**, 1159–1170.

Guilford, T. and Dawkins, M. X. (1991). Receiver psychology and the evolution of animal signals. *Anim. Behav.*, **42**, 1–14.

Hauser, M. D. (1996). *The Evolution of Communication*. Cambridge, MA: MIT Press.

Heinrich, B. and Marzluff, J. M. (1991). Do common ravens yell because they want to attract others? *Behav. Ecol. Sociobiol.*, **28**, 13–21.

Herzel, H. (1993). Bifurcations and chaos in voice signals. *App. Mechan. Rev.*, **46**, 399–413.

Herzel, H., Berry, D., Titze, I. and Steinecke, I. (1995). Nonlinear dynamics of the voice – signal analysis and biomechanical modeling. *Chaos*, **5**, 30–34.

Klein, L. L. (1972). The ecology and social organization of the spider monkey, *Ateles belzebuth*. Unpublished Ph.D. thesis, University of California, Berkeley.

Laska, M., Bauer, V. and Hernandez Salazar, L. T. (2007). Self-annointing behavior in free-ranging spider monkeys (*Ateles geoffroyi*) in Mexico. *Primates*, **48**, 160–163.

Masataka, N. (1986). Rudimentary representational vocal signalling of fellow group members in spider monkeys. *Behaviour*, **96**, 49–61.

May, B., Moody, D. B. and Stebbins, W. C. (1989). Categorical perception of conspecific communication sounds by Japanese macaques, *Macaca fuscata. J. Acoust. Soc. Am.*, **85**, 837–847.

McGregor, P. K. (1993). Signalling in territorial systems: a context for individual identification, ranging and eavesdropping. *Philos. Trans. Roy. Soc. Lond. B*, **340**, 237–244.

McGregor, P. K. and Dabelsteen, T. (1996). Communication networks. In *Ecology and Evolution of Acoustic Communication in Birds*, ed. D. E. Kroodsma and E. H. Miller, Ithaca, NY: Cornell University Press, pp. 409–425.

McGregor, P. K. and Peake, T. M. (2000). Communication networks: social environments for receiving and signalling behaviour. *Acta Ethologica*, **2**, 71–81.

Milton, K. (2000). Quo vadis? Tactics of food search and group movement in primates and other animals. In *On the Move: How and Why Animals Travel in Groups*, ed. S. Boinski and P. A. Garber, Chicago: The University of Chicago Press, pp. 375–417.

Naguib, M. and Wiley, R. H. (2001). Estimating the distance to a source of sound: mechanisms and adaptations for long-range communication. *Anim. Behav.*, **62**, 825–837.

Ramos-Fernández, G. (2000). Patterns of association, feeding competition and vocal communication in spider monkeys, *Ateles geoffroyi*. Unpublished Ph.D. thesis, University of Pennsylvania, Philadelphia.

Ramos-Fernández, G. (2005). Vocal communication in a fission–fusion society: do spider monkeys stay in touch with close associates? *Int. J. Primatol.*, **26**, 1077–1092.

Ramos-Fernández, G., Boyer, D. and Gómez, V. P. (2006). A complex social structure with fission–fusion properties can emerge from a simple foraging model. *Behav. Ecol. Sociobiol.*, **60**, 536–549.

van Roosmalen, M. G. M. and Klein, L. L. (1987). The spider monkeys, genus *Ateles*. In *Ecology and Behavior of Neotropical Primates*, ed. R. A. Mittermeier and A. B. Rylands, Washington DC: World Wide Fund for Nature, pp. 455–537.

Seyfarth, R. M. and Cheney, D. L. (2003). Signalers and receivers in animal communication. *Ann. Rev. Psychol.*, **54**, 145–173.

Seyfarth, R. M., Cheney, D. L. and Marler, P. (1980). Monkey responses to three different alarm calls: evidence of predator classification and semantic communication. *Science*, **210**, 801–803.

Smith, W. J. (1977). *The Behavior of Communicating*. Cambridge, MA: Harvard University Press.

Symington, M. M. (1987). Ecological and social correlates of party size in the black spider monkey, *Ateles paniscus chamek*. Unpublished Ph.D. thesis, Princeton University, NJ.

Symington, M. M. (1990). Fission–fusion social organization in *Ateles* and *Pan. Int. J. Primatol.*, **11**, 47–61.

Teixidor, P. and Byrne, R. W. (1997). Can spider monkeys (*Ateles geoffroyi*) discriminate vocalizations of familiar individuals and strangers? *Folia Primatol.*, **68**, 254–264.

Teixidor, P. and Byrne, R. W. (1999). The 'whinny' of spider monkeys: individual recognition before situational meaning. *Behaviour*, **136**, 279–308.

Wahlberg, M., Ramos-Fernández, G., Ugarte, F., Møhl, B. and Rasch, M. (2002). Recording spider monkeys with a microphone array. *J. Acoust. Soc. Am.*, **112**(5), Pt.2, 2400.

Walters, J. R. and Seyfarth, R. M. (1987). Conflict and cooperation. In *Primate Societies*, ed. B. B. Smuts, D. L. Cheney, R. M. Seyfarth, R. W. Wrangham and T. T. Struhsaker, Chicago: The University of Chicago Press, pp. 306–317.

Waser, P. M. (1977). Sound localization by monkeys: a field experiment. *Behav. Ecol. Sociobiol.*, **2**, 427–431.

9 Social interactions, social relationships and the social system of spider monkeys

FILIPPO AURELI AND COLLEEN M. SCHAFFNER

Introduction

Primates live in a variety of social systems that differ in terms of spacing, grouping and mating patterns as well as the quality of social relationships between individual members (Crook and Gartland, 1966; Clutton-Brock, 1974; van Schaik and van Hooff, 1983; Wrangham, 1987; Dunbar, 1988; Strier, 1994; Janson, 2000; Isbell and Young, 2002; Fuentes, 2007). Three important components of a social system are social organization, social structure and mating system (Kappeler and van Schaik, 2002). In this chapter we focus on the social structure of spider monkeys and its components; social interactions and social relationships. First we start with fission–fusion dynamics, an aspect of social organization that is critical for the understanding of the spider monkeys' social system.

Fission–fusion dynamics

According to Kappeler and van Schaik (2002), social organization describes the size, sexual composition and spatiotemporal cohesion of a social system. The aspect of spatiotemporal cohesion is typically not emphasized in group-living primates because groups are usually viewed as cohesive units. Although variation in spatiotemporal cohesion also occurs in "cohesive" groups, it has mainly been used to characterize species living in so-called "fission–fusion societies."

The term "fission–fusion" was introduced by Hans Kummer (1971) to describe a social system in which group size can be temporarily adjusted to the availability and distribution of resources by means of the fission and fusion of subunits called parties or subgroups. Social systems characterized by such fission–fusion dynamics are considered to be rare among mammals, but they are typical of some primate species (table in Appendix in Smuts *et al.*, 1987; see below) and other mammalian species (e.g. Bechstein's bats, *Myotis bechsteinii*:

Spider Monkeys: Behavior, Ecology and Evolution of the Genus Ateles, ed. Christina J. Campbell. Published by Cambridge University Press. © Cambridge University Press 2008.

Kerth and Konig, 1999; African buffalos, *Syncerus caffer*: Cross *et al.*, 2005; dolphins, *Tursiops* spp: Connor *et al.*, 2000; elephants, *Loxodonta africana*: Wittemyer *et al.*, 2005; spotted hyenas, *Crocuta crocuta*: Holekamp *et al.*, 1997). Fission–fusion dynamics are also typical of modern humans, including hunter-gatherers (Rodseth *et al.*, 1991; Marlowe, 2005).

Among primates several different modal types of social systems characterized by fission–fusion dynamics have been recognized. For example, some prosimians live in dispersed social networks where individuals associate in subgroups of variable size and composition at least some of the time (Bearder, 1987; Müller and Thalmann, 2000; Shülke and Ostner, 2005). Similar social networks are found in orangutans (*Pongo pygmaeus*), where individuals are alone most of the time but associate with others under suitable conditions, especially in food-rich habitats (van Schaik, 1999). Another variation is the social systems of geladas (*Theropithecus gelada*), hamadryas baboons (*Papio h. hamadryas*) and snub-nosed monkeys (*Rhinopithecus bieti* and *R. roxellana*), often called multi-level societies, which are characterized by the rather fixed composition of their basic subunits that typically consist of one adult male, several females and their offspring (Stammbach, 1987; Grüter and Zinner, 2004). Over time "fission–fusion societies" has become the term used to describe social systems in which individuals belonging to the same community are rarely all together, but spend most of the time in temporary subgroups or parties that frequently merge and split again with different compositions. This is the social system of chimpanzees (*Pan troglodytes*) and bonobos (*P. paniscus*) (Nishida and Hiraiwa-Hasegawa, 1987; Hohmann and Fruth, 2002; Stumpf, 2007), and importantly for the focus of our chapter this is also the social system of spider monkeys (*Ateles* spp.: Symington, 1990; Chapman *et al.*, 1995; Di Fiore and Campbell, 2007).

Describing *Pan* and *Ateles* species as living in fission–fusion societies appears to imply that other species do not experience fission–fusion dynamics at all. Flexible spatiotemporal grouping patterns are, however, more common and more complex than generally recognized (Kinzey and Cunningham, 1994; Sussman and Garber, 2007). In fact, there is pronounced variation in cohesion both across and within species, with groups of species not traditionally described as fission–fusion splitting into subgroups (e.g. long-tailed macaques, *Macaca fascicularis*: van Schaik and van Noordwijk, 1988) or experiencing high temporal variation in spatial cohesion (e.g. some baboon, *Papio* spp., populations: Henzi and Barrett, 2003). Thus, it is not appropriate to use the term "fission–fusion" to identify a dichotomy, with species either experiencing fission–fusion dynamics or not (Strier, 1989; Aureli *et al.*, in press).

Although better quantification of the degree of fission–fusion dynamics across groups and species is needed, current understanding is that spider monkeys live in communities that experience a relatively high degree of fission–fusion dynamics (Symington, 1990; Chapman *et al.*, 1995; Di Fiore and Campbell, 2007), in which community members are rarely all together and frequently split and merge into fluid subgroups. The high degree of fission–fusion dynamics is an important aspect of the social system of *Ateles* because it influences the opportunities for community members to interact with one another, and in turn impacts upon their social relationships. Below we explain the conceptual framework that we follow to link social interactions, social relationships and social structure.

A conceptual framework

Following Hinde's (1976, 1979) framework, a social relationship is a concept that links the observable social interactions between group members to the inferred group social structure. An interaction between two individuals implies that A does something to B and possibly B does something else back. An interaction may involve a number of repetitions. Interactions can be described in terms of content (A grooms B) and qualifiers (A grooms B rapidly). When two individuals have a series of interactions over a period of time, any one interaction may affect the subsequent ones. The two individuals can then be said to have a relationship with each other.

A social relationship can be described in terms of the content and quality of the interactions between two individuals and the relative frequencies and patterning of those interactions over time. A relationship is established when two individuals interact on a regular basis, but it is not fixed. Not only is a relationship affected by the individual characteristics of the partners (e.g. age, sex, temperament, relative dominance rank and kinship) and their previous interactions, but also their future interactions are affected by the nature of their relationship (i.e. two individuals interact in a particular way in a given context because they have a particular relationship). A social relationship is also a dynamic concept because any new interaction may steer the relationship in a different direction. Thus, social relationships may change over time.

Each partner in a relationship is also involved in relationships with other group members, so that each relationship is part of a network of relationships or social structure. Such a network influences interactions between individuals as the behavior or even the mere presence of third parties affects the way partners behave with each other (e.g. Kummer, 1967; Chapais, 1988). Thus, the relationships that two individuals have with others affect the relationship that they have with each other. The resulting social structure can therefore be described in terms of the properties and patterns of the constituent relationships. Hinde's (1976, 1979) framework captures therefore the dynamic nature of social interactions, social relationships and social structure and views social relationships and social structure as emergent properties of the observable social interactions.

From a functional point of view, social relationships can be viewed as investments that benefit the individuals involved (Kummer, 1978). In this respect, social interactions are ways of shaping relationships to maximize gain (or minimize loss: Cords, 1997): an individual's interactions with a partner influence the likelihood that the partner behaves in a beneficial way. According to this view, partners are chosen depending on their qualities, behavioral tendencies and availability, and individuals use social interactions to influence the partners' characteristics to their own advantage. Some of the benefits from well-developed and differentiated social relationships of primates include selective tolerance around resources, cooperative hunting, food sharing, mating privileges, agonistic support and protection against harassment (reviewed in Cords, 1997; van Schaik and Aureli, 2000).

Hinde's (1976, 1979) descriptive framework and Kummer's (1978) functional perspective are complementary and useful in understanding social dynamics. It is important to note that by using them we do not imply that nonhuman animals possess a concept of social relationship as described above and similar to ours. This could lead scientists to overestimate the cognitive abilities of primates and other animals (Barrett and Henzi, 2002). The ability to track social interactions with various individuals does not necessarily require that the animals have awareness or high cognitive prowess as such record keeping may be maintained through emotional mediation (Aureli and Schaffner, 2002). Accordingly, emotional mediation may be at the basis of the high flexibility in social interactions depending on relationship quality. Regardless of how animals do it, keeping track of social interactions with their partners clearly facilitates social intercourse if they can predict the actions and responses of their partners with reasonable accuracy (van Schaik and Aureli, 2000).

Social interactions and relationships

Predation avoidance is believed to be the most important and widespread benefit of group living for diurnal primates (van Schaik, 1983; Dunbar, 1988; Janson, 1992). Living in a social system characterized by high fission–fusion dynamics seems to jeopardize such a benefit because group members are often alone or in small subgroups. Socioecological theory suggests that spider monkeys can live in such a social system because they are large-bodied arboreal primates and thus, as adults, face a relatively low predation risk. Being highly frugivorous, spider monkeys, and other large-bodied primates, split into small subgroups to cope with patchily distributed and temporally varying food sources (Wrangham, 1980; Milton, 1984; Dunbar, 1988; Symington, 1988; van Schaik, 1989; Strier, 1992). Under these conditions, fission–fusion dynamics can be viewed as a way to reduce feeding competition among group members by adjusting the number of associated individuals to local resource availability. Individuals split into small subgroups when resources are scarce and are in larger subgroups when resources are plentiful. By doing so, community members also reduce travel time and thus decrease energy demands (Korstjens *et al.*, 2006; Lehmann *et al.*, in press), thereby likely further reducing feeding competition. Fission–fusion dynamics can also provide flexible solutions to the usually contrasting pressures of avoiding predators and minimizing feeding competition when there is temporal or spatial fluctuation in predation pressure and food availability (Boesch and Boesch-Achermann, 2000).

Based on socioecological theory, predictions can be derived about the types of social interactions and relationships between adult spider monkeys (for vocal and sexual behavior see Ramos-Fernández, this volume and Campbell and Gibson, this volume, respectively, and for relationships involving juveniles see Vick, this volume). We focus on the relationships between members of the same or opposite sex. Although some of the predictions for one sex class are relative to another sex class, we present them in separate sections to facilitate clarity. We start with female–female relationships, followed by male–male relationships and finally female–male relationships.

Female–female relationships

According to socioecological theory, female primates that feed primarily on ripe fruit experience strong intragroup contest competition for food as such resources are distributed in discrete, monopolizable patches (van Schaik, 1989; Sterck *et al.*, 1997; Koenig, 2002). These females are expected to have highly differentiated relationships with one another based on varying antagonistic and cooperative interactions. They are expected to display unidirectional agonistic behavior (e.g. an individual attacks another without retaliation) and form clear-cut dominance relationships. They are also expected to form coalitions with other females in order to better compete in contests and predominantly rely on kin for this risky behavior. Thus, female philopatry should be favored under these conditions (see Isbell and Young, 2002 for discussion).

Although ripe fruit is the main source of food for spider monkeys (van Roosmalen and Klein, 1988; Russo *et al.*, 2005; Di Fiore and Campbell, 2007), they do not follow the pattern above because they reduce feeding competition by means of high levels of fission–fusion dynamics. Their main response to feeding competition is to spread out from one another forming small subgroups or foraging alone with dependent offspring. This is a flexible response, and subgroup size depends on local food availability, as spider monkey females are more often in smaller subgroups when food resources are scarce or in smaller patches (Klein and Klein, 1977; Symington, 1988; Chapman *et al.*, 1995; Shimooka, 2003).

Under these conditions, females are not expected to form clear-cut dominance relationships, and the selective pressure to remain in the natal group and rely on strong bonds with kin is low; thus, dispersal from the natal group may even be favored to reduce competition between kin (Wrangham, 1980; Dunbar, 1988). There is evidence that most spider monkey females disperse from their natal group (Di Fiore and Campbell, 2007; Shimooka *et al.*, this volume; Vick, this volume). Thus, the social relationships of female spider monkeys should fall into Sterck *et al.*'s (1997) Dispersal-Egalitarian category, and while they may provide some services to one another (e.g. exchanging information about food location), overall the value of their relationships is expected to be low (van Schaik and Aureli, 2000; but see notable exceptions in bonobos, *Pan paniscus*: Kano, 1992; Hohmann and Fruth, 2002; and West African chimpanzees, *Pan troglodytes*: Lehmann and Boesch, in press).

Females are also expected to spend less time with other community members than males according to Wrangham's (2000) Scramble Competition Hypothesis. Scramble competition is experienced when food patches cannot be monopolized and individuals lose access to resources because others have already used them (Janson and van Schaik, 1988). In general, the intensity of scramble competition is strongly associated with group size; that is, when scramble competition is more intense smaller groups are found (Janson and Goldsmith, 1995). The underlying reason for this association is that individuals in larger groups incur higher travel costs because they need to visit more food patches. Based on the evidence for chimpanzees that females, especially those with dependent offspring, travel more slowly than males (Hunt, 1989), Wrangham (2000) argued that females spend a longer time traveling between patches and have higher relative travel costs than males (given that travel time may predict total energy expenditure: Dunbar, 1988). Thus, females are expected to travel in smaller subgroups than males in order to reduce travel costs (see Lehmann *et al.*, in press for a demonstration of fission–fusion sociality reducing travel costs). Data from chimpanzees seem to support the predictions of the Scramble Competition Hypothesis and suggest that a key ecological feature is the lack of "feed-as-you-go" foraging between fruit trees, which places a premium on rapid travel between food patches (Wrangham, 2000). The data available on spider monkeys suggest that they experience intense scramble competition (Chapman *et al.*, 1995), do little foraging between fruit trees (Symington, 1988; Chapman,

Figure 9.1 Grooming between two adult females (*Ateles geoffroyi*) in Santa
Rosa National Park, Costa Rica. (Photo: Filippo Aureli.)

1990) and females travel slower than males (Shimooka, 2005). Below we review
what is known about social interactions between adult female spider monkeys
to examine whether there is evidence supporting the type of social relationships
and association patterns predicted by the Dispersal-Egalitarian category and the
Scramble Competition Hypothesis.

Spider monkeys are described as living in sex-segregated communities, in
which females are reported to be the "less social" sex (Fedigan and Baxter,
1984). One reason for this characterization is the repeated observation that
females frequently travel alone or in very small subgroups with only their offspring (*A. paniscus*: van Roosmalen and Klein, 1988; *A. geoffroyi*: Chapman,
1990; Fedigan and Baxter, 1984), but females are found in larger subgroups
when food is abundant (*A. belzebuth*: Shimooka, 2003). A second reason for
the characterization of females as the less social sex may relate to the low
frequency of affiliative behavior, such as grooming (Figure 9.1), among adult
females in the same community. In wild *A. geoffroyi*, Ahumada (1992) found
that grooming occurred less often between adult females than between adult
males. Females groomed juveniles, presumably their own offspring, most frequently, but received grooming predominantly from other adult females. In
wild *A. belzebuth chamek* adult female dyads were the most unlikely type of
grooming dyad (Symington, 1990). Furthermore, grooming was so infrequent

Figure 9.2 Mutual embrace with pectoral sniff between two spider monkeys (*Ateles geoffroyi*). (Drawing: Norberto Asensio.)

in one wild population of *A. geoffroyi* that it was not reported as a distinct category, rather it was collapsed together with a variety of other affiliative behaviors (Fedigan and Baxter, 1984). Fedigan and Baxter also reported that females were less affiliative with each other than they were with males. Not surprisingly, studies on captive spider monkeys reveal that more time is spent in affiliative behavior than in wild populations (van Roosmalen and Klein, 1988). In a captive study of *A. geoffroyi*, we found that grooming between females occurred at a rate of about one bout every four hours per dyad (Schaffner and Aureli, 2005), which was within the range of those reported in other captive studies (Rondinelli and Klein, 1976; Anaya-Huertas and Mondragón-Ceballos, 1998; Pastor-Nieto, 2001). However, Pastor-Nieto (2001) found that in one of her study groups, rates were sixfold higher than those we found in our study (Schaffner and Aureli, 2005), although in both groups many individuals were related to each other.

Spider monkeys not only engage in grooming, but they also have a suite of species-specific friendly behaviors, primarily including embraces and pectoral sniffing (Figure 9.2). These two behaviors often occur simultaneously, but they can occur separately and can be mutual or unidirectional (Klein and Klein, 1971; Schaffner and Aureli, 2005). Embraces and pectoral sniffs occur frequently when individuals approach each other (Klein and Klein, 1971) or when individuals of the same community reunite after a period of separation both in captivity and in the wild (Eisenberg and Kuehn, 1966; Klein and Klein, 1971; Izawa *et al.*, 1979; Fedigan and Baxter, 1984; van Roosmalen and Klein, 1988; Schaffner and Aureli, 2005; Aureli and Schaffner, 2007). Exchanges of embraces in this context appear to function as conflict management given that they reduce postfusion aggression (Aureli and Schaffner, 2007). Wild female *A. geoffroyi* exchange embraces less frequently than males, and they do so more often with adult males than adult females (Fedigan and Baxter, 1984). In a captive study on the same species, however, females exchanged embraces exclusively with other adult females (Rondinelli and Klein, 1976). Pastor-Nieto (2001) showed that captive *A. geoffroyi* females in a well-established group exchanged embraces more often than females in a newly formed group. In both groups embraces were much more frequent than grooming. Similar results were found in our study of captive *A. geoffroyi* (Schaffner and Aureli, 2005), but the occurrence of embraces between females was strongly affected by a particular context: the presence of young infants (see below).

Two factors might be critical in determining the quality of female–female social relationships over time. One factor is how long a female has resided in a given community. Since the majority of female spider monkeys emigrate from their natal community at or around the time of reproductive maturity (Symington, 1987; Shimooka *et al.*, this volume; Vick, this volume), they likely enter new communities with little if any established relationships with other individuals. Thus, new immigrants are possibly more vulnerable to attack from resident females, who could view them as competitors for resources. Evidence from *A. paniscus* alludes to this possibility, as although aggression is rare, "leading females," believed to be the oldest, occasionally direct "quite severe aggression toward non-leading females" (van Roosmalen and Klein, 1988: 515). New data suggest this is also the case for wild *A. geoffroyi*, as new immigrant females receive more aggression from other females than long-term resident females do (Asensio *et al.*, in press).

A second factor that might dramatically alter the quality of female social relationships is the presence of infants. Female primates are highly attracted to other females' infants (Maestripieri, 1994). In other species, females increase their rate of grooming with mothers (Seyfarth, 1980; Maestripieri, 1994; Muroyama, 1994; Henzi and Barrett, 2002) and one explanation for the increased grooming is that females provide benefits to the mothers in exchange for access to the infants. This view is in line with the concept of a "biological market" (Nöe *et al.*, 1991; Nöe and Hammerstein, 1994) in which females exchange grooming as a service for access to the infants, which has been clearly demonstrated in baboons (Henzi and Barrett, 2002). In a captive study on *A. geoffroyi*, we found that embraces exchanged between adult females more than doubled in the first six months after infants were born, but no change in grooming occurred (Schaffner and Aureli, 2005). In a study of wild *A. geoffroyi* females with infants also exchanged embraces more than females without infants, but the social partners with whom they exchanged embraces was unclear (Fedigan and Baxter, 1984). In wild *A. belzebuth belzebuth*, it was reported that a special relationship existed between two females with dependent infants, although no mention was made of how this affiliation was manifested (Izawa *et al.*, 1979). Another captive study, largely involving *A. geoffroyi*, revealed that the presence of an infant changed the social status of one female as she experienced a dramatic increase in the rate at which she received embraces and pectoral sniffs from others, but this attention disappeared following the infant's death (Eisenberg and Kuehn, 1966). Furthermore, in a recent study of wild *A. geoffroyi*, there was a dramatic change in the rate at which females received embraces when they had young infants than at other times, and females without infants appeared to give embraces in order to gain access to infants (Slater *et al.*, 2007). There was also evidence for a biological market as the proportion of embraces followed by infant handling was inversely related to the number of infants available; that is, the more infants present in the community the fewer embraces given to each mother in order to gain access to her infant (Slater *et al.*, 2007). Although the information is sparse, the available evidence suggests that the presence of young infants may at least temporarily enhance female–female relationships.

Interactions among female spider monkeys are largely nonantagonistic, with as little as 1% of the overall aggressive interactions observed in *A. geoffroyi* (Fedigan and Baxter, 1984) and an upward ceiling of 15% in *A. belzebuth* (Klein, 1974) being ascribed to female–female aggression. The basis for dominance relationships in spider monkeys is not well understood (Strier, 1999), and the picture emerging from existing studies is unclear. For example, *A. belzebuth* has been described as having high- and low-ranking females based on agonistic displacements at food sources (Symington, 1987), but other researchers have simply discriminated between "leading" or "non-leading females" based on the tendency of one or a few females to determine the foraging paths and general activity of the subgroups (van Roosmalen and Klein, 1988). In *A. geoffroyi*, Chapman (1990) reported that making dominance discriminations was difficult, but he was able to ascribe rank relationships to females, as either high- or low-ranking based on the outcome of aggressive interactions and displacements that predominantly occurred at food resources. However, some field (*A. belzebuth*: Klein, 1971; Izawa *et al.*, 1979; *A. geoffroyi*: Fedigan and Baxter, 1984) and several captive studies (*A. geoffroyi*: Klein and Klein, 1971; Rondinelli and Klein, 1976; Pastor-Nieto, 2001; Schaffner and Aureli, 2005) make no mention of dominance hierarchies among females. Attempts to quantify dominance hierarchies in two captive studies of *A. geoffroyi* revealed that female rank was only loosely apparent. In one study, several adult females were reported to have no clear dominance relationships based on approach/displacement data (Eisenberg and Kuehn, 1966). In another study of a group of hand-reared individuals data from aggression–submission and approach–avoid/retreat patterns indicated little evidence of dominance relationships among the females (Anaya-Huertas and Mondragón-Ceballos, 1998). The latter is the only study in which the actual data for determining dominance rank are presented, and the impression gleaned from the field studies reporting dominance discriminations (Symington, 1987; Chapman, 1990) suggests (through omission rather than admission) that linear dominance hierarchies, in the tradition of the better-studied cercopithecines, do not occur. The most likely differentiation regarding dominance rank, if any, is that long-term resident females are dominant over newly immigrant females (see above).

In summary, the available data, although patchy, suggest that spider monkey female–female relationships conform to Sterck *et al.*'s (1997) Dispersal-Egalitarian category. The relationships are likely low in value (van Schaik and Aureli, 2000), but it is possible that they may be valuable under certain circumstances. One possibility is the arrival of new females into the community, which may serve to promote alliances among long-term residents against the new immigrants. Another circumstance is the presence of young infants and the interest that mothers receive from other females. The available evidence also suggests that females conform nicely to Wrangham's (2000) Scramble Competition Hypothesis that females spend less time with other individuals than males. To further elucidate this picture we turn our attention to male–male social relationships.

Male–male relationships

Given that there is no strong selective pressure for female spider monkeys to remain in their natal group and there are constraints on females to be highly gregarious (see above), selection for males to disperse from the natal group due to inbreeding avoidance is relaxed (Wrangham, 1980; van Hooff and van Schaik, 1994). As a consequence, males can be the philopatric sex (Di Fiore and Campbell, 2007; Shimooka *et al.*, this volume) and develop strong relationships with one another capitalizing on the high degree of familiarity and possibly kinship. Cooperative relationships between males can certainly be used for successful competition with other community males as shown in the case of chimpanzees' coalitionary support to obtain or maintain high dominance rank and mating opportunities (de Waal, 1982; Nishida, 1983; Watts, 1998).

Another consequence of male–male bonding can be at the basis of the main benefit of group living for spider monkeys and chimpanzees. Given that predation avoidance is not a strong selective pressure in these large-bodied primates, one of the main benefits of group living is likely to be the cooperative effort in intergroup competition (Shimooka, 2003; Lehmann *et al.*, in press). Males are then expected to patrol territorial boundaries and defend access to females and food sources from neighboring communities. The longer day-ranges relative to females along with the stronger bonds would allow males to cover a larger area and be more efficient in intergroup interactions (Wrangham, 2000). Thus, male–male relationships are expected to be variable depending on support patterns for intragroup competition, but overall they are expected to be highly valuable because of the male cooperative effort in intergroup competition.

Below we review evidence about social interactions between adult male spider monkeys and evaluate whether their social relationships are stronger and the level of gregariousness higher than those of females. We also examine whether there is evidence for male cooperation in intergroup competition as well as investigate whether male–male interactions within the group reflect positive attraction and competitiveness.

Several behavioral characteristics suggest that males are the more gregarious sex in spider monkeys. For example, they are found more frequently in larger subgroups than females. The average subgroup size for males for *A. belzebuth belzebuth* was nearly double that of females, 5.4 and 2.8 individuals, respectively (Shimooka, 2005). Chapman (1990) indicated that male *A. geoffroyi* are less likely to be encountered alone than females are, although subgroup sizes did not appear to differ between the sexes. In addition, data from several study sites and species indicate that males travel over a greater area of their home range than females (Di Fiore and Campbell, 2007) and males travel more frequently at the boundaries of the home range than females (*A. belzebuth chamek*: Symington, 1988; *A. belzebuth belzebuth*: Shimooka, 2005; *A. geoffroyi*: Chapman, 1990). These ranging patterns are possibly related to the males' defense of females from their community (see below), and males are known to rely on each other for intergroup territorial displays and boundary patrols (*A. belzebuth belzebuth*: Klein 1974; *A. paniscus*: van Roosmalen and Klein, 1988; *A. belzebuth chamek*: Symington, 1990; Wallace, 2001), suggesting there is a need for males to travel together with other males.

Further evidence suggests that adult males have the strongest bonds aside from the mother–offspring bond in spider monkey communities (Fedigan and Baxter, 1984; van Roosmalen and Klein, 1988; Symington, 1990). *Ateles belzebuth chamek* males associate more with each other than they do with females or than females do with other females (Symington, 1990). *Ateles geoffroyi* males are also more likely to be encountered in the company of other males than females are in the company of other females (Chapman, 1990). In the same species adult males are more affiliative than adult females and selectively direct their affiliative efforts toward other males (Fedigan and Baxter, 1984), and grooming is the most frequent in male–male dyads (Ahumada, 1992). Symington (1990) found a similar pattern in *A. belzebuth chamek* where male–male grooming was most common and female–female grooming the least common. Furthermore, van Roosmalen and Klein (1988) indicate that when reuniting *A. paniscus* males regularly exchange embraces and pectoral sniffs with one another and do so much more frequently than females. The higher rate of male–male species-specific friendly behavior, including embraces and pectoral sniffing, has also been identified in *A. geoffroyi* (K. Slater, C. Schaffner and F. Aureli, unpublished data).

There is compelling new evidence showing that the cooperative effort between males in coping with intergroup competition goes well beyond traveling more frequently at territorial boundaries. Aureli *et al.* (2006) observed a series of seven raids in which males of one community walked on the ground, single file, well into the range of their neighbors, reminiscent of similar behavior patterns in chimpanzees (*Pan troglodytes*: Goodall, 1986; Boesch and Boesch-Achermann, 2000; Watts *et al.*, 2006). These raids regularly involved the same 3–4 males and several times culminated in encounters with individuals from the other community. In most cases these encounters resulted in the raiding males chasing encountered residents of the neighboring community. In one case, however, aggression by the raiding males directed toward a resident female was severe and she was "rescued" by her adult son. Provided that such raiding behavior is widespread, if somewhat infrequent, it suggests there is strong selective pressure for secure social bonds among community males in order to garner the level of cooperation and trust to execute such a risky behavior. Indeed, there is a relative dearth of information on aggression occurring between males of the same community, but this picture is also changing.

Males from the same community are rarely if ever reported as being aggressive toward each other and two studies, one of which was a comprehensive review of *Ateles*, make no mention of intracommunity adult male–male aggression (Fedigan and Baxter, 1984; van Roosmalen and Klein, 1988). In addition, in *A. belzebuth belzebuth* out of 52 aggressive events in which the actor and the recipient were known, none was between males (Klein, 1974). However, the seeming lack of male–male aggression is undermined by some recent discoveries. Scouring the captive literature and unpublished data collected from 32 zoos (N. Davis, C. Schaffner and S. Wehnelt, unpublished data) reveals that it is somewhat unusual for captive spider monkey groups to have more than one adult male. For example, Eisenberg and Kuehn (1966) indicated that repeated attempts to introduce more than one adult male into an established group of *A. geoffroyi* inevitably led to such severe aggression that the authors concluded it could not be done. In addition, in cases where there are more than one adult male or adult and maturing males housed in zoo-based groups a disproportionate amount of severe and all of the lethal aggression is done by and directed to males (N. Davis, C. Schaffner and S. Wehnelt, unpublished data). The prevalence of male aggression in captive groups might readily be explained away by the fact that wild spider males are philopatric and attempts to form groups with unrelated or unfamiliar males in captivity could create problems that are not present in wild groups. However, two recent reports from the field muddy the waters even more. Campbell (2006a) and Valero *et al.* (2006) have reported on male–male intragroup coalitionary lethal aggression at two different field sites. At one site repeated coalitionary aggression led to the death of one young adult male (*A. geoffroyi*: Valero *et al.*, 2006) and at the other site two young males have been killed or died as a result of their injuries sustained from coalitionary aggression (*A. geoffroyi*: Campbell, 2006a). These findings are most difficult to reconcile given the philopatric nature of spider monkey males and the reputation of males for having the strongest bonds within spider monkey communities.

Indirect evidence indicates that male–male lethal aggression could be more widespread. Overall, in adult populations, the male to female sex ratio is 1:2.6 (Chapman *et al.*, 1989). This ratio appears to be more dramatic than the sex ratio of males to females in the infant and juvenile population for some sites (Chapman *et al.*, 1989). Thus, the sex ratio bias in favor of the immigrating rather than the philopatric sex is more skewed in the adult than in the juvenile and infant populations. This information makes one wonder what happens to the males, particularly if they are unlikely to emigrate, and suggests that further investigation about the fate of maturing and young adult males is warranted (also see Vick, this volume).

There are observations indicating that relationships between young and older males are filled with uncertainty. At our field sites in Costa Rica and Mexico we have observed several instances in which young adult *A. geoffroyi* males join subgroups with other older males, but keep a "safe distance" (C. Schaffner and F. Aureli, unpublished data). Furthermore, other insights into the relationships between young and older males come from the patterns of grappling behavior that involves a series of approach–retreats by one individual with elements of embracing, tail wrapping, face greeting, face touches and sometimes mutual genital manipulation between the two partners, which are sustained for many minutes. Nearly every account of grappling behavior observed has been exchanged between males of differing age classes (e.g. juveniles to subadult males in *A. belzebuth belzebuth*: Klein, 1971). Our own observations of wild *A. geoffroyi* confirm this relatively rare behavior is largely restricted to male–male dyads and initiated by juvenile and subadult males toward fully adult males. Grappling appears to represent a culmination of strong attraction of younger males to older males as well as high uncertainty, such that the behaviors appear both intense and loaded with tension between partners.

In summary, the reviewed patterns of social interactions between males support the prediction of males having highly valuable relationships for intra- and intergroup competition. The recent discoveries of male raiding parties and intra-community coalitionary lethal aggression also suggest a degree of complexity in male–male social relationships that was heretofore unknown. On the one hand, both behavior patterns require the males to cooperate and form coalitions, behaviors that are commensurate with having strong social bonds. On the other hand, the potential for males to kill each other within a community (at least in *A. geoffroyi*) may create a strong degree of uncertainty in the fabric of their relationships, and this may particularly be the case for subadult and young adult males (Campbell, 2006a; Valero *et al.*, 2006). Thus, males appear to be highly attracted to each other, while simultaneously somewhat repelled.

Female–male relationships

Given that fitness of mammalian females and males is limited by different factors (food for females and fertilizations for males), their social strategies are expected to differ accordingly (Emlen and Oring, 1977). These differences can lead to conflicts of interest between the sexes and to a high variability in female–male relationships within and between primate species (van Schaik and Aureli, 2000; van Schaik *et al.*, 2004; cf. Chapman *et al.*, 2003). At the same time, stable female–male associations are unusually common in primates compared with other mammalian orders (van Schaik and Kappeler, 1997), suggesting that social factors are likely at the basis of female–male interactions. Among these factors, male defense of food resources from neighboring groups and from predators, male policing of intragroup female–female conflicts, sexual harassment and infanticide, and female counterstrategies to them are the most likely (e.g. Smuts and Smuts, 1993; van Schaik and Hoerstermann, 1994; Clutton-Brock and Parker, 1995; Sterck *et al.*, 1997; Watts, 1997; van Schaik and Aureli, 2000; Kappeler and van Schaik, 2004; van Schaik *et al.*, 2004; Williams *et al.*, 2004).

The most varied female–male relationships are expected in situations in which both sexes can choose between multiple partners within the same group (e.g. multimale, multifemale groups) and in which sexual dimorphism is limited, making both sexes suitable allies for one another (van Schaik and Aureli, 2000). These conditions are present in spider monkeys (Di Fiore and Campbell, 2007; Campbell and Gibson, this volume). In addition, in species in which females disperse from the natal group and males are philopatric, both likely in spider monkeys, certain female–male relationships are between mother and adult son. These relationships are expected to be stronger than other female–male relationships as in the case of bonobos (*Pan paniscus*: Furuichi, 1989) and to a lesser extent in chimpanzees (*Pan troglodytes*: Goodall, 1986; Boesch and Boesch-Achermann, 2000), making kinship another source of variation in female–male relationships. Finally, according to biological market theory (Nöe *et al.*, 1991; Nöe and Hammerstein, 1994), the value of potential partners is based on the services and commodities that partners can offer relative to other group members and varies over time and circumstances. In the case of female–male relationships, another source of variation is due to the changes of relative female value over time depending on her reproductive status. Below we review data on social interactions between adult female and adult male spider monkeys. Given the expected high variability, we evaluate whether their social relationships are on average intermediate between the stronger male–male and the weaker female–female relationships.

The evidence for association patterns is mixed. In *A. belzebuth chamek* female–male association indices are intermediate between those of female–female and male–male dyads, although the association levels between female dyads and mixed sex dyads are not significantly different and both are significant lower than those of male dyads (Symington, 1990). In *A. geoffroyi*, males spend most time in proximity with other males, whereas females do not appear to discriminate between same- or opposite-sexed proximity partners (K. Slater, C. Schaffner and F. Aureli, unpublished data). There is also seasonal variation in association patterns between females and males with higher levels during periods of food abundance (*A. belzebuth belzebuth*: Shimooka, 2003).

Information on grooming patterns is similarly mixed, with female–male grooming universally being less common than male–male grooming, but not consistently higher than female–female grooming. For *A. belzebuth chamek*, the pattern identified for association indices is mirrored in the grooming patterns as grooming is most frequently observed in male–male dyads followed by female–male dyads and then female–female dyads (Symington, 1990). A similar pattern of affiliation was identified in a wild population of *A. geoffroyi*, as males predominantly affiliated with other males, while females were more affiliative toward males than they were toward females (Fedigan and Baxter, 1984). A more recent field study of *A. geoffroyi* indicates that males groom same-sex partners and females groom opposite-sex partners, with females almost never grooming same-sexed adults (K. Slater, C. Schaffner and F. Aureli, unpublished data). Grooming patterns differ, however, for another population of *A. geoffroyi* in which the lowest rate of grooming including given and received was between adult males and females. In one study females gave and received about three times more grooming than female–male dyads, whereas male–male grooming was eight times higher than female–male grooming (Ahumada, 1992). However, this particular population of spider monkeys is a closed community in which females are not able to emigrate (Milton and Hopkins, 2006), thus the increased frequency of female–female grooming in this population might be attributed to a greater degree of relatedness and familiarity among females. The mixed pattern of female–male social interactions may be related to another facet of their social relationships: agonistic behavior.

Agonistic patterns appear clear, as all studies that report spider monkey aggression indicate that female-directed male aggression is the most frequent form of aggressive interaction. In one long-term study of *A. belzebuth belzebuth*, 57% of all aggressive interactions in which the actor and the target were known involved adult males targeting adult females (Klein, 1974). Two previous studies suggest the majority of this aggression may be linked to sexual behavior (*A. belzebuth belzebuth*: Symington, 1987 as cited by Campbell, 2003; *A. geoffroyi*: Fedigan and Baxter, 1984), a behavior pattern that is repeatedly reported to occur in secrecy (Klein, 1971; van Roosmalen and Klein, 1988; Campbell, 2006b; Campbell and Gibson, this volume). According to Symington (1987 as cited by Campbell, 2003) female-directed male aggression is strongly associated with periods in which females reproductively cycle. Fedigan and Baxter (1984) noted that males frequently chased females and postulated that this may be a "form of ritualized intimidation display" (p. 291). Two more recent studies have explored female-directed male aggression in much greater detail.

Campbell (2003) examined 107 aggressive interactions in which male *A. geoffroyi* selectively targeted females in order to investigate whether such attacks were more frequent during times when females were cycling. The study focused on three females, and only one received more aggression when she was cycling compared with other times. The data were inconclusive: there was no evidence that males attacked females in order to trigger defecation or urination by the females and there did not appear to be evidence for sexual coercion as much of the aggression took place outside of times when females were likely to conceive. Thus, Campbell (2003) concluded that aggressive behavior by males against females might represent an attempt by males to dominate equally sized females. New data on females from two different study groups of the same species suggest that Fedigan and Baxter (1984) may have been correct in suggesting that female-directed male aggression is a form of ritualized display (Slater *et al.*, 2005; K. Slater, C. Schaffner and F. Aureli, unpublished data). In this study female-directed male aggression was broken into two main categories: prolonged chases with growling and threat faces, and physical aggression involving contact. The distribution of aggressive interactions was examined across different phases of reproduction for the females, including cycling, lactation and pregnancy. The overwhelming majority of prolonged chases occurred during periods when females were cycling, but no pattern for physical aggression was detected across the three female reproductive phases. In addition, males appear to monitor the female reproductive status before or after prolonged chases by consistently sniffing the place where females were sitting. This information coupled with the fact that within the hour following such attacks, females frequently left the subgroup alone with their primary attacker suggests that such aggression is linked to sexual behavior either as a form of sexual coercion (cf. Smuts and Smuts, 1993) or as a display of male quality to impress females (Slater *et al.*, 2005; K. Slater, C. Schaffner and F. Aureli, unpublished data). In fact, when prolonged chases were removed from the data set, no difference was found between the rates of female-directed or male-directed aggression by males. Although more data are warranted before making definitive conclusions about the nature of female-directed male aggression, the evidence from Slater *et al.* (2005) supports the earlier observations of Symington (1987 as cited by Campbell, 2003) that most prolonged chases occur when females cycle. Furthermore, the apparently stereotyped nature of the chases and the female's tendency to leave the subgroup with the primary male aggressor (Slater *et al.*, 2005; K. Slater, C. Schaffner and F. Aureli, unpublished data) suggest it may be a form of ritualized display as originally proposed by Fedigan and Baxter (1984).

In summary, our review finds mixed evidence for female–male relationships being intermediate between male–male and female–female relationships based on data from affiliative behavior. The clear pattern of aggression with prevalence in female–male dyads could be interpreted as evidence against the intermediate relationship quality of female–male dyads. However, careful analysis of its pattern suggests that female-directed male aggression is probably related to reproduction and is not an indication of poor relationship quality. The lack of available kinship data, to date, has not allowed the testing of whether variation in relationship quality between females and males is partially due to the mixture of dyad types, including mother–son as well as unrelated dyads of potential sexual partners. Thus, the emerging picture is that patterns of female–male interactions in spider monkeys are less clear-cut than previously thought with more investigation needed on the various factors affecting their relationship quality.

Conclusions and future research

Our review of what is known about the social interactions of spider monkeys reveals a puzzling picture with many holes to fill. Much still remains to be understood about the social interactions and in turn social relationships of spider monkeys. This reality has limited our attempt to infer variation in the quality of social relationships between group members and to evaluate whether the predictions based on socioecology theory are supported. Thus, a first conclusion is that more empirical data on social interactions, especially from wild populations, are needed before reaching a satisfactory understanding of spider monkeys' social relationships and in turn their social structure.

Although the data are limited, our understanding of female–female social relationships conforms to Sterck *et al.*'s (1997) and Wrangham's (2000) models.

The emerging picture from the data on male–male social relationships suggests that these relationships are valuable but are also variable and at times entail high risks (particularly for maturing males). Such variability calls for a theoretical model that incorporates the dynamic nature of the relationships, characterized by both cooperative and competitive elements. Female–male social relationships also appear to be more ambiguous than previously thought. An integrated framework that incorporates the various aspects of female–female, male–male and female–male social relationships of different ages would be highly beneficial. Long-term studies of multiple groups are the best sources of the needed data for such a framework because the patterns of social interactions can then be interpreted within a rich ecological, demographic and genetic context. There are a number of research issues that our review has highlighted and that can be further addressed when such data become available.

Similarities and differences

Atelines show convergence with some great apes in various aspects of morphology, ecology, life history and social system (Robinson and Janson, 1987; Di Fiore and Campbell, 2007). Our review confirms that convergence in certain social aspects is particularly pronounced between spider monkeys and chimpanzees (previously reviewed in Symington, 1990; Chapman *et al.*, 1995; Di Fiore and Campbell, 2007). Both taxa seem to experience a high degree of fission–fusion dynamics with subgroups that are on average small, but highly variable in size and composition. In both taxa females have smaller ranges than males, and the sexes often move separately. Interactions with neighboring communities are usually hostile with males being mostly involved. Male–male relationships within the community appear to be stronger than female–female or female–male relationships, although there is great variability within each type of relationship that requires further investigation. Such convergent patterns suggest that similar social systems have evolved in response to similar selection pressures in distantly related taxa. Full evaluation of the specific selective pressures will be achieved when direct comparisons are made from detailed socioecological data of both taxa. This is a promising avenue for future collaborative research.

Our review tends to focus on the similarities in the results regarding social interactions across different studies in order to characterize the overall quality of social relationships. The resulting picture is, however, not always clear. This may be partially due to the paucity of detailed studies that forces us to combine information from different species. In fact, the accumulation of more data from multiple groups of different species is likely to reveal interspecific, and possibly intraspecific, variation in spider monkeys' social dynamics. Such variation is well documented in other genera for which more detailed data are available. These examples can provide useful insights about what could be the possible future scenario for spider monkeys.

Squirrel monkeys are one of the most drastic cases of interspecific variation as the three best studied species differ dramatically not only in the rate and quality of social interactions, but also in the dispersal patterns (Boinski *et al.*, 2005). For example, Costa Rican squirrel monkeys (*Saimiri oerstedii*) are characterized by male philopatry, female dispersal, negligible intragroup contest competition for food, and weak female–female relationships; whereas squirrel monkeys living in Peru (*S. boliviensis*) display male dispersal, female philopatry, and well-differentiated female–female relationships based on a strong matrilineal system with females forming coalitions with kin to compete for food; and finally squirrel monkeys living in Suriname (*S. sciureus*) are characterized by dispersal from both sexes, weak female–female relationships, but a stable female dominance hierarchy because females compete directly over small, defensible food patches (Mitchell *et al.*, 1991; Boinski *et al.*, 2002, 2005). A less dramatic, but equally consistent variation has been documented among macaque species. Here the general pattern of male dispersal, female philopatry, and matrilineal structure is common to all species (Thierry *et al.*, 2004). However, there is variation in social interactions within this common theme with certain species consistently having more clear-cut dominance relationships, more kin-biased interactions and a lower degree of tolerance with the average group member than other species (Thierry, 1985, 2000; de Waal and Luttrell, 1989; Aureli *et al.*, 1997). More recent developments have also emphasized that there are sex differences in macaque patterns of covariation of social traits with male–male relationships not always following the pattern found for female–female relationships (Preuschoft *et al.*, 1998; Cooper and Bernstein, 2002). Interspecific and intraspecific variation of social traits within a common theme is also found in baboons (*Papio* populations: Barton *et al.*, 1996; Henzi and Barrett, 2003), langurs (*Semnopithecus entellus*: Koenig, 2000), capuchin monkeys (*Cebus* spp.: Strier, 1999; Fragaszy *et al.*, 2004) and *Pan* (Stumpf, 2007).

From what is known so far, it is more likely that the variation in social dynamics among spider monkey species and groups resembles the macaque case rather than the more drastic case of squirrel monkeys as the available demographic data show consistent dispersal patterns across species (Shimooka *et al.*, this volume). However, more dramatic variation in social interactions associated with different dispersal patterns may emerge from new long-term studies. With more information accumulating on the various spider monkey species and the other members of the Atelinae subfamily (which also includes the genera *Alouatta*, *Lagothrix* and *Brachyteles*) a framework for within-genus, and possibly within-subfamily, variation in social dynamics can be developed building on the existing well-documented cases of other primate taxa.

The challenges of fission–fusion dynamics

Spider monkeys interact with one another within the fluid nature of their social system. As mentioned earlier, the high degree of fission–fusion dynamics experienced by spider monkeys influences the opportunities for community members to interact with one another, and in turn impacts upon their social relationships. This clearly poses challenges for the regulation of dyadic social relationships (i.e. the relationships that one individual has with any other community member) and the possible needed knowledge about third parties' relationships (i.e. the relationships between any other two community members) (Boesch and Boesch-Achermann, 2000; Milton, 2000; Barrett *et al.*, 2003; Dunbar, 2003; Aureli and Schaffner, 2005, 2007). Thus, the patterns of social interactions we reviewed above need to be considered in a context where social information is less complete than in more cohesive groups. We could therefore expect spider monkeys to have enhanced abilities for the extraction of information from subtle social cues and to possibly exchange bond-testing behaviors at fusion when individuals reunite after extended periods of separation. In addition, the exchange of social interactions within biological markets (Nöe and Hammerstein, 1994) is likely to be more challenging in species living in highly fluid social systems because the relative value of partners changes more rapidly depending on the presence or absence of other community members in a given subgroup (Barrett *et al.*, 2003).

Researchers who study species that experience a high degree of fission–fusion dynamics also face a number of challenges. Researchers are usually able to follow only a portion of the community members (usually one subgroup) and thus, like the nonhuman primates they study, need to rely only on partial social information at a given time (see above). Thus, they need to overcome the challenges in inferring social relationships and social structures. For example, several methods have been used to test whether individuals in fluid communities associate in a nonrandom fashion (Whitehead and Dufault, 1999; Lusseau and Newman, 2004; Ramos-Fernández *et al.*, 2006). Another task is to measure the degree of fission–fusion dynamics displayed in a given environment in order to quantify the level of challenge that community members face in their regulation of social relationships. There is therefore a need to develop metrics that capture the degree of variation in spatial cohesion and subgroup membership over time in a number of groups per species (Aureli *et al.*, in press). These data, in turn, will allow comparative analyses of the role that the degree of fission–fusion dynamics may play in the patterning of social interactions and social relationships and in the underlying cognitive abilities. This comparative approach could also lead to new insights on the convergence of social characteristics between *Pan* and species of the Atelinae subfamily, especially *Ateles* and *Brachyteles*.

In conclusion, there are still many challenges on the way toward increasing our knowledge of the social interactions, social relationships and social systems of spider monkeys. New detailed data on social interactions from multiple groups of each species need to be integrated with the accumulating information on their sexual behavior (Campbell and Gibson, this volume) and vocal repertoire (Ramos-Fernández, this volume), as well as with the insights provided by agent-based models (Ramos-Fernández *et al.*, 2006) and time budget models (Korstjens *et al.*, 2006; Lehmann *et al.*, in press). These are exciting challenges for future research. The recent empirical findings regarding the complexity of male–male relationships (Aureli *et al.*, 2006; Campbell, 2006a; Valero *et al.*, 2006; Vick, this volume) and the variability in the quality of female–female relationships (Asensio *et al.*, in press; Slater *et al.*, in press) are promising illustrations of the wealth of discoveries that may lay ahead, which will provide needed information for developing and testing theoretical models.

References

Ahumada, J. A. (1992). Grooming behavior of spider monkeys (*Ateles geoffroyi*) on Barro Colorado Island, Panama. *Int. J. Primatol.*, **13**, 33–49.

Anaya-Huertas, C. and Mondragón-Ceballos, R. (1998). Social behavior of black-handed spider monkeys (*Ateles geoffroyi*) reared as home pets. *Int. J. Primatol.*, **19**, 767–784.

Asensio, N., Korstjens, A. H., Schaffner, C. M. and Aureli, F. (in press). Intragroup aggression, fission–fusion dynamics and feeding competition in spider monkeys. *Behaviour*.

Aureli, F. and Schaffner, C. M. (2002). Relationship assessment through emotional mediation. *Behaviour*, **139**, 393–420.

Aureli, F. and Schaffner, C. M. (2005). Fission–fusion dynamics complicate the regulation of social relationships. *Primate Eye*, **86**, 6.

Aureli, F. and Schaffner, C. M. (2007). Aggression and conflict management at fusion in spider monkeys. *Biol. Lett.*, **3**, 147–149.

Aureli, F., Das, M. and Veenema, H. C. (1997). Differential kinship effect on reconciliation in three species of macaques (*Macaca fascicularis*, *M. fuscata*, and *M. sylvanus*). *J. Comp. Psychol.*, **111**, 91–99.

Aureli, F., Schaffner, C. M., Boesch, C., *et al.* (in press). Fission–fusion dynamics: new research frameworks. *Curr. Anthropol.*

Aureli, F., Schaffner, C. M., Verpooten, J., Slater, K. Y. and Ramos-Fernández, G. (2006). Raiding parties of male spider monkeys: insights into human warfare? *Am. J. Phys. Anthropol.*, **131**, 486–497.

Barrett, L. and Henzi, S. P. (2002). Constraints on relationship formation among female primates. *Behaviour*, **139**, 263–289.

Barrett, L., Henzi, P. and Dunbar, R. (2003). Primate cognition: from 'what now?' to 'what if?'. *Trends Cogn. Sci.*, **7**, 494–497.

Barton, R. A., Byrne, R. W. and Whitten, A. (1996). Ecology, feeding competition and social structure in baboons. *Behav. Ecol. Sociobiol.*, **38**, 321–329.

Bearder, S. K. (1987). Lorises, bushbabies, and tarsiers: diverse societies in solitary foragers. In *Primate Societies*, ed. B. B. Smuts, D. L. Cheney, R. M. Seyfarth, R. W. Wrangham and T. T. Struhsaker, Chicago: University of Chicago Press, pp. 11–24.

Boesch, C. and Boesch-Achermann, H. (2000). *The Chimpanzees of the Taï Forest.* Oxford: Oxford University Press.

Boinski, S., Kauffman, L., Ehmke, E., Schet, S. and Vreedzaam, A. (2005). Dispersal patterns among three species of squirrel monkeys (*Saimiri oerstedii, S. boliviensis* and *S. sciureus*). I. Divergent costs and benefits. *Behaviour*, **142**, 525–632.

Boinski, S., Sughrue, K., Selvaggi, L., *et al.* (2002). An expanded test of the ecological model of primate social evolution: competitive regimes and female bonding in three species of squirrel monkeys (*Saimiri oerstedii, S. boliviensis*, and *S. sciureus*). *Behaviour*, **139**, 227–261.

Campbell, C. J. (2003). Female-directed aggression in free-ranging *Ateles geoffroyi*. *Int. J. Primatol.*, **24**, 223–237.

Campbell, C. J. (2006a). Lethal intragroup aggression by adult male spider monkeys (*Ateles geoffroyi*). *Am. J. Primatol.*, **68**, 1197–1201.

Campbell, C. J. (2006b). Copulation in free-ranging black-handed spider monkeys (*Ateles geoffroyi*). *Am. J. Primatol.*, **68**, 507–511.

Chapais, B. (1988). Experimental matrilineal inheritance of rank in female Japanese macaques. *Anim. Behav.*, **36**, 1025–1037.

Chapman, C. A. (1990). Association patterns of spider monkeys: the influence of ecology and sex on social organization. *Behav. Ecol. Sociobiol.*, **26**, 409–414.

Chapman, C. A., Fedigan, L. M., Fedigan, L. and Chapman, L. J. (1989). Post-weaning resource competition and sex ratios in spider monkeys. *Oikos*, **54**, 315–319.

Chapman, C. A., Wrangham, R. W. and Chapman, L. J. (1995). Ecological constraints on group size: an analysis of spider monkey and chimpanzee subgroups. *Behav. Ecol. Sociobiol.*, **36**, 59–70.

Chapman, T., Arnqvist, G., Bangham, J. and Rowe, L. (2003). Sexual conflict. *Trends Ecol. Evol.*, **18**, 41–47.

Clutton-Brock, T. H. (1974). Primate social organization and ecology. *Nature*, **250**, 539–542.

Clutton-Brock, T. H. and Parker, G. A. (1995). Sexual coercion in animal societies. *Anim. Behav.*, **49**, 1345–1365.

Connor, R. C., Wells, R., Mann, J. and Read, A. (2000). The bottlenose dolphin, *Tursiops* spp.: social relationships in a fission–fusion society. In *Cetacean*

Societies: Field Studies of Whales and Dolphins, ed. J. Mann, R. C. Connor, P. Tyack and H. Whitehead, Chicago: University of Chicago Press, pp. 91–126.

Cooper, M. A. and Bernstein, I. S. (2002). Counter aggression and reconciliation in Assamese macaques (*Macaca assamensis*). *Am. J. Primatol.*, **56**, 215–230.

Cords, M. (1997). Friendships, alliances, reciprocity and repair. In *Machiavellian Intelligence II*, ed. A. Whiten and R. W. Byrne, Cambridge: Cambridge University Press, pp. 24–49.

Crook, J. H. and Gartland, J. S. (1966). Evolution of primate societies. *Nature*, **210**, 1200–1203.

Cross, P. C., Lloyd-Smith, J. O. and Getz, W. M. (2005). Disentangling association patterns in fission–fusion societies using African buffalo as an example. *Anim. Behav.*, **69**, 499–506.

de Waal, F. B. M. (1982). *Chimpanzee Politics*. London: Cape.

de Waal, F. B. M. and Luttrell, L. M. (1989). Toward a comparative socioecology of the genus *Macaca*: different dominance styles in rhesus and stumptail macaques. *Am. J. Primatol.*, **19**, 83–109.

Di Fiore, A. and Campbell, C. J. (2007). The Atelines: variation in ecology, behavior, and social organization. In *Primates in Perspective*, ed. C. J. Campbell, A. Fuentes, K. C. MacKinnon, M. Panger and S. K. Bearder, New York: Oxford University Press, pp. 155–185.

Dunbar, R. I. M. (1988). *Primate Social System*. London: Croom Helm.

Dunbar, R. I. M. (2003). Why are apes so smart? In *Primate Life Histories and Socioecology*, ed. P. M. Kappeler and M. E. Pereira, Chicago: University of Chicago Press, pp. 285–298.

Eisenberg, J. F. and Kuehn, R. (1966). The behavior of *Ateles geoffroyi* and related species. *Smithson. Misc. Coll.*, **151**, 1–63.

Emlen, S. T. and Oring, L. W. (1977). Ecology, sexual selection, and the evolution of mating systems. *Science*, **197**, 215–223.

Fedigan, L. M. and Baxter, M. J. (1984). Sex differences and social organization in free-ranging spider monkeys (*Ateles geoffroyi*). *Primates*, **25**, 279–294.

Fragaszy, D. M., Visalberghi, E. and Fedigan, L. M. (2004). *The Complete Capuchin: The Biology of the Genus Cebus*. Cambridge: Cambridge University Press.

Fuentes, A. (2007). Social organization: social systems and the complexities in understanding the evolution of primate behavior. In *Primates in Perspective*, ed. C. J. Campbell, A. Fuentes, K. C. MacKinnon, M. Panger and S. K. Bearder, New York: Oxford University Press, pp. 592–608.

Furuichi, T. (1989). Social interactions and the life history of female *Pan paniscus* in Wanba, Zaire. *Int. J. Primatol.*, **10**, 173–197.

Goodall, J. (1986). *The Chimpanzees of Gombe: Patterns of Behavior*. Cambridge, MA: Harvard University Press.

Grüter, C. C. and Zinner, D. (2004). Nested societies: convergent adaptations of baboons and snub-nosed monkeys. *Primate Rep.*, **70**, 1–98.

Henzi, S. P. and Barrett, L. (2002). Infants as a commodity in a baboon market. *Anim. Behav.*, **63**, 915–921.

Henzi, S. P. and Barrett, L. (2003). Evolutionary ecology, sexual conflict, and behavioral differentiation among baboon populations. *Evol. Anthropol.*, **12**, 217–230.

Hinde, R. A. (1976). Interactions, relationships and social structure. *Man*, **11**, 1–17.

Hinde, R. A. (1979). *Towards Understanding Relationships*. London: Academic Press.

Hohmann, G and Fruth, B. (2002). Dynamics in social organization of bonobos (*Pan paniscus*). In *Behavioural Diversity in Chimpanzees and Bonobos*, ed. C. Boesch, G. Hohmann and L. F. Marchant, Cambridge: Cambridge University Press, pp. 138–150.

Holekamp, K. E., Cooper, S. M., Katona, C. I., Berry, N. A., Frank, L. G. and Smale, L. (1997). Patterns of association among female spotted hyenas (*Crocuta crocuta*). *J. Mammal.*, **78**, 55–64.

Hunt, K. D. (1989). Positional behavior in *Pan troglodytes* at the Mahale Mountains and the Gombe Stream National Parks, Tanzania. Unpublished Ph.D. thesis, University of Michigan, Ann Arbor, MI.

Isbell, L. A. and Young, T. P. (2002). Ecological models of female social relationships in primates: similarities, disparities, and some directions for future clarity. *Behaviour*, **139**, 177–202.

Izawa, K., Kimura, K. and Nieto, A. S. (1979). Grouping of the wild spider monkey. *Primates*, **20**, 503–512.

Janson, C. H. (1992). Evolutionary ecology of primate social structure. In *Evolutionary Ecology and Human Behavior*, ed. E. A. Smith and B. Winterhalden, New York: Gruyter, pp. 95–130.

Janson, C. H. (2000). Primate socio-ecology: the end of a golden age. *Evol. Anthropol.*, **9**, 73–86.

Janson, C. H. and Goldsmith, M. L. (1995). Predicting group size in primates: foraging costs and predation risks. *Behav. Ecol. Sociobiol.*, **36**, 326–336.

Janson, C. H. and van Schaik, C. P. (1988). Recognizing the many faces of primate food competition: methods. *Behaviour*, **105**, 165–186.

Kano, T. (1992). *The Last Ape: Pygmy Chimpanzee Behavior and Ecology*. Stanford, CA: Stanford University Press.

Kappeler, P. M. and van Schaik, C. P. (2002). Evolution of primate social systems. *Int. J. Primatol.*, **23**, 707–740.

Kappeler, P. M. and van Schaik, C. P. (2004). Sexual selection in primates: review and selective preview. In *Sexual Selection in Primates: New and Comparative Perspectives*, ed. P. M. Kappeler and C. P. van Schaik, Cambridge: Cambridge University Press, pp. 3–23.

Kerth, G. and Konig, B. (1999). Fission, fusion and nonrandom association in female Bechstein's bats (*Myotis bechsteinii*). *Behaviour*, **136**, 1187–1202.

Kinzey, W. G. and Cunningham, E. P. (1994). Variability in platyrrhine social organization. *Am. J. Primatol.*, **34**, 185–198.

Klein, L. L. (1971). Observations on copulation and seasonal reproduction of two species of spider monkeys, *Ateles belzebuth* and *A. geoffroyi*. *Folia Primatol.*, **15**, 233–248.

Klein, L. L. (1974). Agonistic behavior in Neotropical primates. In *Primate Aggression, Territoriality and Xenophobia: A Comparative Perspective*, ed. R. L. Holloway, New York: Academic Press, pp. 77–122.

Klein, L. L. and Klein, D. J. (1971). Aspects of social behavior in a colony of spider monkeys *Ateles geoffroyi*. *Int. Zoo Yrbk.*, **11**, 175–181.

Klein, L. L. and Klein, D. J. (1977). Feeding behavior of the Colombian spider monkey, *Ateles belzebuth*. In *Primate Ecology: Studies of Feeding and Ranging Behaviour in Lemurs, Monkeys and Apes*, ed. T. H. Clutton-Brock, London: Academic Press, pp. 153–181.

Koenig, A. (2000). Competitive regimes in forest-dwelling Hanuman langur females (*Semnopithecus entellus*). *Behav. Ecol. Sociobiol.*, **48**, 93–109.

Koenig, A. (2002). Competition for resources and its behavioral consequences among female primates. *Int. J. Primatol.*, **23**, 759–783.

Korstjens, A. H., Verhoeckx, I. L. and Dunbar, R. I. M. (2006). Time as a constraint on group size in spider monkeys. *Behav. Ecol. Sociobiol.*, **60**, 683–694.

Kummer, H. (1967). Tripartite relations in hamadryas baboons. In *Social Communication among Primates*, ed. S. Altman, Chicago: University of Chicago Press, pp. 63–71.

Kummer, H. (1971). *Primate Societies: Group Techniques of Ecological Adaptation*. Arlington Heights, IL: AHM Publishing Corporation.

Kummer, H. (1978). On the value of social relationships to nonhuman primates: a heuristic scheme. *Soc. Sci. Inform.*, **17**, 687–705.

Lehmann, J. and Boesch, C. (in press). Sex differences in chimpanzee sociality. *Int. J. Primatol.*

Lehmann, J., Korstjens, A. H. and Dunbar, R. I. M. (in press). Fission–fusion social systems as a strategy for coping with ecological constraints: a primate case. *Evol. Ecol.*

Lusseau, D. and Newman, M. E. J. (2004). Identifying the role that animals play in their social networks. *Proc. Roy. Soc. Lond. B* (Suppl.), **271**, S477–S481.

Maestripieri, D. (1994). Influence of infants on females social relationships in monkeys. *Folia Primatol.*, **63**, 192–202.

Marlowe, F. W. (2005). Hunter-gatherers and human evolution. *Evol. Anthropol.*, **14**, 54–67.

Milton, K. (1984). Habitat, diet, and activity patterns of free-ranging woolly spider monkeys (*Brachyteles arachnoides* E Geoffroy 1806). *Int. J. Primatol.*, **5**, 491–514.

Milton, K. (2000). Quo vadis? Tactics of food search and group movement in primates and other animals. In *On the Move: How and Why Animals Travel in Groups*, ed. S. Boinski and P. A. Garber, Chicago: University of Chicago Press, pp. 375–417.

Milton, K. and Hopkins, M. E. (2006). Growth of a reintroduced spider monkey (*Ateles geoffroyi*) population on Barro Colorado Island, Panama. In *New Perspectives in the Study of Mesoamerican Primates: Distribution, Ecology, Behavior and Conservation*, ed. A. Estrada, P. A. Garber, M. S. M. Pavelka and L. Lueke, New York: Springer, pp. 417–435.

Mitchell, C. L., Boinski, S. and van Schaik, C. P. (1991). Competitive regimes and female bonding in two species of squirrel monkeys (*Saimiri oerstedi* and *S. sciurieus*). *Behav. Ecol. Sociobiol.*, **28**, 55–60.

Müller, A. E. and Thalmann, U. (2000). Origin and evolution of primate social organisation: a reconstruction. *Biol. Rev.*, **75**, 405–435.

Muroyama, Y. (1994). Exchange of grooming for allomothering in female patas monkeys (*Erythrocebus patas*). *Behaviour*, **128**, 103–119.

Nishida, T. (1983). Alpha status and agonistic alliance in wild chimpanzees (*Pan troglodytes schweinfurthii*). *Primates*, **24**, 318–336.

Nishida, T. and Hiraiwa-Hasegawa, M. (1987). Chimpanzees and bonobos: cooperative relationships among males. In *Primate Societies*, ed. B. B. Smuts, D. L. Cheney, R. M. Seyfarth, R. W. Wrangham and T. T. Struhsaker, Chicago: University of Chicago Press, pp. 165–177.

Noë, R. and Hammerstein, P. (1994). Biological markets: supply and demand determine the effect of partner choice in cooperation, mutualism and mating. *Behav. Ecol. Sociobiol.*, **35**, 1–11.

Noë, R., van Schaik, C. P. and van Hooff, J. A. R. A M. (1991). The market effect: an explanation for pay-off asymmetries among collaborating animals. *Ethology*, **87**, 97–118.

Pastor-Nieto, R. (2001). Grooming, kinship, and co-feeding in captive spider monkeys (*Ateles geoffroyi*). *Zoo Biol.*, **20**, 293–303.

Preuschoft, S., Paul, A. and Kuester, J. (1998). Dominance styles of female and male Barbary macaques (*Macaca sylvanus*). *Behaviour*, **135**, 731–755.

Ramos-Fernández, G., Boyer, D. and Gómez, V. P. (2006). A complex social structure with fission–fusion properties can emerge from a simple foraging model. *Behav. Ecol. Sociobiol.*, **60**, 536–549.

Robinson, J. G. and Janson, C. H. (1987). Capuchins, squirrel monkeys, and atelines: socioecological convergence with Old World primates. In *Primate Societies*, ed. B. B. Smuts, D. L. Cheney, R. M. Seyfarth, R. W. Wrangham and T. T. Struhsaker, Chicago: University of Chicago Press, pp. 69–82.

Rodseth, L., Wrangham, R. W., Harrigan, A. M. and Smuts, B. B. (1991). The human community as a primate society. *Curr. Anthropol.*, **32**, 221–254.

Rondinelli, R. and Klein, L. L. (1976). Analysis of adult social spacing tendencies and related social interactions in a colony of spider monkeys (*Ateles geoffroyi*) at San Francisco Zoo. *Folia Primatol.*, **25**, 122–142.

Russo, S. E., Campbell, C. J., Dew, J. L., Stevenson, P. R. and Suarez, S. A. (2005). A multi-forest comparison of dietary preferences and seed dispersal by *Ateles* spp. *Int. J. Primatol.*, **26**, 1017–1037.

Schaffner, C. M. and Aureli, F. (2005). Embraces and grooming in captive spider monkeys. *Int. J. Primatol.*, **26**, 1093–1106.

Seyfarth, R. M. (1980). The distribution of grooming and related behaviours among adult female vervet monkeys. *Anim. Behav.*, **28**, 798–813.

Shimooka, Y. (2003). Seasonal variation in association patterns of wild spider monkeys (*Ateles belzebuth belzebuth*) at La Macarena, Colombia. *Primates*, **44**, 83–90.

Shimooka, Y. (2005). Sexual differences in ranging of *Ateles belzebuth belzebuth* at La Macarena, Colombia. *Int. J. Primatol.*, **26**, 385–406.

Shülke, O. and Ostner, J. (2005). Big times for dwarfs: social organization, sexual selection, and cooperation in the Cheirogaleidae. *Evol. Anthropol.*, **14**, 170–185.

Slater, K. Y., Schaffner, C. M. and Aureli, F. (2005). Female-directed aggression in wild spider monkeys: male display and female mate choice. *Primate Rep.*, **72**, 89–90.

Slater, K. Y., Schaffner, C. M. and Aureli, F. (2007). Embraces for infant handling in spider monkeys: evidence for a biological market? *Anim. Behav.*, **74**, 455–461.

Smuts, B. B. and Smuts, R. W. (1993). Male aggression and sexual coercion of females in nonhuman primates and other mammals: evidence and theoretical implications. *Adv. Stud. Behav.*, **22**, 1–63.

Smuts, B. B., Cheney, D. L., Seyfarth, R. M., Wrangham, R. W. and Struhsaker, T. T. (1987). *Primate Societies*. Chicago: University of Chicago Press.

Stammbach, E. (1987). Desert, forest, and montane baboons: multilevel societies. In *Primate Societies*, ed. B. B. Smuts, D. L. Cheney, R. M. Seyfarth, R. W. Wrangham and T. T. Struhsaker, Chicago: University of Chicago Press, pp. 112–120.

Sterck, E. H. M., Watts, D. P. and van Schaik, C. P. (1997). The evolution of female social relationships in nonhuman primates. *Behav. Ecol. Sociobiol.*, **41**, 291–309.

Strier, K. B. (1989). Effects of patch size on feeding associations in muriquis (*Brachyteles arachnoids*). *Folia Primatol.*, **52**, 70–77.

Strier, K. B. (1992). Atelinae adaptations: behavioural strategies and ecological constraints. *Am. J. Phys. Anthropol.*, **88**, 515–524.

Strier, K. B. (1994). Myth of the typical primate. *Yrbk. Phys. Anthropol.*, **37**, 233–271.

Strier, K. B. (1999). Why is female kin bonding so rare? Comparative sociality of neotropical primates. In *Comparative Primate Socioecology*, ed. P. C. Lee, Cambridge: Cambridge University Press, pp. 300–319.

Stumpf, R. (2007). Chimpanzees and bonobos – diversity within and between species. In *Primates in Perspective*, ed. C. J. Campbell, A. Fuentes, K. C. MacKinnon, M. Panger and S. K. Bearder, New York: Oxford University Press, pp. 321–344.

Sussman, R. W. and Garber, P. A. (2007). Cooperation and competition in primate social interactions. In *Primates in Perspective*, ed. C. J. Campbell, A. Fuentes, K. C. MacKinnon, M. Panger and S. K. Bearder, New York: Oxford University Press, pp. 636–651.

Symington, M. M. (1987). Sex ratio and maternal rank in wild spider monkeys: when daughters disperse. *Behav. Ecol. Sociobiol.*, **20**, 421–425.

Symington, M. M. (1988). Food competition and foraging party size in the black spider monkey (*Ateles paniscus chamek*). *Behaviour*, **105**, 117–132.

Symington, M. M. (1990). Fission–fusion social organization in *Ateles* and *Pan. Int. J. Primatol.*, **11**, 47–61.

Thierry, B. (1985). Patterns of agonistic interactions in three species of macaque (*Macaca mulatta, M. fascicularis, M. tonkeana*). *Aggress. Behav.*, **11**, 223–233.

Thierry, B. (2000). Covariation of conflict management patterns across macaque species. In *Natural Conflict Resolution*, ed. F. Aureli and F. B. M. de Waal, Berkeley: University of California Press, pp. 106–128.

Thierry, B., Singh, M. and Kaumanns, W. (2004). *Macaque Societies: A Model for the Study of Social Organization*. Cambridge: Cambridge University Press.

Valero, A., Schaffner, C. M., Vick, L. G., Aureli, F. and Ramos-Fernández, G. (2006). Intragroup lethal aggression in wild spider monkeys. *Am. J. Primatol.*, **68**, 732–737.

van Hooff, J. A. R. A. M. and van Schaik, C. P. (1994). Male bonds: affiliative relationships among nonhuman primate males. *Behaviour*, **130**, 309–337.

van Roosmalen, M. G. M. and Klein, L. L. (1988). The spider monkeys, genus *Ateles*. In *Ecology and Behavior of Neotropical Primates*, Vol. 2, ed. R. A. Mittermeier, A. F. Coimbra-Filho and G. A. B. de Fonseca, Washington, DC: World Wildlife Fund, pp. 455–537.

van Schaik, C. P. (1983). Why are diurnal primates living in groups? *Behaviour*, **87**, 120–144.

van Schaik, C. P. (1989). The ecology of social relationships amongst female primates. In *Comparative Socioecology: The Behavioural Ecology of Humans and Other Mammals*, ed. V. Standen and R. A. Foley, Oxford: Blackwell, pp. 195–218.

van Schaik, C. P. (1999). The socioecology of fission–fusion sociality in orangutans. *Primates*, **40**, 69–86.

van Schaik, C. P. and Aureli, F. (2000). The natural history of valuable relationships in primates. In *Natural Conflict Resolution*, ed. F. Aureli and F. B. M. de Waal, Berkeley: University of California Press, pp. 307–333.

van Schaik, C. P. and Hoerstermann, M. (1994). Predation risk and the number of adult males in a primate group: a comparative test. *Behav. Ecol. Sociobiol.*, **35**, 261–272.

van Schaik, C. P. and Kappeler, P. M. (1997). Infanticide risk and the evolution of male-female association in primates. *Proc. Roy. Soc. Lond. B*, **264**, 1687–1694.

van Schaik, C. P. and van Hooff, J. A. R. A. M. (1983). On the ultimate causes of primate social systems. *Behaviour*, **85**, 91–117.

van Schaik, C. P. and van Noordwijk, M. A. (1988). Scramble and contest feeding competition among female long-tailed macaques (*Macaca fascicularis*). *Behaviour*, **105**, 77–98.

van Schaik, C. P., Pradhan, G. R. and Noordwijk, M. A. (2004). Mating conflict in primates: infanticide, sexual harassment and female sexuality. In *Sexual Selection in Primates: New and Comparative Perspectives*, ed. P. M. Kappeler and C. P. van Schaik, Cambridge: Cambridge University Press, pp. 131–150.

Wallace, R. B. (2001). Diurnal activity budgets of black spider monkeys, *Ateles chamek*, in a southern Amazonian tropical forest. *Neotrop. Primates*, **9**, 101–107.

Watts, D. P. (1997). Agonistic interventions in wild mountain gorilla groups. *Behaviour*, **134**, 23–57.

Watts, D. P. (1998). Coalitionary mate guarding by male chimpanzees at Ngogo, Kibale National Park, Uganda. *Behav. Ecol. Sociobiol.*, **44**, 43–55.

Watts, D. P., Muller, M., Amsler, S. J., Mbabazi, G. and Mitani, J. C. (2006). Lethal intergroup aggression by chimpanzees in the Kibale National Park, Uganda. *Am. J. Primatol.*, **68**, 161–180.

Whitehead, H. and Dufault, S. (1999). Techniques for analysing vertebrate social structure using identified individuals: review and recommendations. *Adv. Stud. Behav.*, **28**, 33–74.

Williams, J. M., Oehlert, G. W., Carlis, J. V. and Pusey, A. E. (2004). Why do male chimpanzees defend a group range? *Anim. Behav.*, **68**, 523–532.

Wittemyer, G., Douglas-Hamilton, I. and Getz, W. M. (2005). The socioecolgy of elephants: analysis of the process creating multitiered social structures. *Anim. Behav.*, **69**, 1357–1371.

Wrangham, R. W. (1980). An ecological model of female bonded primate groups. *Behaviour*, **75**, 262–300.

Wrangham, R. W. (1987). Evolution of social structure. In *Primate Societies*, ed. B. B. Smuts, D. L. Cheney, R. M. Seyfarth, R. W. Wrangham and T. T. Struhsaker, Chicago: University of Chicago Press, pp. 282–296.

Wrangham, R. W. (2000). Why are male chimpanzees more gregarious than mothers? A scramble competition hypothesis. In *Primate Males: Causes and Consequences of Variation in Group Composition*, ed. P. M. Kappeler, Cambridge: Cambridge University Press, pp. 248–258.

10 Spider monkey reproduction and sexual behavior

CHRISTINA J. CAMPBELL AND K. NICOLE GIBSON

Introduction

Although studies on the behavioral ecology of spider monkeys are becoming more common (see Campbell, this volume), few data are currently available concerning their reproductive biology and sexual behavior. Early investigations provided information largely limited to life history parameters, indicating that spider monkeys show an extended gestation length (7–7.5 months; Eisenberg, 1973; Milton, 1981; Nunes and Chapman, 1997) and interbirth interval relative to most other monkeys (approximately three years, Eisenberg, 1973; Milton, 1981; Chapman and Chapman, 1990; C. J. Campbell, unpublished data). However, specific data pertaining to the reproductive biology and behavior of *Ateles* have continued to be underrepresented in the literature. The major hindrance limiting these types of investigations is the sheer difficulty of studying spider monkeys in the wild. In addition, spider monkey sexual behavior is rarely observed. Until the 1970s, copulation had never been reported, even in captivity (Klein, 1971) and thus was thought to occur at night (Eisenberg and Kuehn, 1966). Researchers carrying out long-term studies of *Ateles* in free-ranging conditions have consistently observed relatively low numbers of copulations (Table 10.1) reiterating this fact. One of the difficulties in observing copulations is related to the behavior of spider monkeys when they pair off to mate. Spider monkeys mate promiscuously, but copulations occur almost exclusively during a "consortship." We define the term consortship as a specific social interaction when a male–female pair coordinate to stay apart from others in their social community for a few minutes, hours, or full days, and copulate (Symington, 1987; Campbell, 2000, 2006; K. N. Gibson, unpublished data). Consortships are extremely secretive; both the male and female avoid and sneak away from other spider monkeys and thus frequently lose observers (Campbell, 2006; K. N. Gibson, unpublished data). Spider monkeys almost certainly copulate with some frequency, but researchers generally have not been attuned to the subtle cues of mating behavior.

Spider Monkeys: Behavior, Ecology and Evolution of the Genus Ateles, ed. Christina J. Campbell. Published by Cambridge University Press. © Cambridge University Press 2008.

Table 10.1 *Number of spider monkey copulations observed at various study sites*

Study site	Species	Approximate # of contact hours	# of copulations observed	Reference(s)
Barro Colorado Island, Panama	*A. geoffroyi ornatus*	1200	18	Campbell (2000, 2006)
Punta Laguna, Mexico	*A. geoffroyi yucatanensis*	1896 (2 groups)	5	Ramos-Fernández (2001)
Cocha Cashu, Manu National Park, Peru	*A. belzebuth chamek*[a]	>1360 (2 groups)	22	Symington (1987)
		500	2	S. Russo (unpublished data)
		2252 (4 groups)	43[b]	K. N. Gibson (unpublished data)
Lago Caiman, Noel Kempff Mercado National Park, Bolivia	*A. belzebuth chamek*	2700	2	Wallace (1998)
La Chonta, Concesion Forestal, Bolivia	*A. belzebuth chamek*	>1400	3	A. Felton (unpublished data)
Yasuní National Park, Ecuador	*A. belzebuth belzebuth*	1273	1	S. Suarez (unpublished data)
		500	0	J. L. Dew (unpublished data)
La Macarena, Colombia	*A. belzebuth belzebuth*	1263	5	Shimooka (2000 and unpublished data)
		965	2	Matsushita and Nishimura (2000)
Raleighvallen-Voltzberg, Suriname	*A. paniscus paniscus*	865	27	van Roosmalen (1985)

[a] Previously published as *A. paniscus chamek*.
[b] Of these 43, 39 were considered "complete." Four of the 43 copulations observed were considered "interrupted" – by males arriving; by the male stopping copulation, perhaps because males were arriving.

The information reviewed in this chapter draws largely from an ongoing study of the reproductive biology and behavior of free-ranging spider monkeys (*Ateles geoffroyi*) on Barro Colorado Island (BCI), Panama (Campbell, 2000, 2003, 2004, 2006; Campbell *et al.*, 2001) and a 22-month study (2003–2006) on the mating strategies and socioecology of male spider monkeys (*Ateles belzebuth chamek*) at the Cocha Cashu Biological Station in Manu National Park, Peru (K. N. Gibson, unpublished data). While there are few pertinent publications from other study sites, future long-term studies of well-habituated populations with known individuals and demographic histories will almost certainly improve our understanding of the reproduction and sexuality of this taxon.

Female morphology and physiology

Ovarian and menstrual cycles

Cycle lengths of both captive and free-ranging female spider monkeys fall consistently between 20 and 24 days whether from hormonal or vaginal swab data (Goodman and Wislocki, 1935; Hodges *et al.*, 1981; McDaniel *et al.*, 1993; Hernández-López *et al.*, 1998; Campbell *et al.*, 2001). Data also suggest that menstrual bleeding is present over 2–4 days, although it may not always be externally visible (Goodman and Wislocki, 1935; Campbell *et al.*, 2001), and as such does not provide a reliable visual indicator of female reproductive status to researchers. Campbell *et al.* (2001) describe the ovarian cycle for both captive and free-ranging females using enzyme immunoassays to measure estrogen (E1C) and progesterone (PdG) metabolites (Figure 10.1). E1C excretion increases dramatically from follicular phase levels to peak during the peri-ovulatory period and these high levels are sustained throughout the luteal phase. PdG concentrations also peak during the peri-ovulatory period, and these high levels are maintained until the late luteal phase. The cycle profiles reported by Campbell *et al.* (2001) were similar for both free-ranging and captive females. However, in the free-ranging animals they were 2–3 days longer on average, follicular phases tended to be a little longer in duration (approximately 4–8 days), and E1C and PdG levels did not peak until midway through the luteal phase of the cycle. In addition, nonpregnant PdG levels were significantly higher for the free-ranging animals. The authors suggest dietary differences between the captive and free-ranging females, and the advanced age of the captive female subjects to be the most likely explanation for these disparities (Campbell *et al.*, 2001).

(a)

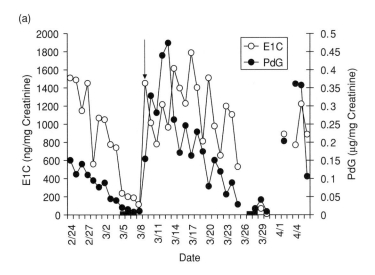

(b)

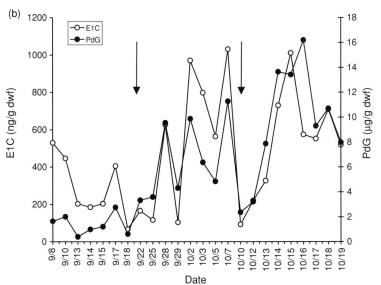

Figure 10.1 Ovarian cycle of *Ateles geoffroyi*. (a) Captive female (from urine), black bars represent blood present in the urine. (b) Free-ranging female (from feces). Arrows in both figures represent presumed days of ovulation, second arrow in (b) represents beginning of pregnancy. (After Campbell *et al*., 2001.)

Figure 10.2 A juvenile female black-handed spider monkey (*Ateles geoffroyi*) showing the hypertrophied, pendulous clitoris. (Photo: C. J. Campbell.)

Clitoris morphology

A hypertrophied clitoris is present in many species of New World monkey and in *Ateles* spp. it is pendulous (Figure 10.2) and larger than the flaccid penis of the male (Wislocki, 1936). At birth, the clitoris is pendulous and both of us have used this characteristic to successfully determine the sex of newborns within days of birth. Individual variation in clitoris length and shape occurs among females within a social group (C. J. Campbell and K. N. Gibson, personal observations) and can be successfully used by researchers to identify individual females. Variation in clitoral length appears to be governed by genetics. On BCI, one matriline of females (determined by observational data and mitochonidrial DNA) all have visually much longer clitorides than other females in the group (C. J. Campbell, unpublished data). In addition to intraspecific variation, there appears to be phylogenetic variation at the species level in at least the color of the clitoris, with *A. geoffroyi panamensis* and *A. g. yucatanensis* showing less pigmentation and a more uniform pink coloration (C. J. Campbell and K. N. Gibson, personal observations; C. M. Schaffner, personal communication) than two South American subspecies, *A. belzebuth belzebuth* (J. L. Dew, Y. Shimooka, S. Suarez, personal communication) and *A. b. chamek* (K. N. Gibson, personal observation; R. B. Wallace, personal communication). In these two subspecies a portion of the clitoris is a dark gray/black color. The location and degree of the darker pigmentation is highly variable among females, making it a useful trait for individual identification (K. N. Gibson, personal observation; Y. Shimooka and S. Suarez, personal communication).

Unlike the large sexual swellings seen in some members of the Cercopithecoidea and Hominoidea superfamilies (Dixson, 1983), extremely limited swelling of the spider monkey clitoris can occur midcycle, but this is highly variable within cycles and between individuals (Goodman and Wislocki, 1935; C. J. Campbell, personal observation). The relatively inconspicuous and inconsistent nature of swelling, along with the fact that its timing in relation to ovarian cycle stage has not been examined, makes it a highly unreliable indicator of ovulation. As such, like most New World primates, female spider monkeys have no obvious visual cue to reproductive status and/or ovarian cycle stage.

The most commonly cited function of the spider monkey pendulous clitoris is that it aids in the communication of sexual state via chemical cues in urine and/or other vaginal secretions (Klein, 1971). Chemical or olfactory cues may be more important signals of female reproductive status for New World monkeys and prosimians than the Old World anthropoids (Epple, 1976; Dixson, 1983). Adult male spider monkeys and less frequently other adult females, and immature males and females, smell the branches where females have previously been sitting ("place sniffing"), grasp the female's clitoris and then sniff their hand, and/or smell or lick urine that has been deposited on branches and leaves. Klein (1971) postulated that the dorsal groove of the clitoris (Wislocki, 1936) might channel and/or collect urinary and vaginal fluids, allowing them to be utilized as olfactory cues. At both of our study sites the clitoris does appear to be used for olfaction to aid in the male assessment of a female's reproductive state (Campbell, 2004; K. N. Gibson, unpublished data). The most noted example comes from male behavior during mating consorts. Males place sniff before and after copulation and touch the clitoris and smell their fingers multiple times during copulation (see also Pastor-Nieto, 2000). These behaviors suggest that males are able to identify olfactory cues in the female's urine or other vaginal excretions (only previously suggested in primates for cottontop tamarins *Saguinus oedipus*; see Ziegler *et al.*, 1993).

Dixson (1998) suggested that the dispersed fission–fusion social organization of spider monkeys provides the key to the evolution of the extreme clitoral morphology in these monkeys. When males and females do not always associate in stable groups, males could effectively keep track of a female's reproductive status if the female deposits olfactory cues in a wide-ranging area (Dixson, 1998). A major problem with this idea is that all of the behaviors thought to be related to this function (e.g. place sniffing and urine sniffing and licking) take place when the female and male are in close vicinity. There is no evidence that olfactory cues are available to males outside of the immediate temporal and spatial association with the female in question, again leading us to ask the question – why is the clitoris so enlarged?

Spider monkeys are monomorphic in body size (Di Fiore and Campbell, 2007) and it is only the clitoris that clearly distinguishes females from males (Eisenberg and Kuehn, 1966; C. J. Campbell and K. N. Gibson, personal observations). The clitoris therefore almost certainly aids in the identification of adult females from a distance, which could be necessary given the dispersed nature of the spider monkey community. Indeed this is something we use on a frequent basis to make an initial assessment of the identity of an animal from a distance. In the closely related *Brachyteles* and *Lagothrix*, the clitoris is hypertrophied, but is not nearly as large as in *Ateles*. If indeed sexual discrimination is a function of the size of the clitoris in *Ateles*, two possible factors may help to explain this difference. First, the fission–fusion social organization in *Brachyteles* and *Lagothrix* is not as prevalent as it is in *Ateles* (Di Fiore and Campbell, 2007), thus males and females are probably in closer visual contact than they are in *Ateles*. Second, *Brachyteles* and *Lagothrix* males are significantly larger than the females (Di Fiore and Campbell, 2007) and have relatively large testicles (Dixson, 1998), making the sexes visually more distinctive than in *Ateles* – even from a distance.

Dixson (1998) states that there is no evidence that the clitoris provides tactile feedback over and above what there is in other primate species. However, one of us (C. J. Campbell) has observed that in *A. geoffroyi* the grooved surface of the clitoris continuously rubs along the inferior surface of the erect penis during intromission attempts and copulation, suggesting that the clitoris may act as a "guide" to direct the large penis of this species into the vaginal canal (Campbell, 2006), or may indeed provide further sensory feedback to the male and/or female during prolonged intromission.

Thus, even though its utility in scent marking is the most cited and most conspicuous function of the extreme spider monkey clitoris, it is probable that there are multiple functions, and it is difficult to determine if any of these actually fully explain the extreme amplification in size.

Male morphology and physiology

Penile morphology

In its flaccid condition, the penis of the male spider monkey is barely visible because it is retracted within the prepuce. All references to the flaccid penis of male spider monkeys (*Ateles* spp.) describe it as being relatively elongated (Wislocki, 1936; Hill, 1962) with a simple morphology (Hill, 1962; Harrison and Lewis, 1986). From our observations of the erect penis of spider monkeys, it is clear that it is relatively large for their body size and that the distal end is

Figure 10.3 An adult male, immediately following a copulation, showing the relative size and shape of the erect penis. (Photo: C. J. Campbell.)

actually mushroom shaped (Figure 10.3), as was described by Hill (1962) for the closely related muriquis (*Brachyteles* spp.) and woolly monkeys (*Lagothrix* spp.). In *A. belzebuth belzebuth* and *A. geoffroyi* at least, the penis is darkly pigmented, being almost black along its entire length (Wislocki, 1936; C. J. Campbell and K. N. Gibson, personal observations).

The penis of most primates contains a bone (*os penis*, *os priapi*, or baculum), as it does in several other orders of mammals. However, this bone is completely absent in spider monkeys and the closely related *Lagothrix* and *Alouatta* (Wislocki, 1936; Hill, 1962; Dixson, 1987b), although it is present in *Brachyteles* (Dixson *et al.*, 2004). The only other anthropoid that completely lacks this bone is our own species, *Homo sapiens*. In species where the baculum is present, one of its potential roles is the maintenance of an erection during prolonged copulations (Romer, 1962). Alternatively, it has been suggested that bacula may assist in copulation when initial penetration is difficult either as a result of the male being much larger than the female, or because of a complex penile morphology (Long and Frank, 1968). The absence of this bone in spider monkeys is particularly interesting because they have prolonged single mount intromissions which are more commonly associated with longer baculum lengths across the primate order (Dixson, 1987a, 1987b).

The surface of the penis is covered with cornified spines that are directed toward the base of the penis (Wislocki, 1936). Dixson (1998) includes the penile spines of *Ateles* in his second category of "robust simple spines" and suggests that they could function in any number of ways, including tactile stimulation of the male and/or female during intromission attempts and copulation, the removal of coagulated semen during copulation, or possibly aiding in the production of a "genital lock" (see below for evidence of this in *A. geoffroyi*).

Testes and scrotum

Relative to their body size, the testicles of spider monkeys are smaller than their closest relatives, weighing approximately 0.17% of their total body weight (Schultz, 1938). Unlike many primates, the testes of spider monkeys are descended at birth (Wislocki, 1936). Nevertheless, the testes of immature males are relatively small and remain close to the body until maturity. Scrotal sac color in mature (subadult and adult) males varies individually, and like the clitoris of the female can be used to reliably identify individual animals (C. J. Campbell and K. N. Gibson, personal observations).

Sexual behavior

Copulatory behavior

All studies of spider monkey sexual behavior have reported prolonged single mount copulations in a dorsoventral position (Klein, 1971; van Roosmalen, 1985; Symington, 1987; Campbell, 2006). Detailed published accounts of the behaviors leading up to and included in a spider monkey sexual encounter are rare and much of what follows comes from our own research (Campbell, 2006; K. N. Gibson, unpublished data). We have found that precopulation and copulation behavior varies between our study populations. Sexual encounters between spider monkeys on BCI are characterized by a total lack of vocalizations by the copulating pair until separation. Spider monkeys at Cocha Cashu on the other hand do occasionally make specific precopulation vocalizations when they are communicating their readiness to start a consort. Once the male and female join each other in the same tree they are quiet. However, at both study sites the female's associated offspring may vocalize before and while the pair copulate. When the copulation is finished, one or both of the adults may respond to long calls of other group members, and often initiate long calls before joining another group. This only occurs when the pair ends their consortship. If the consortship continues, the pair maintain their silence – not making or responding to long calls or other vocalizations. Copulation is performed in a dorsoventral position (Figure 10.4) with the male behind the female either sitting and placing his legs over her thighs in a "leg-lock" (originally described by Klein, 1971) or squatting behind the female. The male usually holds onto the female by wrapping his

Figure 10.4 An adult male and female engaging in copulation. The male is seated behind the female. (Photo: C. J. Campbell.)

arms (one or both) around the female's torso while she grasps branches or vines above them. Positional changes do occur during the sequence. The "leg-lock" is not always in place and both the male and female may reposition themselves for better stability.

Although there is some variation in the average copulation length reported, all studies have consistently reported average lengths of close to or well over 10 minutes (van Roosmalen, 1985; Symington, 1987; Campbell, 2006; K. N. Gibson, unpublished data). In *A. geoffroyi*, attempts to insert the penis into the vaginal canal are made approximately every 2–5 seconds, and the female and male often change positions during this time. Difficulties in attaining intromission appear to relate to the large glans penis of *A. geoffroyi* (Figure 10.3). In *A. belzebuth*, penis insertion is usually rapid, but when not, appears to be related to the poor angle of insertion, when the male is not directly behind the female (K. N. Gibson, unpublished data). The average length of copulation from intromission to presumed ejaculation (as indicated by the cessation of thrusting) was over 17 minutes for *A. geoffroyi* (Campbell, 2006) and close to 14 minutes for *A. belzebuth* (K. N. Gibson, unpublished data). Upon termination of thrusting, the male may remain intromitted, usually pausing for just less than 2 minutes (Campbell, 2006; K. N. Gibson, unpublished data). A white viscous substance, possibly semen from a previous copulation, is often seen dripping from the vaginal canal and also coagulating around the penis soon after intromission and throughout copulation, as well as streaming from the vaginal canal after thrusting stops, although the exact source of this substance has yet to be verified. Nevertheless, the white fluid is not always present, even when a female has mated earlier on the same day (K. N. Gibson, unpublished data). Urine and blood may also be seen streaming or dripping from the female's genitalia during copulation (Campbell, 2006; K. N. Gibson, unpublished data). If females urinate they generally do so towards the middle or end of the copulation, before thrusting has stopped.

Thrusting occurs continuously at a constant rate of one thrust every 3 seconds for the entire copulation with the thrust sequence involving three stages (thrust in, full insertion, and pull back) each lasting almost exactly 1 second. A double thrust may also occur where the first of the two thrusts is much shallower than the second. In *A. geoffroyi* thrusting does not increase in frequency toward the end of the copulation, and termination of thrusting may be instigated by the female who starts moving in the tree, often swinging between branches with the male still intromitted for approximately 30 seconds before coming to a rest (Campbell, 2006). At this point, only the penis glans is still inserted into the vagina and the pair are still until it slides out (suggesting that there may be a "genital lock" in *A. geoffroyi*). After separation, both the male and female immediately rub their genitalia on a branch. There is variation between species in these behaviors. In *A. belzebuth* the female usually separates from the male after cessation of thrusting or after a brief pause following the end of thrusting and the pair does not swing between branches as seen in *A. geoffroyi*. There is also no behavioral indication of a genital lock for *A. belzebuth*. In this species, after separation, usually only the male or the female rub their genitals on branches, and they do so only half the time (K. N. Gibson, unpublished data).

The female's associated juvenile and/or subadult offspring are usually present during copulation; they tend to remain quiet, sitting in contact with and watching the copulating pair. When the offspring (mostly young juveniles) interact with the pair they may crawl on one or both adults, sit next to and in contact with them, nurse, touch, smell, or lick the genitals of the adults, lick urine or fluid that is deposited on leaves below the pair, and solicit one or both adults to play. Juvenile offspring may also "harass" the pair by lunging or grabbing at them or by pushing to get in between the pair at any time throughout the copulation (van Roosmalen, 1985; Campbell, 2006; K. N. Gibson, unpublished data). However, in *A. geoffroyi* this usually does not begin until the end of the copulation when the female has started moving with the male still intromitted (Campbell, 2006).

At BCI and Cocha Cashu, copulation usually takes place during consortships (defined above) and nearly always at least 50 m away from other adult and subadult males. The majority of cases that take place outside of a consort involve the presence of additional adult females and their immature offspring. In one known case where an additional adult male was present, the copulating male was the dominant male in the group (Campbell, 2000). Isolation from other adult males is achieved by one of three ways depending on what group the male and female were associated with prior to copulation. When the male and female are part of the same larger subgroup, the male and female each leave the subgroup at different times or follow each other to meet in a new location, or they remain in the location of the larger subgroup after all the other individuals have traveled away. When the male and female are not part of the same subgroup, they may leave their respective subgroups and meet each other at a new location. Once separation from other males is achieved the pair may remain in a consortship (making up their own subgroup) and still be intermittently joined by other males and females. Despite the possibility of being discovered in a consort, pairs eventually separate again from all other adult and subadult males and copulations only take place when the pair is likely not to be observed (K. N. Gibson, unpublished data). While consortships are usually the rule in spider monkey sexual pairings, there is one case at Cocha Cashu of a female that was uncooperative and forced to copulate (K. N. Gibson, unpublished data).

During consorts, the male and female frequently scan the canopy and do not make or respond to vocalizations (even when they hear long calls given by familiar individuals). This behavioral repertoire suggests that males and females – that intend to copulate – avoid other group members. Campbell (2006) proposed that consorting males and females specifically avoid other adult males because the presence of additional adult females is not uncommon. Data from Cocha Cashu also support this conclusion (K. N. Gibson, unpublished data). Sexual behavior in the closely related muriqui (*Brachyteles* spp.) contrasts strongly with the data for spider monkeys – in muriquis, copulation may take place in "aggregations," where females copulate with multiple males in quick succession (Milton, 1985; Strier, 1996).

Van Roosmalen (1985) suggested that if the copulating pair sought seclusion it was to avoid harassment by the female's associated juvenile. We disagree with this assessment for two reasons. First, and most importantly, in the copulations we have witnessed, females with juvenile offspring never copulated in seclusion from them or attempted to travel away from or without them. Second, while offspring do interact with the copulating pair – by sitting in close proximity or contact and in some cases harassing them – interference to the point of cutting short a copulation is often avoided by the adult male threatening the juvenile with open mouth threats, pushing and slapping out at or cuffing the juvenile, as well as the mother giving open mouth threats (K. N. Gibson, unpublished data). Only in rare cases has the unrestrained behavior of a juvenile stopped a copulation that was in progress (K. N. Gibson, unpublished data). While we recognize that juveniles can disrupt copulations involving their mothers, our data do not support the idea that pairs avoid associated juvenile offspring. Even in the event of a failed copulation, pairs do eventually (successfully) copulate and always in the presence of the poorly behaved juvenile (K. N. Gibson, unpublished data).

Patterns of behavior across the female ovarian cycle

In spite of the paucity of information on spider monkey sexual behavior, previous authors have suggested that a number of rarely observed behaviors are indicative of "estrus" in adult females such as copulation (Symington, 1987), female directed male aggression (Symington, 1987), clitoral manipulation by males and females (van Roosmalen, 1985), and place sniffing (Klein, 1971). Much of the work by one of us (Campbell, 2000, 2003, 2004; Campbell *et al.*, 2001) has integrated behavioral and endocrinological data (from the study of hormones in fecal samples) in an attempt to determine whether any of these behaviors are strongly associated with any one reproductive condition or stage of the ovarian cycle.

As suggested by Klein (1971), "place sniffing," when a male sniffs the place where a female was sitting, is observed more often when the female is reproductively cycling. However, males also place sniff when females are not reproductively cycling (Campbell, 2004). Likewise, urine sniffing, clitoral hold, and clitoral sniff behaviors are also directed at females in varying reproductive states and thus cannot reliably indicate to researchers what reproductive state the female is in (Campbell, 2004). While these four behaviors appear to occur during all stages of the ovarian cycle it is unclear at this point whether any are more strongly associated with ovulation. Nevertheless, these are investigative behaviors directed towards females, usually by adult or subadult males (Campbell, 2000; K. N. Gibson, unpublished data). They more than likely function as a means to gauge the reproductive condition of a female even though they may not perfectly relate to any one female reproductive condition or stage of the ovarian cycle. Males almost certainly investigate females of all reproductive states in order to gather information about changes in their condition.

Copulations appear to be more likely to take place when the female is reproductively cycling (Campbell, 2004). However, as with most anthropoid primates, this behavior is not limited to reproductively cycling females and occurs when females are acyclic, pregnant or lactating (Campbell, 2004; K. N. Gibson, unpublished data). Indeed one of us observed one female to copulate during two months when hormonal data indicated anovulatory cycles (Campbell, 2004), strengthening the conclusion that this behavior is not strictly tied to ovulation in spider monkeys.

Male spider monkeys are known to be aggressive towards females (Fedigan and Baxter, 1984; Symington, 1987; Campbell, 2003; Slater *et al.*, 2005) and aggression may be linked to conflicts of reproductive interests between males and females and, in turn, mating strategies (Aureli and Schaffner, this volume; K. N. Gibson, unpublished data). Male aggression towards females can be a form of "sexual coercion," where the use or threat of force by a male influences females – at some cost to the female – to mate with one male over another either now or in the future (Smuts and Smuts, 1993). Symington (1987) postulated that male spider monkeys direct aggression only at cycling females, to determine the reproductive state of a female by inducing defecation or urination. Data from *A. geoffroyi* on BCI – where female reproductive state was known through fecal derived hormonal data – do not support these hypotheses because there was no correlation between acts of aggression and the target females' reproductive state (Campbell, 2003). Females are targets not only when cycling, but also when pregnant and lactating. Moreover, aggression towards females on BCI rarely induced defecation or urination and there was no association between aggression and observed copulations (Campbell, 2003; but see Aureli and Schaffner, this volume). Data do tentatively support the notion that male aggression may be the only way for males to maintain dominance over equally sized females (Strier, 1994; Campbell, 2003).

Males appear to bias their aggressive behaviors and copulation attempts towards "mature" adult females, rather than younger females that have just reached sexual maturity. On BCI, a subadult female that was reproductively cycling was not observed to copulate and was never observed to be the target of any of the behaviors discussed above. At this time she was about 6 years old, but at about 7 years of age she was seen copulating and the following year she gave birth (C. J. Campbell, personal observation).

Van Roosmalen (1985) proposed that self-clitoral manipulation is indicative of the estrous period. However, the relationship between this behavior and reproductive condition is unclear (Campbell, 2004). Rates of self-clitoral hold and self-clitoral rub do not consistently rise in association with observed copulations, or with any particular stage in the ovarian cycle (Campbell, 2004). We suggest instead that these behaviors are more likely related to clitoral self-grooming and insect repulsion because the clitoris is large and mostly hairless.

Of all the behaviors suggested to indicate estrus in female spider monkeys, not one has been shown to occur only in reproductively cycling females or to cluster exclusively with copulations and/or any particular hormonal profile. In addition, while certain behaviors (e.g. copulation, place sniffing, and self-clitoral rub) appear to have some association with female reproductive state, the pattern is not always consistent among females. Researchers that study spider monkeys must be careful when they attribute various reproductive states to females based on behavioral observations alone. They should consider changes in the frequency of the behaviors mentioned above, rather than simply presence/absence. Ultimately, these behaviors are not good indicators of where a female is in her ovarian cycle because they are not only associated with the peri-ovulatory period. The concept of "estrus," as originally defined by Heape (1900), is not applicable to spider monkey sexual behavior (or the sexual behavior of any anthropoid primate) because copulations occur at all times, even when conception is impossible (Campbell, 2007; see also Dixson, 1998; Loy, 1987). However, this does not exclude the possibility that there are more subtle behaviors that may be strongly associated with female reproductive state and cycle stage. We suggest that more research – focused on female hormone profiles and behavior both in the wild and in captivity – will continue to improve our understanding of female spider monkey sexual behavior.

Reproduction

Conception

Young female primates typically go through a period of adolescent sterility or adolescent subfecundity (Knott, 2001) where females do not ovulate during ovarian cycling, or where seemingly normal hormonal cycling coincides with a complete lack of sexual behavior, a lack of interest by adult and subadult males, and a lack of conception. This also appears to be the case in spider monkeys (e.g. *A. geoffroyi*, Campbell, 2004). Data show that multiparous females (those who have previously reproduced) usually cycle for a few months before conception occurs (e.g. *A. geoffroyi*, 3–6 cycles, Campbell *et al.*, 2001; *A. belzebuth*, K. N. Gibson, unpublished data) suggesting a cycling-to-conception delay similar to muriquis and woolly monkeys (Nishimura, 2003; Strier *et al.*, 2003).

Birth

To our knowledge there are no published reports of the spider monkey birth process – suggesting that births take place at night. Both wild and captive females have been observed carrying new babies the morning after they were "put to bed" without an infant (C. J. Campbell and K. N. Gibson, personal observations; C. M. Schaffner, personal communication). Like most anthropoid primates, spider monkeys usually give birth to one offspring at a time. However, twinning is also possible, but not common. Two sets of twins have been born in different populations of spider monkeys in South America. Link *et al.* (2006) report a case of twins (*A. belzebuth belzebuth*) born at Tinigua National Park, Colombia. In Peru, one set of female twins (*A. b. chamek*) was born and monitored at Cocha Cashu (K. N. Gibson, unpublished data). Twins appear to hinder their mother's ability to feed and show slower physical and social development compared with their peers (Link *et al.*, 2006). The female twins at Cocha Cashu showed greater independence from their mother compared with others their same age, and were often seen feeding and traveling without their mother or any other adults (K. N. Gibson, unpublished data). This set of twins often relied on each other for support, instead of relying on their mother, which is very unusual in spider monkey mother–offspring social relationships.

Paternity

Mating in spider monkeys is not limited to any particular male within a community (van Roosmalen, 1985; Symington, 1987; Campbell, 2006). All of the adult males in the BCI study group were observed to copulate at least once within a 14-month period (Campbell, 2006). Nevertheless, dominant males appear to copulate more often than lower-ranking males (Symington, 1987; Campbell, 2000), and to be more successful at maintaining consortships with females (Symington, 1987). In the past we have not been able to address properly whether or not mating is skewed towards a dominant male because males have not been the focus of study. However, one of us (K. N. Gibson) has undertaken a study on male social relationships and the results should allow us to understand better the dynamics of male mating strategies. Even though mating success may be skewed towards a dominant male in spider monkeys this may not translate into higher paternity success (cf. Alberts *et al.*, 2006). Genetic data concerning paternity have not been published for *Ateles* although genetic data from two different communities at Yasuní (Ecuador) indicate that different males do sire similarly aged individuals in a community (A. Di Fiore, A. Link and S. Spehar, unpublished data).

Seasonality

Although spider monkeys are not considered to be seasonal breeders, data from various sites throughout their distribution show birth peaks during certain months of the year (Klein, 1972; Milton, 1981; Symington, 1987; Chapman and Chapman, 1990; Campbell, 2000). As predicted by Di Bitetti and Janson (2000) this pattern appears to correlate with fruit availability across sites (i.e. births begin when fruit availability increases). What seems to be the most striking

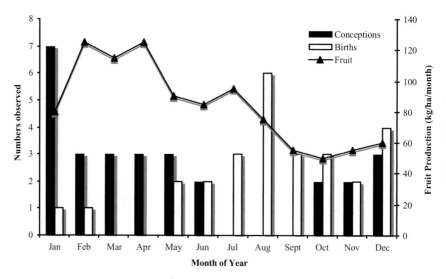

Figure 10.5 Combined spider monkey births and calculated conceptions on Barro Colorado Island, Panama, from two time periods (1968–1980, from Milton, 1981, and October 1997–December 1998 from Campbell, 2000) in relation to fruit production. (Modified from Wright et al., 1999.)

feature of spider monkey birthing patterns is the 2–4 months each year when births do not occur. Symington (1987) first noted this and suggested that rather than a birth season, spider monkeys appear to experience a "non-birth" season. Why births do not occur evenly throughout the year is unclear. Thus, rather than considering births, Symington (1987) looked at the timing of conceptions and noted that spider monkey females experience a 3–4 month period of "non-conception." The absence of conception during these months may relate to a decrease in female reproductive condition during times of reduced fruit availability (Symington, 1987). Indeed when one examines reproductive seasonality patterns in *Ateles* spp., there is a consistent pattern of females not conceiving during months when ripe fruit (their principal food) is declining and approaching its lowest production (see Figure 10.5 for data from BCI). To date, no study has used physiological data on spider monkey females to test the hypothesis that low fruit availability negatively impacts the female ovarian cycle, causing a reduction in the likelihood of conception.

Hernández-López et al. (2002) provide data that suggest that male sperm quality may vary according to season regardless of food availability. In the captive population they studied (where food availability was consistent) the number of live sperm was higher in the dry season than in the rainy season. Though their sample size was small it appears that seasonal changes may influence spider monkey sperm quality and, in turn, reduce conceptions.

Summary

As authors that study two species of spider monkey found in different regions of the Neotropics, we have been able to show that there is both consistency and variation within this genus concerning their reproduction and sexuality. At this stage, however, interpretation of the differences we have uncovered is hampered by the lack of data from other study sites where demographies and individuals are known. We hope that future research and analyses will remedy the ambiguities we have identified. Consistent findings include both morphological and behavioral characteristics of reproduction and sexual behavior. Both males and females have uncommon external genitalia – males lack a baculum and have a large penis relative to their body size, and females have an extreme hypertrophied and pendulous clitoris. Sexual encounters are generally long and private, yet the mating system is promiscuous and mating takes place at times when conception is not likely or impossible. Lastly, although sexual behavior is not seasonal, there appears to be a "non-conception" season most likely related to the onset and continuation of low fruit production.

Acknowledgements

Fieldwork by C. J. Campbell on Barro Colorado Island was supported by grants from the National Science Foundation (Dissertation Improvement Grant # SBR-9711161), the L.S.B. Leakey Foundation, the Department of Anthropology, University of California, Berkeley, and the Smithsonian Tropical Research Institute. Drs. Phyllis Dolhinow and Katharine Milton provided invaluable advice during her project and she would like to thank them both for their help. C. J. Campbell is deeply appreciative of the assistance of Antigone Thomas in the field that made the research presented in this chapter and elsewhere possible. K. N. Gibson thanks the Peruvian government Ministerio de Agricultura Instituto Nacional de Recursos Naturales (INRENA) and the INRENA Parque Nacional del Manu for authorization to work in Manu National Park. She especially appreciates the assistance given by Jessica Espinoza. K. N. Gibson expresses her gratitude to Linda and David McDonough for the quiet retreat where the manuscript was written. K. N. Gibson's research was funded by the Fulbright Institute of International Education, the L.S.B. Leakey Foundation,

Organization for Tropical Studies Francis J. Bossuyt and David and Deborah Clark Fellowship, National Science Foundation Graduate Research Fellowship, John Perry Miller Fund, Williams Fund, and Yale University.

References

Alberts, S. C., Buchan, J. C. and Altmann, J. (2006). Sexual selection in wild baboons: from mating opportunities to paternity success. *Anim. Behav.*, **72**, 1177–1196.

Campbell, C. J. (2000). The reproductive biology of black-handed spider monkeys (*Ateles geoffroyi*): integrating behavior and endocrinology. Unpublished Ph.D. thesis, University of California, Berkeley.

Campbell, C. J. (2003). Female directed aggression in free-ranging *Ateles geoffroyi*. *Int. J. Primatol.*, **24**(2), 223–238.

Campbell, C. J. (2004). Patterns of behavior across reproductive states of free-ranging female black-handed spider monkeys (*Ateles geoffroyi*). *Am. J. Phys. Anthropol.*, **124**(2), 166–176.

Campbell, C. J. (2006). Copulation in free-ranging black-handed spider monkeys (*Ateles geoffroyi*). *Am. J. Primatol.*, **68**, 507–511.

Campbell, C. J. (2007). Primate sexuality and reproduction. In *Primates in Perspective*, ed. C. J. Campbell, A. F. Fuentes, K. C. MacKinnon, M. Panger and S. Bearder, New York: Oxford University Press, pp. 423–437.

Campbell, C. J., Shideler, S. E., Todd, H. E. and Lasley, B. L. (2001). Fecal analysis of ovarian cycles in female black-handed spider monkeys (*Ateles geoffroyi*). *Am. J. Primatol.*, **54**, 79–89.

Chapman, C. A. and Chapman, L. J. (1990). Reproductive biology of captive and free-ranging spider monkeys. *Zoo Biol.*, **9**, 1–9.

Di Bitetti, M. S. and Janson, C. H. (2000). When will the stork arrive? Patterns of birth seasonality in Neotropical primates. *Am. J. Primatol.*, **50**, 109–130.

Di Fiore, A. and Campbell, C. J. (2007). The atelines: variation in ecology, behavior, and social organization. In *Primates in Perspective*, ed. C. J. Campbell, A. F. Fuentes, K. C. MacKinnon, M. Panger and S. Bearder, New York: Oxford University Press, pp.155–185.

Dixson, A. F. (1983). Observations on the evolution and behavioral significance of "sexual skin" in female primates. *Adv. Stud. Behav.*, **13**, 63–106.

Dixson, A. F. (1987a). Observations on the evolution of the genitalia and copulatory behaviour in male primates. *J. Zool. Lond.*, **213**, 423–443.

Dixson, A. F. (1987b). Baculum length and copulatory behavior in primates. *Am. J. Primatol.*, **13**, 51–60.

Dixson, A. F. (1998). *Primate Sexuality: Comparative Studies of the Prosimians, Monkeys, Apes, and Human Beings*. Oxford: Oxford University Press.

Dixson, A. F., Pissinatti, A. and Anderson, M. J. (2004). Observations on genital morphology and anatomy of a hybrid male muriqui (genus *Brachyteles*). *Folia Primatol.*, **75**, 61–69.

Eisenberg, J. F. (1973). Reproduction in two species of spider monkeys, *Ateles fusciceps* and *Ateles geoffroyi*. *J. Mammal.*, **54**, 955–957.

Eisenberg, J. F. and Kuehn, R. E. (1966). The behavior of *Ateles geoffroyi* and related species. *Smithson. Misc. Coll.*, **151**(8), 1–63.

Epple, G. (1976). Chemical communication and reproductive processes in nonhuman primates. In *Mammalian Olfaction, Reproductive Processes, and Behavior*, ed. R. L. Doty, New York: Academic Press, pp. 257–282.

Fedigan, L. M. and Baxter, M. J. (1984). Sex differences and social organization in free-ranging spider monkeys (*Ateles geoffroyi*). *Primates*, **25**(3), 279–295.

Goodman, L. M. and Wislocki, G. B. (1935). Cyclical uterine bleeding in a New World monkey (*Ateles geoffroyi*). *Anat. Rec.*, **61**(4), 379–387.

Harrison, R. M. and Lewis, R. W. (1986). The male reproductive tract and its fluids. In *Comparative Primate Biology*, Vol. 3, ed. W. R. Dukelow and J. Erwin, New York: Alan R. Liss, pp. 101–148.

Heape, W. (1900). The "sexual season" of mammals and the relation of the "pro-oestrum" to menstruation. *Q. J. Microscop. Sci.*, **44**, 1–70.

Hernández-López, L., Mayagoita, L., Esquivel-Lacroix, C., Rojas-Maya, S. and Mondragón-Ceballos, R. (1998). The menstrual cycle of the spider monkey (*Ateles geoffroyi*). *Am. J. Primatol.*, **44**, 183–195.

Hernández-López, L., Parra, G. C., Cerda-Molina, A. L., *et al.* (2002). Sperm quality differences between the rainy and dry seasons in captive black-handed spider monkeys (*Ateles geoffroyi*). *Am. J. Primatol.*, **57**(1), 35–42.

Hill, W. C. O. (1962). *Primates, Comparative Anatomy and Taxonomy*. Vol. 4: *Cebidae, Part B*. Edinburgh: Edinburgh University Press.

Hodges, J. K., Gulick, B. A., Czekala, N. M. and Lasley, B. L. (1981). Comparison of urinary oestrogen excretion in South American primates. *J. Reprod. Fertil.*, **61**, 83–90.

Klein, L. L. (1971). Observations on copulation and seasonal reproduction of two species of spider monkeys, *Ateles belzebuth and A. geoffroyi*. *Folia Primatol.*, **15**, 233–248.

Klein, L. L. (1972). The ecology and social organization of the spider monkey, *Ateles belzebuth*. Unpublished Ph.D. thesis, University of California, Berkeley.

Knott, C. (2001). Female reproductive ecology of the apes. In *Reproductive Ecology and Human Evolution*, ed. P. T. Ellison. New York: Aldine de Gruyter, pp. 429–464.

Link, A., Palma, A. C., Velez, A. and de Luna, A. G. (2006). Costs of twins in free-ranging white-bellied spider monkeys (*Ateles belzebuth belzebuth*) at Tinigua National Park, Colombia. *Primates*, **47**(2), 131–139.

Long, C. A. and Frank, T. (1968). Morphometric function and variation in the baculum with comments on the correlation of parts. *J. Mammal.*, **49**, 32–43.

Loy, J. (1987). The sexual behavior of African monkeys and the question of estrus. In *Comparative Behavior of African Monkeys*, ed. E. L. Zucker, New York: Alan R. Liss, pp. 175–195.

Matsushita, K. and Nishimura, A. (2000). Colombia Macarena chousachi ni seisoku suru yasei kumozaru MB-3 gun no seitai-shakai gakuteki kenkyu. (English title: Ecological and social studies on wild spider monkey MB-3 group at La Macarena,

Colombia). In *Adaptive Significance of Fission-Fusion Society in* Ateles, ed. K. Izawa, Sendai: Miyagi University of Education, pp. 57–76 (in Japanese).

McDaniel, P. S., Janzow, F. T., Porton, I. and Asa, C. S. (1993). The reproductive and social dynamics of captive *Ateles geoffroyi* (black-handed spider monkey). *Am. Zool.*, **33**, 173–179.

Milton, K. (1981). Estimates of reproductive parameters for free-ranging *Ateles geoffroyi*. *Primates*, **22**(4), 574–579.

Milton, K. (1985). Mating patterns of woolly spider monkeys, *Brachyteles arachnoides*: implications for female choice. *Behav. Ecol. Sociobiol.*, **17**, 53–59.

Nishimura, A. (2003). Reproductive parameters of wild female *Lagothrix lagotricha*. *Int. J. Primatol.*, **24**(4), 707–722.

Nunes, A. and Chapman, C. A. (1997). A re-evaluation of factors influencing the sex ratio of spider monkey populations with new data from Maracá Island, Brazil. *Folia Primatol.*, **68**(1), 31–33.

Pastor-Nieto, R. (2000). Female reproductive advertisement and social factors affecting the sexual behavior of captive spider monkeys. *Lab. Primate Newsl.*, **39**, 5–9.

Ramos-Fernández, G. (2001). Association patterns, feeding competition and vocal communication in spider monkeys (*Ateles geoffroyi*). Unpublished Ph.D. thesis, University of Pennsylvania, Philadelphia.

Romer, A. S. (1962). *The Vertebrate Body*, 3rd edn. Philadelphia: Saunders.

Schultz, A. H. (1938). The relative weights of the testes in primates. *Anat. Rec.*, **72**, 387–394.

Shimooka, Y. (2000). Colombia Macarena chousachi ni seisoku suru yasei kumozaru MB-2 gun no seitai-shakai gakuteki kenkyu. (English title: Ecological and social studies on wild spider monkey MB-2 group at La Macarena, Colombia). In *Adaptive Significance of Fission-fusion Society in* Ateles, ed. K. Izawa, Sendai: Miyagi University of Education, pp. 21–36 (in Japanese).

Slater, K. Y., Schaffner, C. M. and Aureli, F. (2005). Female-directed aggression in wild spider monkeys: male display and female mate choice. *Primate Rep.*, **72**, 89–90.

Smuts, B. B. and Smuts, R. W. (1993). Male aggression and sexual coercion of females in nonhuman primates and other mammals: evidence and theoretical implications. *Adv. Study Behav.*, **22**, 1–63.

Strier, K. B. (1994). Brotherhoods among atelines: kinship, affiliation and competition. *Behaviour*, **130**, 151–167.

Strier, K. B. (1996). Reproductive ecology of female muriquis. In *Adaptive Radiations of Neotropical Primates*, ed. M. A. Norconk, A. L. Rosenberger and P. E. Garber, New York: Plenum Press, pp. 511–532.

Strier, K. B., Lynch, J. W. and Ziegler, T. E. (2003). Hormonal changes during the mating and conception seasons of wild northern muriquis (*Brachyteles arachnoides hypoxanthus*). *Am. J. Primatol.*, **61**, 85–99.

Symington, M. M. (1987). Ecological and social correlates of party size in the black spider monkey, *Ateles paniscus chamek*. Unpublished Ph.D. thesis, Princeton University, Princeton, NJ.

van Roosmalen, M. G. M. (1985). Habitat preferences, diet, feeding strategy and social organization of the black spider monkey (*Ateles paniscus paniscus* Linnaeus 1758) in Surinam. *Acta Amazonica*, **15**, 3–238.

Wallace, R. B. (1998). The behavioural ecology of black spider monkeys in north-eastern Bolivia. Unpublished Ph.D. thesis, Liverpool University, UK.

Wislocki, G. B. (1936). The external genitalia of the simian primates. *Hum. Biol.*, **8**, 309–347.

Wright, S. J., Carrasco. C., Calderón, O. and Paton, S. (1999). The El Niño southern oscillation, variable fruit production and famine in a tropical forest. *Ecology*, **80**(5), 1632–1647.

Ziegler, T. E., Epple, G., Snowdon, C. T., Porter, T. A., Belcher, A. M. and Küderling, I. (1993). Detection of the chemical signals of ovulation in the cotton-top tamarin, *Saguinus oedipus*. *Anim. Behav.*, **45**, 313–322.

11 Immaturity in spider monkeys: a risky business

L A U R A   G R E E R   V I C K

Introduction

Socialization is an important and complicated process in primates, involving morphological, ecological and social factors, which together serve to channel individuals into age- and gender-specific roles that help determine the population biology of each species. Despite its extreme importance, however, the socialization process has received little attention relative to primatological studies of adult nonhuman primates. Most studies have focused chiefly on social structure, as mediated by adults through dominance or affiliative interactions; mating strategies and their function in mediating the reproductive success of the mature individuals utilizing them; and socioecology, again primarily examining male and female adults' differential need for and utilization of resources and the resultant impact on social structure. This chapter investigates immature spider monkeys, that is, individuals from birth until emigration or attainment of social adulthood. The chapter focuses mainly on the Yucatecan spider monkey, *Ateles geoffroyi*, a relatively monomorphic and large-brained cebid with a complex social organization and social roles clearly differentiated by gender. Socialization and the immature phase in these spider monkeys will be considered in the broader context of the adult roles they will eventually exhibit and with regard to what is known about immaturity in primates in general.

Background

Much of the data available on immatures is based on infancy and parent–offspring relationships (usually mother–infant) (Gubernick and Klopfer, 1981; Taub, 1984), starting with and building on Harlow's pioneering work (Harlow, 1959; Harlow and Zimmermann, 1959; Harlow, 1971) on attachment and the role of maternal attributes in infant development. Immatures have often been seen as rather passive "objects" of adult behavior, rather than as complex

Spider Monkeys: Behavior, Ecology and Evolution of the Genus Ateles, ed. Christina J. Campbell.
Published by Cambridge University Press. © Cambridge University Press 2008.

288

individuals with distinct needs and individualized agendas that sometimes differ from those of adults (Pereira and Fairbanks, 1993b). Indicative of the dearth of research on immature primates, Pereira and Leigh (2003) noted that between 1993 and 2003, only four studies of play appeared in the principal primatological journals, despite the fact that play is such an important characteristic of juvenile primates and alleged to be a critical factor in the socialization process. Today, research on primate socialization that includes a more detailed examination of juveniles and subadult animals, augmenting our understanding of this critical postinfant developmental period, is still relatively rare (Cheney, 1977, 1978a, 1978b; Clark, 1977; Kraemer *et al.*, 1982; Hayaki, 1983; Pusey, 1983, 1990; Pereira and Altmann, 1985; Pereira, 1988a, 1988b, 1989; Pereira and Fairbanks, 1993a).

Despite the lack of attention previously paid to the immature phase of nonhuman primate behavioral development, it is an exceedingly important period in the life of primates. It is not only a period of preparation for adult life, but also a time filled with distinct challenges, including, but not limited to, finding food and avoiding predation. Immature animals are less efficient foragers than adults and are therefore particularly vulnerable to starvation in times of food shortage. Their small body size makes them not only easier targets for predators, but they also do not appear to be as vigilant (Janson, 1990; Janson and van Schaik, 1993) or as able to detect predators as are adults (van Schaik and van Noordwijk, 1989; Janson and van Schaik, 1993). They sometimes give alarms to inappropriate stimuli and/or fail to react to adult-generated alarms (Seyfarth and Cheney, 1980, 1986; Macedonia, 1991; Janson and van Schaik, 1993). Moreover, some predators appear to direct more of their predation strategies toward obtaining infant prey (e.g. Boinski, 1988; Stanford *et al.*, 1994; Stanford, 1995). Countering this risk, living in groups and even the social structure itself can serve to buffer the risk of predation, as when juveniles stay in subgroups at the group's protected center (Rhine, 1975; Robinson, 1981; Janson, 1990) or when adults remain vigilant during infant play (Goldizen, 1987). To compare the effect of predation risk versus access to food on juveniles' spatial locations within a group, Janson (1990) measured predation rates (based on rates of vigilance behaviors) and potential food intake in different locations within a group of brown capuchins (*Cebus apella*). He determined that juveniles preferred locations with minimum risk of predation, even when food availability was not as high. Nonhuman primate juveniles are in fact usually found in larger foraging parties, despite competition for food being higher in such associations (van Schaik and van Noordwijk, 1985; Janson, 1990; Janson and van Schaik, 1993). On the other hand, juvenile red colobus monkeys (*Colobus badius tephrosceles*) exhibit no consistent association pattern as a countermeasure in the face of predation by chimpanzees at Gombe. Immature colobus are consumed most by chimpanzees (Stanford *et al.*, 1994; Stanford, 1995) and their tendency toward peripheralization, no matter the context, possibly contributes to their frequent capture (Stanford, 1995).

Immatures are also at risk from conspecific aggression, directed either intentionally or unintentionally to them. Infanticide directed toward dependent young by immigrant males is perhaps the most noteworthy example of the risk to immatures from intragroup aggression (see review in van Schaik and Janson, 2000). But, immatures can also be injured or even die as the result of being "in the wrong place at the wrong time," as when aggression directed toward a mother also injures her dependent young. As well as becoming targets of aggression, immatures are often involved in alliances and conflicts that involve adults as well as other juveniles (Silk *et al.*, 1981; Pereira, 1988b, 1995; Fairbanks, 1993; Pereira and Leigh, 2003). These conflicts and alliances serve to affect not only survival during the immature phase but can also influence future survival and reproductive success. In short, immature nonhuman primates play much more than a passive role in their societies.

Primates are noted for their long period of immaturity; in fact, the length of this period relative to the total life-span is the most unusual aspect of growth and development in nonhuman primates (Pereira and Fairbanks, 1993a). Primates have been characterized as having "slow" life histories, growing slowly and attaining reproductive maturity late; however, over the trajectory of the immature phase, parts of it can be characterized by "fast" periods, as for example, the developmental growth spurts characterizing some primate species. Slow growth has been explained in various ways: from constraints on overall growth due to the costs involved in the development and maintenance of the characteristically large primate brain (Sacher and Staffeldt, 1974; Sacher, 1975; Armstrong, 1983; Harvey *et al.*, 1989; Allman *et al.*, 1993; Charnov and Berrigan, 1993; Martin, 1996; Deaner *et al.*, 2003); to an "ecological risk aversion" theory which holds that mortality due to predation or starvation is minimized during the juvenile period through a slow growth strategy (Janson and van Schaik, 1993); to a link between arboreality and slow life histories, whereby arboreality promulgates the evolution of this feature (Ross and Jones, 1999). Whatever the proximate or ultimate explanation, life histories are the result of adaptive evolutionary processes, and the timing of maturation and length of the juvenile period can be modified as a result of socioecological variables affecting survival and successful reproduction (Pagel and Harvey, 1993). In primates, it is thought that generally "slow" life histories enable the immature primate to gain valuable experience and to fine tune behavior while delaying the point at which they can engage in successful reproduction. Those primates with the largest relative brain sizes appear to have the longest juvenile periods (Harvey *et al.*, 1987; Janson and van Schaik, 1993), and spider monkeys exemplify this trend.

As is true for research on nonhuman primates generally, most studies of immatures have been conducted on cercopithecines and the apes. In *Juvenile Primates: Life History, Development and Behavior*, Pereira and Fairbanks (1993a) attempted to redress this imbalance: four of nine chapters devoted to issues of juvenility in specific primate taxa focus on Neotropical species. For spider monkeys (*Ateles* spp.), little is known about infancy, postinfant development, or the socialization process in general. Milton's research on the diet, social organization and population growth of the Barro Colorado Island (BCI) spider monkey population, made up of individuals reintroduced as immatures, is a notable exception (Milton, 1981, 1993; Milton and Hopkins, 2005). Her research supports the idea that some species- and gender-typical behaviors in spider monkeys, such as feeding on ripe fruit and the ability to discriminate between types of such fruit in terms of dietary quality, apparently occur as the result of trial and error and/or inherited tendencies. According to Milton, "the feeding behavior of the BCI spider monkeys suggests that some chemical constituents of fruits interact with chemosensory receptors to permit them to gauge fruit quality and modify their behavior in response to this information" (1993, p. 180). The social structure of the BCI spider monkeys is the typical fission–fusion type of society characteristic of *Ateles* spp. Males are more gregarious and preferentially seek out the company of other males as they mature, as with spider monkeys observed in other locales (Milton, 1993). Some differences are evident in the behavior of the BCI spider monkey group, however. Campbell's research on BCI has demonstrated that females do not have core areas. Instead, the entire group ranges together, foraging in various parts of the island, dependent on food availability, but they then split into subgroups within the different areas. Campbell (2000, 2003) attributes this pattern to the fact that there is only one group on the island, so that reduced competition among females counteracts the need to maintain core areas.

Although previous research has pinpointed important gender differences among adult spider monkeys, including greater sociability among males and a tendency for males to direct most of their aggression toward females (Fedigan and Baxter, 1984; Campbell, 2003), little is known about the ontogeny of what appear to be gender-based differences. If the assumption of adult roles is heavily dependent on or even fine-tuned through learning, one should be able to find early gender-related differences in social phenomena such as time spent in proximity to like-sexed adults or attentiveness displayed toward adult male versus adult female activities. Investigators have examined such factors in other primates; Pereira (1984), for instance, used measures of spatial proximity to quantify differential attraction toward like-sexed adults in juvenile male and female yellow baboons (*Papio cynocephalus*).

Symington (1987) reported that spider monkeys (*Ateles belzebuth chamek*) at Cocha Cashu Biological Station, Manu National Park, Peru, exhibit a larger preweaning investment in male offspring. In her study there was a longer inter-birth interval after the birth of males; infant males were carried longer than infant females; and there was a longer interval between the initiation of nursing and attempts by the mother to reject nursing. She observed no gender-based difference in postweaning investment, with juveniles of both sexes traveling and sleeping with their mothers for their first four years, after which males began associating more with the community males and females began visits to neighboring communities.

McDaniel's study of *A. geoffroyi* in troop MR at Hacienda Los Inocentes Private Reserve in northwest Costa Rica suggests that age is a better predictor of activity budgets than sex class (McDaniel, 1994). McDaniel's study also demonstrates that differences in sex roles begin as early as the juvenile phase. Juvenile males, for example, were involved in agonistic interactions approximately as often as were juvenile females but initiated more of these incidents than they received, the opposite of the situation for juvenile females. Like their adult counterparts, juvenile males directed more aggression to females and the only instance of aggression directed by a juvenile female was toward an adult female. Males engaged in most instances of olfactory investigations and marking (e.g. urine lick/sniff, genital/anal inspection, ventral rub, place sniff), sometimes exceeding the adult males in the frequency of these activities. Greetings were similar to other forms of affiliation, with adult and juvenile males exhibiting this behavior more within their own class. Grooming most often involved adult and juvenile females. With the exception of the adult female with infant class, the age/sex class receiving most grooming, all other age–sex classes initiated and received grooming approximately equally; this finding contrasted with what Ahumada (1992) found in his study: that juvenile females received grooming the most. As further evidence that higher levels of affiliation in males begin early, juvenile males played twice as frequently with members of their own age/sex classes and engaged more in social locomotion with them than was the case with juvenile females.

Contrary to the norm, McDaniel found that immatures do not always exhibit patterns reflecting those of their adult like-sexed counterparts. Unlike adult males, for example, juvenile males touched adult females most often, and juvenile females had the lowest frequencies for social resting, in direct contrast to the patterns of adult females and adult females with infants, age/sex classes that had higher levels of social rest than did adult males. On the other hand, juvenile males exhibited levels of social resting similar to that of adult males. McDaniel's data give us important information about the ontogeny of adult roles. However, the length and scope of her study did not allow examination of interindividual differences, whether based on ecology, troop structure, or personal life histories (McDaniel, 1994).

Some of Chapman's findings regarding *Ateles geoffroyi* at Santa Rosa National Park, Costa Rica, have particular relevance to understanding the development of immatures as well as risks to their survival. Adult males traveled more widely over their home ranges and were frequently seen near a community boundary, whereas adult females spent more time solitarily and seemed to modulate their use of their home range based on whether or not they had dependent infants. Females with dependent young were found in significantly smaller subgroups than females without infants. Mothers of young infants also restricted their range use and seemed to avoid community boundaries. Chapman (1990) suggests that intratroop aggression is possibly a major source of injury for young spider monkeys and, based on six occurrences of serious injury to immatures, hypothesizes that through restricted range use and avoidance of large associations, mothers with dependent young are employing a strategy aimed at promoting their immatures' survival. In contrast, Fedigan and Baxter (1984) compared females with and without infants and found females without infants to be more solitary, as did McDaniel (1994), for *A. geoffroyi* in Costa Rica, and Shimooka (2003) in La Macarena, Colombia. McDaniel (1994) also observed no aggression directed toward dependent infants. As with Fedigan and Baxter (1984), McDaniel (1994) and Shimooka (2003), our preliminary study on wild Yucatecan spider monkeys found that average subgroup size for females with dependent infants is larger than is the size of subgroups containing females with no dependent infants (Vick and Taub, 1995).

Based on Eisenberg's (1976) report of close relationships between adult females and their offspring in captivity and Izawa and colleague's (Izawa *et al.*, 1979) suggestion that juveniles seen in frequent association with adult females are their offspring, Fedigan and Baxter (1984) hypothesized that some of the subadult and adult males seen in *Ateles geoffroyi* groups at Tikal, Guatemala, are grown sons of the adult females. These authors further point out that, as with common chimpanzees, where an adult male sometimes travels with other males or with his mother and siblings, subadult and grown spider monkey males at Tikal can be found traveling or otherwise engaged in peaceful associations with particular adult females and their young. However, they were unable to assess the biological relatedness of these males and females. Shimooka (2003) found that associations between individual pairs of spider monkeys at La Macarena, Colombia, are inconsistent over time; most notably, in two cases of presumed mothers and sons, these kin-related dyads did not always associate in a predictable manner after the son became independent of his mother.

Although juvenile and subadult primates often harass copulating adults (Walters, 1987), among spider monkeys such harassment has been noted only toward the end of copulatory events. On BCI, Campbell (2006a) recorded nine cases of harassment during the 18 copulations she observed. These occurred only when thrusting had ended and the mother began locomoting with the male still intromitted. Harassment consisted of the immature lunging at the copulating pair; in all cases the copulating male retaliated by threatening the immature. Based on her observations, Campbell disagrees with van Roosmalen's assertion that if spider monkey copulatory pairs sought to mate in seclusion, they were probably attempting to avoid harassment that could be directed toward the pair by the female's associated offspring (van Roosmalen, 1985). In 16 of 18 copulations observed by Campbell on BCI, associated immature offspring (juvenile or subadult) remained in the company of their copulating mothers. Behaviors of the immatures varied from sitting in contact, to nursing, crawling on one or both members of the copulating pair, or engaging in what appeared to be play solicitation. In all but two cases, associated young were quiet throughout the copulation (Campbell, 2006a; see also Campbell and Gibson, this volume).

Due to the long socialization period for *Ateles* spp. – five or more years for each individual – long-term studies are necessary in order to follow several cohorts throughout this phase of their lives. Longitudinal studies can describe not just the "average" immature at any age point, but can also identify individual responses to different circumstances that affect development, which is very important in understanding primates, who are known for their behavioral plasticity (Pereira and Leigh, 2003). For example, spider monkey females are known to emigrate around the time of puberty (Fedigan and Baxter, 1984; Symington, 1987; Chapman and Chapman, 1990; Shimooka et al., this volume). Previous studies have not given us a clear picture of just how this happens, how much variability there is in the timing of emigration, or, indeed, if all females need to emigrate, and, if not all do, why some emigrate and others do not. We know that adult males tend to affiliate chiefly with other males (Fedigan and Baxter, 1984; Symington, 1987; Chapman, 1990; Shimooka, 2003) but have no clear understanding of the process whereby young males cease to socialize primarily with their mothers and instead seek the company of adult males nor whether preferential relations remain between these males and their mothers into adulthood, as is the case among chimpanzees (Goodall, 1968, 1971, 1986, 1990; Pusey, 1983). In fact, we still know very little about how gender-differentiated social roles develop in spider monkeys. The current study attempts to satisfy these requirements: several cohorts of known individuals with known genealogical relationships are being followed, longitudinally, in two social groups. For that reason, results presented in this chapter are necessarily preliminary, and observations are ongoing. This chapter focuses primarily on life history events, maternal investment, and association patterns.

Methods

Study site and community

The study site is located near the Maya village of Punta Laguna, in a deciduous lowland tropical forest in the Yucatecan state of Quintana Roo, about 50 kilometers inland from the Caribbean coast Maya site of Tulum (described in Ramos-Fernández and Ayala-Orozco, 2003; Ramos-Fernández, Ayala-Orozco *et al.*, 2003, 2005). Two spider monkey communities, East Group and West Group, living in the Otoch Ma'ax Yetel Kooh (OMYK) Protected Area of 5367 hectares, have been under periodic observation since 1994. Since early 1997, trained local observers have carried out continuous observations averaging about six hours per day during weekdays. The smaller East Group is better habituated, and, as a result, birth dates are known for all individuals born after late 1995. Because it is harder to find members of West Group consistently, fewer birth dates are known from this group.

Behavioral observations

Distinguishing males from females in spider monkeys is relatively easy; the enlarged clitoris is a prominent marker even in newborn females (see Campbell and Gibson, this volume). Testicular size helps to distinguish immature from mature males, and nulliparous subadult females can be recognized by the lack of prominence of the nipples and the nonflattened pelage in the nipple area. Individual recognition in this study was based on facial characteristics, pelage color, and the pattern of spots on and/or coloration of genitalia. Following Walters (1981), who determined that mothers of juvenile yellow baboons (*Papio cynocephalus*) could be identified by behavioral data, immature individuals in frequent, close, nonaggressive association with certain females, even if not observed as infants, were considered to be offspring of those females. In the field, individuals were labeled as infants if they were still nursing and/or being carried by their mothers, as juveniles if a subsequent young had been born, and as subadult or young adult if the mother had given birth to at least two infants subsequent to their own births or when males began associating primarily with adult males. However, these lines of demarcation are arbitrary and quite individually variable. Therefore, for purposes of this discussion, the term "immatures" will be used to describe females who have not yet emigrated (natal females) or given birth (immigrant females or those who remained in the group) or males who have not yet joined the adult male hierarchy or started associating chiefly with adult (natal) males.

To obtain information on association and behavioral interactions of immatures, three primary data collection techniques were used. The primary method was the "focal interval" technique (Altmann, 1974) and data were collected at 10-second intervals (Balau and Redmond, 1978). Focal subjects included all immatures and young adults in both groups, from birth up to emigration (females) or assumption of a stable place in the dominance hierarchy (males). Four trained local field assistants and myself were the principal data collectors, and we had periodic help from student assistants.

When an immature was encountered, a focal session would begin. However, if there were several immatures in a subgroup, an effort was made to observe animals that had not recently been sampled. At the end of one session, another animal would be selected. Repeat samples on an individual in any given day only occurred if an hour had elapsed since the previous sample, but few animals were sampled twice in the same day. Each focal session lasted 15 minutes or until the animal had been out of sight for at least four minutes. At the onset of each focal session, date, hour initiated, focal identity, locality according to sector or major landmark, altitude, and the identity of other animals within the subgroup (if any) were recorded, along with any other unusual observation, such as individuals traveling outside their normal range. Beginning with the onset of the focal session and at each 10-second interval thereafter, the behavior of the focal individual was noted. For social interactions, the identity of the participant as well as the directionality of the behavior were recorded. If an object was being manipulated, the identity of the object was indicated; when feeding, the identity of food type and part eaten were also recorded. At intervals of one minute, observers recorded linear distances of the immature focal subject from the mother, other sibling (if present), or if neither was present, the distance to and identity of the nearest neighbor. The basic ethogram included the following behaviors: feed, locomote, rest, groom, attack, nurse, dorsal carry, ventral carry, play nonsocial, play social, approach, leave.

Additionally, *ad libitum* diary type entries were made of infrequent or "special events" such as births, aggression, emigration or immigration, intertroop encounters, developmental milestones, or any unusual occurrence, and these data serve as a source of information on trends of associations within spider monkey groups. These "all occurrences" data collected between 2003 and early 2006 have been used in the statistical analyses, although *ad libitum* data recorded between 1998 and 2005 are used for descriptive purposes.

Data analysis

To date, it has only been possible to analyze a preliminary subset of the very large set of focal data collected as part of our longitudinal study. The part of the

Table 11.1 *Sex ratios at OMYK (1997–2006)*

East Group	01/97	12/99	12/02	1/06	West Group	01/97	12/99	12/02	1/06
Males	6	6	6	11		12	13	15	13
Females	13	10	14	13		22	28	25	27
Sex ratio, M:F	1:2.17	1:1.7	1:2.33	1:1.18		1:1.83	1:2.15	1:1.67	1:2.08

Table based on Ramos-Fernández *et al.* (2003) and more recent project data.

data set that has been analyzed includes both male and female focal samples as well as focal samples representing all age groups from the newborn period to young adulthood (1618 samples). Complete longitudinal data from birth to emigration or entry into the adult hierarchy do not yet exist for individual animals in our study population. Therefore, association patterns for immatures necessarily are based on a cross-sectional methodology. Primary nonparametric statistical techniques included the Mann–Whitney U Test and one-sample and 2×2 Chi Square Tests. All tests were carried out using the computer software SPSS version 14.0 for Windows.

Results

Community and group size

At OMYK, community size has shown a slight increase over time, ranging from 19 in East Group and 34 in West Group in January 1997, to 16 and 41 in December 1999, to 20 and 40 in December 2002, to 24 and 40 in January 2006, respectively (Ramos-Fernández and Ayala-Orozco, 2003). Based on line-transect surveys conducted between September–December 1995 and March–October 1997, 21 subgroups were sexed at OMYK, yielding a 1:2.6 sex ratio (Gonzalez-Kirchner, 1999). However, based on counts of known individuals, our study yields a sex ratio less biased toward females as compared with other locales (Carpenter, 1935; Coehlo *et al.*, 1976; Klein and Klein, 1977; Cant, 1978; van Roosmalen, 1985; Symington, 1987; Chapman, 1990; Ramos-Fernández *et al.*, 2003). The sex ratio in the two groups has also varied slightly over time, ranging from 1:2.2 in January 1997 in East Group to 1:1.2 in East Group as of January 2006 (Table 11.1).

Interbirth intervals

Symington (1988) found that the interbirth interval for *Ateles belzebuth chamek* at Cocha Cashu varied according to infant's gender, with a significantly longer interbirth interval following the birth of males: 36 versus 29 months. We have little evidence to support the notion of a gender bias in interbirth interval at OMYK. In our initial report, based on eight females who had at least two births between January 1997 and December 2002, we calculated a 32 month mean interbirth interval (Ramos-Fernández *et al.*, 2003). We now have been able to increase this pool of females who have had two births to 20, and the interbirth interval calculated on these yields an overall mean of 36.6 months, with a range of 24 to 48 months. This interbirth interval at OMYK is consistent with those reported from other sites, which approximates 36 months (Eisenberg, 1973; Milton, 1981; Chapman and Chapman, 1990; Campbell, 2000). Considering all births, there is little difference between interbirth intervals following the birth of female versus male offspring: 37 months versus 36 months, respectively. There does seem to be a gender-bias difference between the two communities, although this could possibly be an artifact of small sample size. Considering only East Group, the disparity in interbirth intervals is greater following male births: 36 months versus 31 months following the birth of females (this is the same trend observed by Symington, 1988, although the disparity is not as great). Because it is difficult to observe/record a stillbirth or neonatal death, especially in the less habituated West Group, it is possible that East Group interbirth interval data (where perinatal deaths would be much harder to miss) more accurately reflect the trend for this population; these data suggest a slight bias toward maternal investment in male offspring. However, the range in interbirth intervals overall, utilizing data from both groups, appears to be within the normal range for the species. Moreover, the two shortest interbirth intervals (which followed births of females) come from a single female in East Group. Thus, the shortness of these intervals may be as much a reproductive characteristic particular to one female, as any indication of heavier maternal investment in males. If this female's data are omitted, the interval is still slightly longer in East Group after the birth of males: 36 versus 33.6 months.

Age at weaning, traveling independently

The exact time when an individual is "weaned," that is when the immature no longer nurses, is difficult to determine since it is not always possible to ascertain if an infant whose head is at its mother's breast is in fact nursing. As pointed out by Symington (1987), attempts at rejection are easier to observe. Although most offspring at Cocha Cashu are weaned by age two years (Symington, 1987), very preliminary data from OMYK suggest not only that there is considerable variation in ages at completed weaning but also that there is no clear pattern relative to greater maternal investment in one sex. For example, the shortest interbirth interval at OMYK, which is also the earliest weaning time (since mothers only nurse their newborn), is slightly less than 24 months. In contrast, a male from East Group, TN, was observed nursing as late as 31 months. TN's mother (VE) directed no aggression at TN despite the fact that she gave birth to a subsequent sibling seven months and eight days later. Since the gestation length for spider monkeys is 7 to 7.5 months (Eisenberg, 1973; Milton, 1981; Chapman and Chapman, 1990; Nunes and Chapman, 1995), VE may have already been pregnant while TN was still nursing. A West Group female infant, RS, attempted to nurse when she was 31 months old, but her mother prevented this by chasing her twice, and then moving off. RS's mother did not give birth until 17.5 months later. Maternal rejection of infant attempts to nurse begins during the second year and is observed as early as 14 months. For example, KA, older sister of the late-weaned TN, was observed jumping up and down and squealing when denied access to the breast at 14 months, although her mother did not give birth again until 17 months later.

Infants move away from their mothers and explore the nearby environment, sometimes including other individuals, before they are weaned. For example, in March 2001 in East Group, CH's four-day-old infant reached out to an adult male (BE) who was sitting nearby, then began climbing on him. After several months, infants are often off their mothers as they rest or feed. During these initial independent forays, infants appear to be sampling the environment, learning about foods, other animals, and even the humans observing them. For example, the 7.5-month-old TN was observed more than a meter from his mother, who was feeding in a subgroup that contained adult males. After returning to his mother's back, he reached out and chewed on leaf tips. At 6.5 months, infant female LC (West Group) spent several minutes off her mother in solitary play. As she swung back and forth, holding onto a vine, she appeared to be trying to threaten observers (orienting toward us, with what appeared to be an open mouth threat).

Although infants appear to initiate independent locomotion at an early age as they explore their immediate environment, mothers initially carry them whenever they move; later, mothers carry infants during long-distance travel or when open spaces need to be navigated. As young spider monkeys begin to locomote independently and spend more time off their mothers, they increasingly need to communicate with their mothers in order to be carried long distances or over open spaces. Mothers emit "an invitation to carriage" by giving a series of high, soft twitters, which brings babies back to them so that they can climb aboard to ride or to bridge. At other times mothers approach infants and posture or touch the infants, so that the infants can climb on.

Infants are carried into their second year and there appears to be considerable variability with regard to when mothers refuse to carry their infants, or the time

Table 11.2 *Age at emigration for natal females*

Group	Monkey	Emigration date	Age in yrs. at emigration	Age in months at emigration	Emigrated before 2nd subsequent sibling born	Still in natal group?	Gave birth in natal group?
E	KA	3/24/2003	4.97	59.7	Yes	No	No
E	LA	8/15/2003	5.3	64.1	Yes	No	No
E	EQ	10/30/2003	4.9	58.5	Yes	No	No
E	PI	12/30/2003	5.8	69.8	Yes	No	No
W	LC	4/30/2004	6.3	75.5	Yes	No	No
W	MN	8/28/2004	5.7	68.4	Yes	No	No
W	RS	5/31/2004	5.5	66.5	Yes	No	No
W	FC	9/20/2004	5.9	70.3	Yes	No	No
W	OC	–	–	–	No	Yes	Yes
W	LH	–	–	–	No	Yes	Yes
W	CQ	–	–	–	No	No[a]	Yes

[a]Female disappeared later.

when infants attempt to cross open spaces alone. For example, 16.5-month-old infant male LI was observed crossing over a dirt road; although he crossed by means of a continuous arboreal pathway, he was hesitant at first, but then crossed by himself. In contrast, 15-month-old female infant EQ was observed making "gemidos" (lost infant calls); her mother CH made a bridge for her over a large gap in the canopy and waited for EQ to approach. To do so, EQ had to make several smaller bridges on her own. This infant was still riding the mother for long-distance travel at 18 months.

Usually, when a mother pauses in her locomotion to make a bridge, the offspring uses the bridge; but at other times, youngsters take an alternative route, ignoring the maternal bridge. Mothers traveling with their offspring (infants and/or juveniles) appear to "steer" the group by sometimes forging ahead of the independently locomoting youngsters, and, by taking an alternate direction, influence the youngsters to follow.

Age at emigration

Female spider monkeys are thought to emigrate from their natal group near puberty, which for spider monkeys is about four to five years of age (Fedigan and Baxter, 1984; Symington, 1987; Shimooka *et al.*, this volume). Based on data from eight females of known age who have emigrated, ages at the time of emigration at OMYK range from 58.5 months to 75.5 months (4.9 to 6.3 years of age; Table 11.2). Emigration occurs at a slightly younger age on average in

East Group (5.2 years) than in West Group (5.9 years), although not significantly so ($U_{(4,4)} = 2.0, p > 0.05$). Similar to the situation among chimpanzees, some spider monkey females do not emigrate. In West Group, for example, three natal females did not emigrate from their natal group before their first pregnancy/birth. One of these individuals and her female offspring have since disappeared from the natal group. One adult female in East Group and two other adult females in West Group, whose origins and ages are not known, have disappeared as adults (with dependent young) from their resident group, and while death cannot be excluded, these females may have emigrated. Therefore, the possibility exists that although spider monkey females typically tend to emigrate around puberty, some females at OMYK may emigrate when older or perhaps not at all. Resolution of the variability of the phenomenon of female emigration in this population awaits more life history data.

Association patterns of immature animals

Spider monkeys are noted for their fission–fusion type of organization (see Aureli and Schaffner, this volume), which presumably represents an adaptation for successful foraging for mature fruit (Fedigan and Baxter, 1984; Symington, 1987, 1988; Chapman, 1990; Chapman *et al.*, 1995). It is a social organization quite similar to that characterizing the common chimpanzee (*Pan troglodytes*). Prevailing socioecological theory posits that subgroup size will be determined primarily by predation risk and food availability (Crook, 1972; Alexander, 1974; Clutton-Brock and Harvey, 1977; Terborgh, 1983; van Schaik, 1983; Terborgh and Janson, 1986; Chapman *et al.*, 1995). Larger subgroups offer better predator detection and protection; however, a larger subgroup requires more food to support all its members, which usually means more travel time and energy expenditure. When travel costs become too high or subordinate members are not meeting their food requirements, individuals will seek smaller subgroups.

Subgroup size of spider monkeys at Santa Rosa appears to be correlated with resource availability; however, the considerable variation observed at this site needs additional explanation (Chapman, 1990). Similarly, at OMYK subgroup size increases during times of high food availability; however, individuals in these larger groupings do not travel further than those in smaller subgroups (Ramos-Fernández, 2001). Thus the composition rather than the size of subgroups might determine whether or not particular individuals join or leave subgroups (Ramos-Fernández, 2001).

To determine if any differences based on age or gender exist in associational patterns for immatures, subgroup size and individual identification at the beginning of each focal sample were examined. Since it was not possible to

Table 11.3 *Percentage of samples where immatures aged 42–60 months were in subgroups with and without their mothers*

	Males	Females
With mothers	14	56
Without mothers	37	3

$\chi^2 = 42.538$, df $= 1$, $p < 0.01$.

collect equal numbers of focal samples on each animal, average subgroup size was calculated for each age/sex category. Across all ages, there is a small, but nevertheless statistically significant difference in average subgroup size between male and female immatures ($U_{(698,883)} = 287\ 353.0$, $p < 0.05$), when all subgroups are considered; focal males are found in slightly larger subgroups: 4.8 versus 4.5. When immatures (across all ages) were sampled in subgroups that contained or did not contain their mothers, for males, subgroups are significantly larger if mothers are also present, 5.0 when mothers are there versus 3.5 when they are not ($U_{(606,89)} = 18\ 885.5$, $p < 0.01$). For females, the difference in subgroup size when mothers are present is not significantly larger: 4.6 for subgroups with mothers versus 4.4 for subgroups without mothers ($U_{(837,34)} = 13\ 858.5$, $p = 0.794$). When only those individuals under five years of age are considered, the trend for immature males to be found in slightly larger subgroups than immature females is still visible, although not statistically significant ($U_{(539,726)} = 185\ 113.0$, $p = 0.097$), with average subgroup sizes of 5.0 for male focals and 4.8 for female focals respectively. Subgroups containing immatures under age five and their mothers are larger than those without mothers (for males: 5.1 with mothers versus 3.5 without mothers; $U_{(508,28)} = 4790.0$, $p < 0.01$; for females: 4.8 with mothers versus 3.3 without mothers; $U_{(696,25)} = 7528.0$, $p = 0.247$).

Based on the knowledge that spider monkey adult males are often separate from females (forming what Fedigan and Baxter called "two parallel 'societies'" [1984: 289]), relationships between mothers and offspring were examined by comparing the percentage of all focal samples when immatures were with and without their mothers. For males less than 42 months of age there are no samples in which young males were not with their mothers. Symington (1987) reported that at about four years old individuals of both sexes begin to frequently leave their mothers' company, males to travel with males in the group and females to visit neighboring communities. Data from OMYK indicate that male immatures exhibit the same trend ($\chi^2 = 42.538$, df $= 1$, $p < 0.01$; Table 11.3). On the

Table 11.4 *Average distance, in*
meters, from mother for immatures
under 5 years of age

Age in months	Females	Males
0–6	1.13	0.44
6–12	1.81	1.96
12–18	2.43	2.85
18–24	3.35	4.02
24–30	3.64	3.46
30–36	4.49	4
36–42	5.62	4.03
42–48	5.71	6.7
48–54	5.19	7.46
54–60	6.24	4
Average distance from mother, 0–60 months	4.29	3.62

$U = 154\ 299.5, p < 0.01.$

other hand, when immatures under five years of age are in subgroups containing their mothers, females maintain greater interindividual distances from their mothers ($U_{(497,681)} = 154\ 299.5$, $p < 0.01$), although the pattern is variable (Table 11.4).

Immature males appear to be attracted to adult males and often take the initiative in establishing contact with them, even at early ages. For example, 20-month-old HU (East Group) approached the dominant adult male PA and ventrally embraced him, and at 30 months, HU approached adult male DA, embraced him, made vocalizations, and then sat next to him. Similarly, CC and adult male AL (West Group) were observed embracing when CC was older but still juvenile (approximately three years old). Yet, when immature males and females between birth and age five years are compared, using samples where they were found in subgroups containing either no male or at least one adult male, there is no significant difference between them ($\chi^2 = 0.076$, df $= 1$, $p = 0.782$; Table 11.5). However, comparing only those data for individuals between 42 and 60 months of age, there is a significant trend for immature males to spend more time with adult males, as measured by number of samples immatures appear in groups containing them ($\chi^2 = 21.613$, df $= 1$, $p < 0.01$; Table 11.6, Figure 11.1). The trend for maturing males to gravitate toward other group males is even more noticeable when only those subgroups not containing the mother are considered ($\chi^2 = 28.450$, df $= 1$, $p < 0.01$; Table 11.7). Anecdotally, a two and a half year old juvenile male

Table 11.5 *Number of samples in which immatures under age 5 were in subgroups containing at least one adult male*

Number of adult males in subgroup	0	1 or more
Immature males	440	146
Immature females	624	200

$n = 1410$, $\chi^2 = 0.076$, df $= 1$, $p = 0.782$.

Table 11.6 *Number of samples in which immatures 42–60 months of age were in subgroups containing at least one adult male*

Number of adult males in subgroup	0	1 or more
Immature males	44	42
Immature females	205	60

$n = 351$, $\chi^2 = 21.613$, df $= 1$, $p < 0.01$.

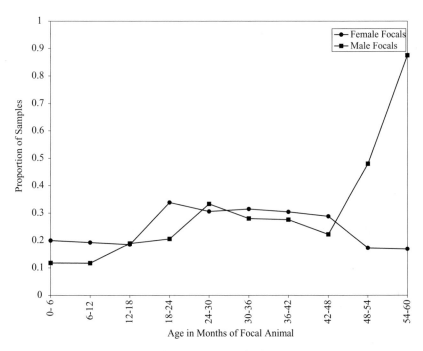

Figure 11.1 Proportion of all samples where focal animals were in subgroups containing at least one adult male.

Table 11.7 *Percentage of samples where immatures 42–60 months of age were in subgroups not containing their mothers and containing at least one adult male*

Number of adult males in subgroup	0	1 or more
Immature males	2	26
Immature females	16	3

$n = 47$, $\chi^2 = 28.450$, df $= 1$, $p < 0.01$.

PE, whose mother disappeared, frequently associated with adult males, sometimes sleeping with them at night, for several months before he himself disappeared.

Based on the analysis of a very large sample of *ad libitum* data taken by field assistants between 2003 and 2006, there appears to be a trend for immature males to be observed in more affiliative interactions with all other animals than is the case for immature females ($\chi^2 = 46.278$, df $= 1$, $p < 0.01$). In fact, the number of affiliative interactions involving immature females is less than one quarter that of episodes involving males (Figure 11.2), despite the fact that there were more immature females than immature males during this time period (10 males: 8 females in East Group and 9:24 in West Group).

Immature males also appear to direct more affiliative interactions to adult males than do immature females, mirroring the adult pattern ($\chi^2 = 19.174$, df $= 1$, $p < 0.01$; Figure 11.3). There are no observed cases of adult males directing affiliative behavior toward immature females and only four incidences of males directing affiliation toward immature males out of hundreds of hours of total observations. Consistent with the gender-based preference for affiliative associations in adults, immature females are involved in more affiliative interactions with other females of all ages ($n = 11$) than they are with males ($n = 3$) whereas male immatures are involved in more affiliative interactions with other males of all ages ($n = 45$) than they are with females ($n = 33$) ($\chi^2 = 6.255$, df $= 1$, $p < 0.05$). Although the sample size is very small for cases where direction of affiliation could be observed, the trend for interactions with adult females suggests that immature females direct more affiliative interactions toward adult females ($n = 7$) than they receive from them ($n = 4$). In contrast, for immature males, the number of affiliative interactions given to adult females ($n = 16$) and received from them ($n = 14$) is more nearly equal; however, these differences are not statistically significant ($\chi^2 = 0.347$, df $= 1$, $p = 0.556$).

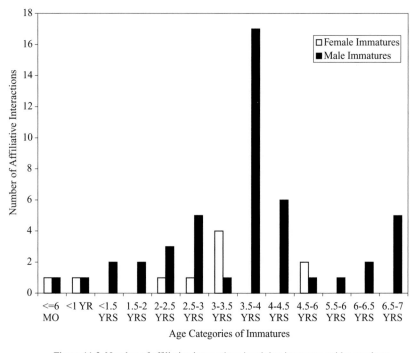

Figure 11.2 Number of affiliative interactions involving immature spider monkeys.

Affiliation between immatures and nonmothers

As noted in the previous section, mothers at OMYK typically bridge only for their own infants; however, other troop members sometimes bridge for them as well. For example, when PE was 22 months of age, a juvenile female not believed to be PE's sibling helped PE to bridge an open space. Other affiliative behaviors may be aimed at comforting a distressed youngster and/or helping it find its mother: at age 26.5 months LI was emitting "gemidos" as he followed in the general direction that his mother had taken a short while earlier. At first an unidentified juvenile stayed close by LI as he vocalized; then the juvenile began traveling and LI followed. Eventually LI stopped vocalizing.

Nonmaternal group members sometimes carry infants for short distances. Usually the infant approaches the troop member and climbs on that individual's back or shoulders. Often the individual is a sibling; at other times nonrelatives perform the carrying. For example, while his mother was about two or three meters away, 7.5-month-old infant male TN was observed being carried by, first, AR (his 68-month-old brother; at this age, AR would be considered a subadult or young adult since he spent much of his time with the older adult males) and then

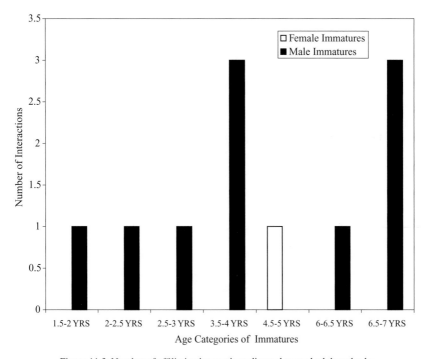

Figure 11.3 Number of affiliative interactions directed toward adult males by immature spider monkeys.

by BE, an unrelated adult male. Between 2003 and early 2006, nine episodes of infant carrying by nonmothers were recorded, of which five involved females as carriers (3 carriers were juveniles and 2 were adults). One of these adult female episodes involved the resident adult female CL who carried male HU, who was almost 36 months old. HU was also carried by subadult/young adult female HE when HU was 25 months old; HE was in the process of gaining acceptance as an immigrant into East Group. The other four "allomother"-carrying episodes involved males as carriers. Two were juvenile males; two were adults. The adult males only carried immature males, although one immature was carried at an unusually advanced age: 51 months. One juvenile male carried an infant female during two episodes.

Interestingly, of the five episodes when juveniles were carriers, none carried their own siblings. Two of these five juvenile-initiated carrying episodes are particularly interesting. In the first, female immature LA, at almost five years of age, approached 17-month-old HU, who was approximately eight meters from his mother and was persistently screaming. LA allowed HU to climb on her back; then she carried him to his mother. A second juvenile carrying episode is interesting because of the closeness in age of the two participants: the juvenile female CO, at 28.5 months, carried the 18-month-old female infant JS for a short distance.

Play

Subgroups in which play is observed are often of larger than average size (Vick and Taub, 1995). Episodes of play involve immatures of various ages, as in the playgroup observed on March 22, 2000, which contained 36-month-old male LI, 54-month-old female MI, 2-year-old female KA and 15-month-old female EQ. While playing, youngsters usually make soft and throaty "chuckle" vocalizations although sometimes play is silent. Play most often involves play chases and fights. While playing, individuals often dash along broad semihorizontal limbs making leaps into leafy branches below. Sometimes a hollow filled with water is utilized, the monkeys dashing back and forth and splashing water. Although play usually occurs arboreally while adults are resting nearby, at OMYK, play also occurs on the ground, sometimes involving several individuals. One ground-based "game" involves several individuals in turn climbing rapidly down one tree, racing across the open ground, and then rapidly climbing up another tree, then locomoting arboreally to the first tree and beginning the cycle again. To date this play has only been carried out by juveniles. Despite the fact that terrestrial behavior in spider monkeys is generally associated with anxiety, characterized by frequent scans and taking long periods of time to descend to the ground, terrestrial players at OMYK seemed relatively unwary. Terrestrial play at OMYK is often accompanied by adult males acting as sentries, as is the case on Barro Colorado Island, Panama (Campbell *et al.*, 2005). At OMYK, adult females also appear vigilant during episodes of terrestrial play and sometimes threaten observers by branch shaking. Play invitations can be quite casual, as when a mother grabs her infant by its leg and then plays with it, causing the infant to make the "chuckle" vocalization. Another invitation consists of slapping the tree, then backing up, lying down, then racing forward, backing up, etc., until the invitee reciprocates. Although play usually occurs during the day, the play "chuckle" is very distinctive and is sometimes heard at night, particularly when the moon is full.

Intratroop aggression and the risk to immatures

Injury as the result of intratroop aggression appears to be a risk for maturing young at OMYK. From 1996 to 2000, all offspring born into East Group survived. In June 2000, PA, the dominant adult male of East Group, attacked a mother–infant pair, when the infant was two weeks of age. The infant sustained lethal wounds, although the mother was uninjured (Vick *et al.*, 2001). From subsequent known birthdates, it is clear that this mother subsequently conceived, approximately six weeks after the attack and death of her infant. Also, in early 2001 two other infants, sex undetermined, were observed with severe wounds, and they subsequently died within days. The research team thinks it likely these deaths too were the result of intratroop aggression since both were very young infants, dependent on mothers, and unlikely to travel far from them. Whether or not infants were targeted for attack is uncertain (Vick *et al.*, 2001). However, our data support Chapman's (1990) conclusion that intratroop aggression can cause injury, and even death, to immatures and poses a threat to a spider monkey mother's successful rearing of her offspring. Fedigan and Baxter (1984) have also reported unprovoked attacks on mothers with infants (see also Shimooka *et al.*, this volume).

Analysis of *ad libitum* data suggests a significant intergender difference wherein immature males, over all ages, are involved in more aggressive interactions than are immature females ($\chi^2 = 24.923$, df $= 1$, $p < 0.01$); this difference is also true when only those individuals under five years of age are considered ($\chi^2 = 10.667$, df $= 1$, $p < 0.01$) (Figure 11.4). Males also initiate more aggression than do immature females, beginning at young ages (Figure 11.5). However, there is no difference in the amount of aggression directed by adult males toward or submissive behavior received from immature males and females at young ages ($\chi^2 = 0.000$, df $= 1$, $p = 1.000$; Figure 11.6). Immature males direct little aggression toward adults, and, then, only toward adult females. There were seven such episodes for animals younger than six years; in two of these cases the young males were in the company of an adult female. Only seven episodes were recorded where adult females directed aggression to or received submission from immature animals (females [5] versus males [2]); only three involved immatures known to be under age five years (females [2] versus males [1]) (Figure 11.7).

Maternal defense and long-term associations between mothers and offspring

As with most primates, spider monkey mothers come to the defense of their immature offspring (Symington, 1987; Chapman, 1990). At OMYK, mothers intervene on behalf of both their male and female offspring, frequently triggered by infant squeals in the context of rough play. In such cases, maternal

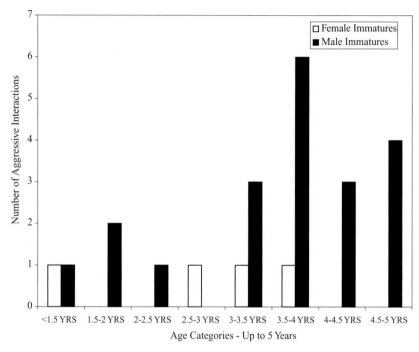

Figure 11.4 Number of aggressive interactions including immature spider monkeys of varying age and sex.

interventional aggression is usually directed toward an older juvenile who is playing roughly with her infant. Aggression sometimes escalates if the targeted juvenile is unrelated and the recipient's own mother in turn comes to its defense. These defensive and counteraggressive episodes are usually of short duration and consist of harsh vocalizations, open mouth threats, lunges, and sometimes cuffs or slaps. Defense of young sometimes also occurs in feeding contexts, when an unrelated adult female threatens an immature and its mother comes to its defense. Females most often support their young against attack by other adult females but will also sometimes challenge aggressive adult males in defense of their offspring.

Intuitively it is expected that mothers will defend their young offspring. But how long does intervention by mothers on behalf of their older immatures continue? At OMYK mothers have intervened on behalf of subadult and young adult offspring. For example, two adult males were attacking VE's adult son DA when VE inserted herself into the middle of the fray, interrupting the attack. Another female, CH, tried to defend her subadult son JO when he was almost

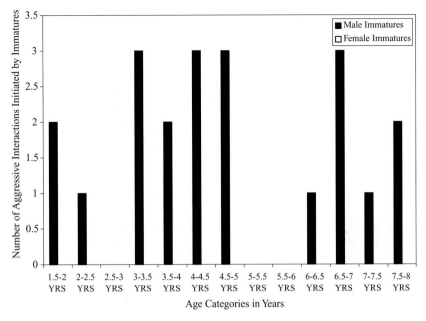

Figure 11.5 Number of aggressive interactions initiated by immature spider monkeys of varying age and sex.

six years of age during a lethal attack by adult males (Valero *et al.*, 2006). It appears that adult females on occasion will risk serious injury in order to protect their offspring even after their offspring reach maturity.

Conversely, when under attack by adult males, both juvenile males and females at OMYK have been observed defending their mothers. When he was 45 months old, TN tried to defend his mother VE when adult male PA attacked her; PA's associate, adult male BE, then attacked TN. When the female immature LO was 40 months old, she tried to defend her mother CH who was being attacked by adult male LI, resulting in LO being attacked in retaliation. During that same episode, the unrelated immature male TL, who was 43 months old, tried to defend both LO and CH.

OMYK data support Fedigan and Baxter's contention that some of the subadult and adult males observed in close association with females are probably their own offspring (Fedigan and Baxter, 1984). Even after subadult/young adult males begin traveling and interacting primarily with adult males, they still are often seen traveling or foraging with their mothers. In June 2004, for example, LI joined his mother FL and rested close to her although he was over seven years old at the time.

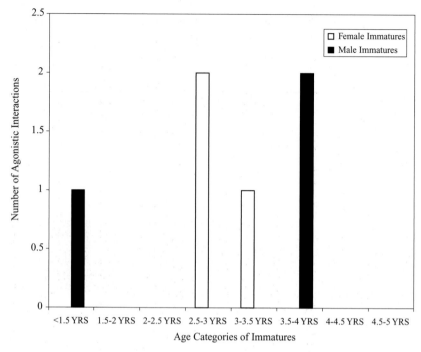

Figure 11.6 Number of agonistic interactions between immature spider monkeys and adult males.

Attaining a position in the male hierarchy

In East Group, only one young male has firmly established himself in the adult hierarchy; this individual (BE) appeared to be a subadult in 1995, and, for months, persistently tried to approach and follow the adult males while making high pitched infant-like vocalizations. Notably, BE's presumed mother, BI, sometimes followed her son, also vocalizing, as he followed after the adult males. Three males were born in East Group between the time our research began and 2000, and none of these succeeded in becoming established in this group. One of these unsuccessful males, JO, was repeatedly attacked by at least two of the resident adult males over a period of several months and was ultimately killed by troop males when he was almost six years of age, in 2002 (Valero *et al.*, 2006). Several months later, the partially dismembered body of AR, a young adult male almost seven years old, was discovered when it was dragged by village dogs into the outskirts of Punta Laguna, located nearest the core area of East Group. AR was one of the individuals who came to the ground most often; however, most of the researchers consider it unlikely that he would

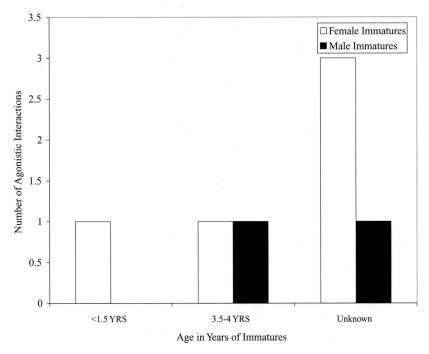

Figure 11.7 Number of agonistic interactions between immature spider monkeys and adult females.

have been unwary enough to suffer an attack by dogs. Moreover, since village dogs seldom range far from the village, it is considered probable that the male spider monkey died close to the village, making it less likely that his death was the result of predation by a large felid or other native predator. Death due to aggression by adult males is therefore a strong possibility in both cases, and male reproductive competition has been suggested, since the operational sex ratio (number of males: number of fertilizable females) at the time of the attacks was 5:0 (Valero *et al.*, 2006). Before and after the demise of AR, adult male DA (a subadult in 1995) suffered repeated attacks by adult males, sometimes sustaining severe wounding, and in the spring of 2005, he too disappeared. When he disappeared, the operational sex ratio was 4:1. Finally, LI, at age eight years, disappeared after Hurricane Emily in 2005, the only individual who was not observed again after the hurricane; due to DA's disappearance, the operational sex ratio at the time of LI's disappearance was 3:1. Although his death may have been hurricane-related, prior to this event, he had been the recipient of attacks by the only two surviving adult males in East Group, PA and BE.

In contrast to the situation in East Group, in West Group five young males appear to have established themselves in the hierarchy: EN, FI, CD, BC, and CC. Although their birthdates are unknown, at least three were considered to be infants in 1997. Beginning in 1997, the resident adult males of West Group disappeared over time, perhaps allowing easier access to the adult hierarchy by this coterie of young males. It is not possible to accurately determine the operational sex ratio for West Group when these males joined the adult male hierarchy, since it was not always possible to determine exact dates of birth or disappearance; moreover, this ratio has varied over time. However, when one compares the average sex ratio in the two groups, based on four sampling points between January 1997 and January 2006 (Table 11.1), there were slightly more females as compared with males in West group, overall (1:1.8, East Group; 1:1.9, West Group).

Discussion

Community and group size

At OMYK, the sex ratio, which has ranged over time from 1:2.3 to 1:1.2 (Table 11.1), is slightly less biased toward females than has been reported at many spider monkey sites (see review in Chapman *et al.*, 1989). Chapman compared sex ratios of spider monkeys in various locales, noting that the bias toward females was smaller in areas typified by a ready supply of water. In accord with the resource competition hypothesis, wherein the number of males should be limited in order to avoid or reduce intraspecific competition in those species in which females disperse, these authors further suggested that the greater the competition for postweaning resources, the stronger the bias would be toward females (Clark, 1978; Silk, 1983; Chapman *et al.*, 1989). At OMYK, the home ranges of both East and West groups contain areas of "monte alto" or semi-evergreen old growth forest, with trees up to 25 meters in height, located chiefly near the large lakes, as well as areas of "kelenché," 30–50 year old successional forest, with trees up to 10 meters and containing some valuable food sources (Ramos-Fernández and Ayala-Orozco, 2003; Ramos-Fernández *et al.*, 2003). Access to the "monte alto," especially, may aid in supporting the more balanced sex ratio observed at OMYK. As has been demonstrated in a number of primate species, social behavior can vary depending on environmental conditions, including not only ecology, but also social structure. The relative reduction in number of breeding females, as evidenced by a more even sex ratio, such as exhibited in East Group, may have set the stage for what appears to be more intrasexual competition in that group.

Interbirth intervals, age at weaning, traveling independently

The theory that there should be greater parental investment in the sex with greater variance in reproductive success (in our case, spider monkey males) is not supported by data from OMYK. Despite the slight trend in East Group toward a longer interbirth interval after the birth of males, the mean interbirth intervals, after the birth of males versus females (both groups combined) are very similar: 36 months after the birth of males, 37 months after the birth of females. Symington (1987) reported that at Cocha Cashu, high-ranking spider monkey mothers carry infant males three months longer than they carry female infants; they discourage nursing attempts by infant daughters four months earlier than for male infants. At OMYK, a dominance hierarchy among females is hard to discern and does not seem to be an important phenomenon. Moreover, preliminary data on age at weaning and for traveling independently show interindividual variability, rather than according to sex (also see Shimooka *et al.*, this volume).

Age at emigration

In both East and West Group, there is much interindividual variation in age at emigration. Notwithstanding this variability, at OMYK, all emigrating females of known age remained in the natal group until several years after the birth of the emigrant's first younger sibling, but all subsequently emigrated before the birth of the second sibling. What might account for this timing-of-emigration phenomenon?

Social factors have been proposed as proximate causes of emigration (Harcourt and Stewart, 1981; Pusey, 1983, 1990; Crockett and Janson, 2000), although for OMYK females, there are not yet sufficient data to determine if characteristics of the mother or of mother–daughter relationships make a difference in whether and/or when to emigrate. For example, among Gombe chimpanzees, some daughters of high-ranking females have not emigrated from their natal groups. Yet, even in this well-studied population, it is unclear what factors enable a female to stay within the natal group. Factors such as mother's rank, mother's range within the group, ecological variables, the possibility of finding a group into which to immigrate, the density of resident adult females in the natal and neighboring communities, and the aggression likely to be received upon staying or transferal have all been suggested (Pusey, 1983, 1990).

However, other factors may also be at work in spider monkeys. If females profit in terms of future reproductive success by observing mothering skills prior to emigration, staying with their mothers during the entire course of at least one sibling's birth and subsequent postbirth periods of nursing and maternal transport would be beneficial, a phenomenon well documented to occur among marmosets (Epple, 1975; Hoage, 1978; Goldizen, 1987). Moreover, these young females theoretically would hone additional life skills, such as more efficient foraging, predator avoidance and interpersonal skills, than if they emigrated earlier. OMYK spider monkey females appear to spend most of their time near their own mothers prior to emigration. Could the impending second birth trigger more feeding competition in the natal unit? At OMYK both male and female immatures exhibit an increase, over time, in distance from mother while in the same subgroup, and perhaps this is a consequence of more feeding competition. But does increased feeding competition trigger increased separation and/or emigration? Pusey examined this possibility in Gombe chimpanzees. Although she saw more aggressive interactions over food among mothers and offspring than among adults, she thought that food competition was unlikely to have triggered separation since this usually started during the mother's consortships or when daughters began breeding (Pusey, 1983, 1990).

West Group females emigrate slightly later than females of East Group, and some West Group females did not emigrate at all (Table 11.2). Data on ecological differences between the two groups are not yet sufficient to assess their effects on emigration patterns, although the home range for West Group is larger: 1.66 km^2 (0.29 km^2 medium forest and 1.37 km^2 minimum successional forest) compared with East Group's 0.95 km^2 (0.29 km^2 medium forest and 0.66 km^2 minimum successional forest) (Ramos-Fernández and Ayala-Orozco, 2003, p. 8). A consequence of emigration is that it serves to avoid inbreeding: a larger number of adult males in West Group may have lessened the potential for inbreeding in that group (also see Shimooka *et al.*, this volume).

Association patterns, travel, play

During the time immatures are dependent, their mothers primarily determine their associations with other group members. The composition and size of the subgroups in which they are found, for example, are usually presumed to be due to factors such as food availability and the degree of predator protection afforded. Although it does not appear to be the case at OMYK (Vick and Taub, 1995), Chapman (1990) reported that in the populations he studied, mothers avoided large subgroups, possibly in order to safeguard their young. It is probable that multiple factors determine how mothers choose whether or not to join or leave a subgroup, not simply based on its size or in order to promote the safety of their offspring. Important social factors such as the need for male offspring to establish relationships with adult males probably also play some role. For example, the fact that OMYK immature males, while in subgroups containing their mothers, are found in slightly larger subgroups than is the case for immature females may be a consequence of such a choice. Moreover, dependent immature individuals might actually play a more active role than is generally thought, especially as they become older. Immature male chimpanzees appear to be attracted to other troop members and often lead their mothers toward other subgroups (Pusey, 1983). At OMYK, a similar phenomenon occurs, as juveniles sometimes appear to be "leading" their mothers as they progress ahead of them during travel.

Data from OMYK indicate that the pattern for strong male–male bonds and greater affiliative rates for males begins early in spider monkeys, with young males displaying more affiliative behaviors overall, directing more of their affiliative interactions toward adult males, and sometimes taking the initiative in establishing contact with adult males. Since spider monkey males remain in their natal group and must join the established hierarchy therein, such interactions probably pave the way for future success.

That immature females are involved in fewer affiliative interactions with others, including adult males, also reflects the species-typical adult pattern. Since females usually emigrate, the degree of genetic relatedness between group females is thought to be less than that among resident males. Moreover, adult females must compete for scattered resources, and, like female chimpanzees, often feed in small subgroups or alone. Interactions between adult males and females are often not friendly, with males directing most of their aggression against adult females (Fedigan and Baxter, 1984; Campbell, 2003; see also Aureli and Schaffner, this volume).

Although immature females direct more affiliative interactions toward adult females than they receive, the interaction is more nearly equal for immature males. This gender bias in affiliative dynamics may reflect the fact that these same males will play a large role in natal-troop interactions in the future. Eventually, they will dominate the adult females and, in some cases, even serve as potential mates, whereas young females will eventually leave the troop. Some young males have also been observed coming to the aid of adult females. In line with the fact that immature males stay in the troop and must join the hierarchy, whereas females usually emigrate, the tendency for young males to associate with adult males increased over time. Association with adult males probably serves not only as a training ground for the future but also provides protection for the juveniles; in the case of the juvenile male whose mother disappeared, association with the adult males may have permitted his survival for the short time he lived in the troop after his mother's disappearance. Since adult males are thought to cooperate to defend their females from the males of other groups, association with and protection of younger males may help build future coalitions. For young females, the pattern of association with adult males over time is more variable but eventually decreases, as would be expected since females generally leave the group before breeding.

Allomaternal care, such as bridging, carrying and sitting with infants, is not common among spider monkeys at OMYK, but our limited set of data ($n = 9$ cases of carrying) does not show a gender-based bias in the distribution of allomaternal care by nonmothers. Both female and male juveniles play with and sometimes carry infants, although no juveniles were observed carrying their own siblings. At least one case, where the juvenile female CO at 28.5 months carried the 18-month-old infant JS, can perhaps be interpreted as practice for adulthood. The incident in which the adolescent or young adult female HE carried the 25-month-old HU can be interpreted as practice or as an attempt at bonding with the resident females, since HE was in the process of immigrating into the group, or even with a potential mate (HU). Adult males carried infants twice, both times male immatures. One of these incidents involved a younger sibling. Instances involving nonsiblings may represent attempts to establish relationships with the mothers of the immatures or even as a way of establishing relationships with future colleagues.

Although much has been written concerning the potential functions of play (Poirier and Smith, 1974; Martin and Caro, 1985; Meaney et al., 1985; Fairbanks, 2000), ranging from improving motor and/or neuromotor development, to practice of social skills, it is also true that under extreme ecological conditions play can drop out of the social repertoire (Baldwin and Baldwin, 1974; Smith, 1978; Lee, 1984). Data to date from OMYK indicate that the habitat provides sufficient food and safety from predators for frequent play, large playgroups, and even occasional terrestrial play to occur. We are fortunate to have a "natural experiment" regarding the role of the habitat on immature play: play data collected after the two major hurricanes in 2005, Emily and Wilma, will allow comparison of play behavior before and after the devastating effects to the OMYK forest of these storms.

Intratroop aggression and the risk to immatures, especially males

The species-typical pattern of aggression, with adult males exhibiting most of it, begins early. Immature spider monkey males appear not only to be involved in more aggressive interactions, but also to initiate more aggression than do immature females. The limited data to date do not indicate a differentiation by adult males in the aggression they direct toward immature males versus females, although adult females do appear to direct more aggression toward immature females. These young females – at least until emigration – will be likely competitors for resources, and the young males will soon be dominant to them. At Hacienda Los Inocentes in Costa Rica, McDaniel (1994) was able to discern a dominance hierarchy based on the frequency of mixed-class dyadic agonistic interactions. In her study, juvenile males ranked below adult females with infants and above the adult female class, whereas juvenile females ranked last.

Proximate explanations of known and possible cases of infanticide at OMYK must remain tentative due to small sample size. Within the span of about one year, three out of five infants in East Group died as the result of wounding. The perpetrator in the observed case was the most dominant male and may well have been the infant's father. What might account for the increase in violent deaths in such a short period at OMYK? Even if the infants' deaths were only an accidental result of aggression aimed at the mothers, the spike in such serious aggression, especially after a four-year interval with no such deaths, requires some scrutiny. It is possible that, in the observed case of infanticide, the infant male was not sired by the dominant male since a brief "invasion" by adult males from West Group occurred approximately seven to seven and a half months prior to the infant's birth. Infanticide by chimpanzee males has sometimes been associated with infants of females living on the group's periphery, infants who were likely to have been fathered by males from a neighboring group (Goodall, 1983; Nishida and Kawanaka, 1985; Fruth *et al.*, 1999). On the other hand, the increased number of late adolescent/young adult males in East Group at the time may have stimulated more competition for fertilizable females. Certainly, one of these young males – later killed by the adult males (Valero *et al.*, 2006) – was observed mating with troop females both before and after the observed case of infant killing.

It is possible that the two presumed cases of infant mortality consequent to wounding were the result of aggression by females and served to improve their own infants' survival through elimination of potential competitors. Chapman (Chapman *et al.*, 1989, p. 317) suggested that "Females can limit the number of males in a community by either adjusting the sex ratio before birth (Clark, 1978) or by decreasing the probability that other females in the community can successfully raise male offspring (Silk, 1983)." Citing Symington's (1987) study at Cocha Cashu Biological Station, Manu National Park, where the sex ratio at birth was female biased, and his own research at Santa Rosa as well as that of Fedigan and Baxter's study (1984) at Tikal National Park, where immature males were approximately twice as likely as immature females to receive aggression, Chapman *et al.* (1989) hypothesized that both phenomena may be important in reducing the number of adult males compared with females in spider monkeys. Female aggression does not seem a likely explanation for the infant mortality seen at OMYK. OMYK females, the most intensely observed adult animals in this study, have not been observed inflicting serious wounds. Moreover, at the time of the fatal wounding incidents involving infants, neither the troop size nor the number of potentially competing infants was larger than in the past.

Reproductive competition may also be implicated in the death or disappearance of young adult males in East Group; all of the males born between 1995 and 1997 died or disappeared, and all were sometime victims of aggression directed toward them by resident adult males. Similarly, Campbell (2006b) reported three cases of coalitionary aggression directed toward subadult males by adult male spider monkeys (*A. geoffroyi*) on Barro Colorado Island, Panama. These cases are thought to have led to the deaths of two of these young males.

It is generally thought that individuals of the emigrating sex face greater challenges to survival than individuals of the philopatric sex. At OMYK, immigrant females receive aggression from resident adult females. However, OMYK data suggest that although adult male spider monkeys affiliate more closely than do members of other age/sex classes, cooperatively guard troop females and territory, and are thought to be closely related (Fedigan and Baxter, 1984; Symington, 1990; van Roosmalen and Klein, 1988), young immature males still face many perilous challenges.

Maternal defense and long-term associations between mothers and offspring

As appears to be the case in Santa Rosa (Chapman *et al.*, 1989), injury poses a substantial risk for immatures at OMYK. Moreover, adult females sometimes risk serious injury to themselves in order to protect their offspring, even those offspring who have reached maturity, which would have the consequence of protecting these mothers' own reproductive interests. However, these females may be doing more than that, since both juvenile males and females have been observed defending their mothers when under attack. A protective mother may also be safeguarding a coalition, which can help her in feeding competition with other individuals, especially other resident-group adult females. Such mother–adolescent and mother–adult daughter interactions have been observed in Gombe chimpanzees (Pusey, 1983). Pusey also suggests other benefits of staying together among chimpanzees, which might also be important for spider monkey mothers: finding food for the mother; providing grooming; playing with younger siblings and thus allowing mothers more time for feeding or other social interactions; and possibly even adoption, in the case of a mother's death. In spider monkeys, the presence of older offspring may also help deter aggression directed toward younger offspring or provide other aid and thus be a factor in promoting a mother's overall reproductive success. In addition to coalitionary support during agonistic encounters, frequent association with mother and siblings may provide palliative aid to wounded offspring. Chapman and Chapman (1987) for example, report a case where a spider monkey mother provided care to a severely wounded, already weaned juvenile male, resuming nursing and carrying him over an extended period of time. At OMYK, the adult male DA sustained a severe wound to his knee, presumably as a result of male–male aggression; in this case, his 38-month-old sister KA was observed sitting with him and grooming his face.

Conclusions

Despite the work that has been conducted on spider monkeys in recent years, there is much yet to learn about the period of immaturity in this genus and how developing individuals are channeled into their appropriate adult social roles. Data from OMYK and other sites, however, do allow us to draw two important conclusions. First, sex role differentiation begins at an early age, and, throughout their immature years, spider monkeys are not just preparing for future roles. They are hard at work, negotiating a difficult ecological and sometimes perilous social environment. Second, immaturity for spider monkeys is a time of risk, especially as individuals make the transition into adolescence and young adulthood. Such risks prevail, no matter whether the individual must leave its natal group, as with females, and make the difficult transition into a new group or whether the maturing male must try to establish a secure place in the adult male hierarchy of his natal group. In fact, data from OMYK suggest that the transition from subadult to early adult status appears to present special risks to maturing males from resident adult males, especially when reproductive competition is high.

References

Ahumada, J. A. (1992). Grooming behavior of spider monkeys (*Ateles geoffroyi*) on Barro Colorado Island, Panama. *Int. J. Primatol.*, **13**, 39–49.

Alexander, R. D. (1974). The evolution of social behavior. *Annu. Rev. Ecol. Syst.*, **5**, 325–383.

Allman, J., McLaughlin, T. and Hakeem, A. (1993). Brain weight and life-span in primates. *Proc. Natl. Acad. Sci. USA*, **90**, 118–122.

Altmann, J. (1974). Observational study of behavior: sampling methods. *Behaviour*, **49**, 227–267.

Armstrong, E. (1983). Relative brain size and metabolism in mammals. *Science*, **220**, 1302–1304.

Balau, J. and Redmond, D. E., Jr. (1978). Some sampling considerations in the quantification of monkey behavior under field and captive conditions. *Primates*, **19**, 391–400.

Baldwin, J. D. and Baldwin, J. I. (1974). Exploration and social play in squirrel monkeys (*Saimiri*). *Am. Zool.*, **14**, 303–315.

Boinski, S. (1988). Sex differences in the foraging behavior of squirrel monkeys in a seasonal habital. *Behav. Ecol. Sociobiol.*, **23**, 177–186.

Campbell, C. J. (2000). The reproductive biology of black-handed spider monkeys (*Ateles geoffroyi*): integrating behavior and endocrinology. Unpublished Ph.D. thesis, University of California, Berkeley.

Campbell, C. J. (2003). Female directed aggression in free-ranging *Ateles geoffroyi*. *Int. J. Primatol.*, **24**, 223–238.

Campbell, C. J. (2006a). Copulation in free-ranging black-handed spider monkeys (*Ateles geoffroyi*). *Am. J. Primatol.*, **68**, 507–511.

Campbell, C. J. (2006b). Lethal intragroup aggression by adult male spider monkeys (*Ateles geoffroyi*). *Am. J. Primatol.*, **68**, 1197–1201.

Campbell, C. J., Aureli, F., Chapman, C. A., *et al.* (2005). Terrestrial behavior of *Ateles* spp. *Int. J. Primatol.*, **26**, 1039–1051.

Cant, J. (1978). Population survey of the spider monkey *Ateles geoffroyi* at Tikal, Guatemala. *Primates*, **19**, 525–535.

Carpenter, C. R. (1935). Behavior of red spider monkeys in Panama. *J. Mammal.*, **16**, 171–180.

Chapman, C. A. (1990). Association patterns of spider monkeys: the influence of ecology and sex on social organization. *Behav. Ecol. Sociobiol.*, **26**, 409–414.

Chapman, C. A. and Chapman, L. J. (1987). Social responses to the traumatic injury of a juvenile spider monkey (*Ateles geoffroyi*). *Primates*, **28**, 271–275.

Chapman, C. A. and Chapman, L. J. (1990). Reproductive biology of captive and free-ranging spider monkeys. *Zoo Biology*, **9**, 1–9.

Chapman, C. A., Fedigan, L. M., Fedigan, L. and Chapman, L. J. (1989). Post-weaning resource competition and sex ratios in spider monkeys. *Oikos*, **54**, 315–319.

Chapman, C. A., Wrangham, R. W. and Chapman, L. J. (1995). Ecological constraints on group size: an analysis of spider monkey and chimpanzee subgroups. *Behav. Ecol. Sociobiol.*, **36**, 59–70.

Charnov, E. and Berrigan, D. (1993). Why do female primates have such long life spans and so few babies? Or life in the slow lane. *Evol. Anthropol.*, **2**, 191–194.

Cheney, D. (1977). The acquisition of rank and the development of reciprocal alliances among free-ranging immature baboons. *Behav. Ecol. Sociobiol.*, **2**, 303–318.

Cheney, D. (1978a). Interactions of immature male and female baboons with adult females. *Anim. Behav.*, **26**, 389–408.

Cheney, D. L. (1978b). The play partners of immature baboons. *Anim. Behav.*, **26**, 1038–1050.

Clark, A. B. (1978). Sex ratio and local resource competition in a prosimian primate. *Science*, **201**, 163–165.

Clark, C. B. (1977). A preliminary report on weaning among chimpanzees of the Gombe National Park, Tanzania. In *Primate Biosocial Development*, ed. S. Chevalier-Skolnikoff and F. E. Poirier, New York: Garland, pp. 235–260.

Clutton-Brock, T. H. and Harvey, P. H. (1977). Primate ecology and social organization. *J. Zool*, **183**, 1–39.

Coehlo, A. M., Jr., Bramblett, C. A., Quick, L. B. and Bramblett, S. S. (1976). Resource availablity and population density in primates: a sociobioenergetic analysis of the energy budgets of Guatemalan howler and spider monkeys. *Primates*, **17**, 63–80.

Crockett, C. M. and Janson, C. H. (2000). Infanticide in red howlers: female group size, male membership, and a possible link to folivory. In *Infanticide by Males and its Implications*, ed. C. P. van Schaik and C. H. Janson, Cambridge: Cambridge University Press, pp. 75–98.

Crook, J. H. (1972). Sexual selection, dimorphism, and social organization in the primates. In *Sexual Selection and the Descent of Man, 1871–1971*, ed. B. G. Campbell, Chicago: Aldine, pp. 231–281.

Deaner, R. O., Barton, R. A. and van Schaik, C. P. (2003). Primate brains and life histories: renewing the connection. In *Primate Life Histories and Socioecology*, ed. P. Kappeler and M. Pereira, Chicago: University of Chicago Press, pp. 233–265.

Eisenberg, J. F. (1973). Reproduction in two species of spider monkeys, *Ateles fusciceps* and *Ateles geoffroyi*. *J. Mammal.*, **54**, 955–957.

Eisenberg, J. F. (1976). Communication mechanisms and social integration in the black spider monkey, *Ateles fusciceps robustus* and related species. *Smithson. Contrib. Zool.*, **213**, 1–108.

Epple, G. (1975). Parental care in *Saguinus fuscicollis* spp. (*Callithricidae*). *Folia Primatol.*, **24**, 221–238.

Fairbanks, L. A. (1993). Juvenile vervet monkeys: establishing relationships and practicing skills for the future. In *Juvenile Primates: Life History, Development and Behavior*, ed. M. E. Pereira and L. A. Fairbanks, Oxford: Oxford University Press, pp. 211–227.

Fairbanks, L. A. (2000). The developmental timing of primate play: a neural selection model. In *Biology, Brains, and Behavior: The Evolution of Human Development*, ed. S. T. Parker, J. Langer and M. L. McKinney, Santa Fe, CA: School of American Research Press, pp. 131–158.

Fedigan, L. M. and Baxter, M. J. (1984). Sex differences and social organization in free-ranging spider monkeys (*Ateles geoffroyi*). *Primates*, **25**, 279–294.

Fruth, B., Hohmann, G. and McGrew, W. C. (1999). The *Pan* species. In *The Nonhuman Primates*, ed. P. Dolhinow and A. Fuentes, Mayfield, CA: Mayfield Publishing Company, pp. 64–72.

Goldizen, A. W. (1987). Tamarins and marmosets: communal care of offspring. In *Primate Societies*, ed. B. B. Smuts, D. L. Cheney, R. M. Seyfarth, R. W. Wrangham and T. T. Struhsaker, Chicago: University of Chicago Press, pp. 34–43.

Gonzalez-Kirchner, J. P. (1999). Habitat use, population density and subgrouping pattern of the Yucatan spider monkey (*Ateles geoffroyi yucatanensis*) in Quintana Roo, Mexico. *Folia Primatol.*, **70**, 55–60.

Goodall, J. (1968). The behaviour of free-living chimpanzees in the Gombe Stream Reserve. *Anim. Behav. Monogr.*, **1**, 161–311.

Goodall, J. (1971). *In the Shadow of Man*. London: Collins.

324   *L. G. Vick*

Goodall, J. (1983). Population dynamics during a 15 year period in one community of free-living chimpanzees in the Gombe National Park, Tanzania. *Z. Tierpsychol.*, **61**, 1–60.

Goodall, J. (1986). *The Chimpanzees of Gombe: Patterns of Behavior*. Cambridge, MA: Belknap Press.

Gubernick, D. J. and Klopfer, P. H., eds. (1981). *Parental Care in Mammals*. New York: Plenum Press.

Harcourt, A. H. and Stewart, K. J. (1981). Gorilla male relationships: can differences during immaturity lead to contrasting reproductive tactics in adulthood? *Anim. Behav.*, **29**, 206–210.

Harlow, H. F. (1959). Love in infant monkeys. *Sci. Am.*, **200**, 68–74.

Harlow, H. F. (1971). *Learning to Love*. New York: Ballentine Books.

Harlow, H. F. and Zimmermann, R. R. (1959). Affectional responses in the infant monkey. *Science*, **130**, 421–432.

Harvey, P. H., Martin, R. D. and Clutton-Brock, T. H. (1987). Life histories in comparative perspective. In *Primate Societies*, ed. B. B. Smuts, D. L. Cheney, R. M. Seyfarth, R. W. Wrangham and T. T. Struhsaker, Chicago: University of Chicago Press, pp. 181–196.

Harvey, P. H., Read, A. F. and Promislow, D. E. L. (1989). Life history variation in placental mammals: unifying the data with theory. *Oxford Surv. Evol. Biol.*, **6**, 13–31.

Hayaki, H. (1983). The social interactions of juvenile Japanese monkeys on Koshima islet. *Primates*, **24**, 139–153.

Hoage, R. J. (1978). Parental care in *Leontopithecus rosalia rosalia*: sex and age differences in carrying behavior and the role of prior experience. In *The Biology and Conservation of the Callitrichidae*, ed. D. G. Kleiman, Washington, DC: Smithsonian Institution Press, pp. 293–305.

Izawa, K., Kimura, K. and Nieto, A. S. (1979). Grouping of the wild spider monkey. *Primates*, **20**, 503–512.

Janson, C. H. (1990). Ecological consequences of individual spatial choice in foraging groups of brown capuchin monkeys, *Cebus apella*. *Anim. Behav.*, **40**, 922–934.

Janson, C. H. and van Schaik, C. P. (1993). Ecological risk aversion in juvenile primates: slow and steady wins the race. In *Juvenile Primates: Life History, Development, and Behavior*, ed. M. Pereira and L. A. Fairbanks, New York: Oxford University Press, pp. 57–74.

Klein, L. L. and Klein, D. B. (1977). Feeding behaviour of the Colombian spider monkey. In *Primate Ecology: Studies of Feeding and Ranging Behavior in Lemurs, Monkeys and Apes*, ed. T. H. Clutton-Brock, London: Academic Press, pp. 153–182.

Kraemer, H. C., Horvat, J. R., Doering, C. and McGinnis, P. R. (1982). Male chimpanzee development focusing on adolescence: integration of behavioral with physiological changes. *Primates*, **23**, 393–405.

Lee, P. C. (1984). Ecological constraints on the social development of vervet monkeys. *Behaviour*, **91**, 245–262.

Macedonia, J. M. (1991). Vocal communication and anti-predator behavior in the ring-tailed lemur (*Lemur catta*) with a comparison to the ruffed lemur (*Varecia variagatta*). Unpublished Ph.D. thesis, Duke University, Durham, NC.

Martin, P. and Caro, T. (1985). On the functions of play and its role in behavioral development. In *Advances in the Study of Behavior*, ed. J. S. Rosenblatt, C. Beer, M. C. Busnel and P. J. B. Slater, Orlando, FL: Academic Press, pp. 59–103.

Martin, R. D. (1996). Scaling of the mammalian brain: the maternal energy hypothesis. *News Physiol. Sci.*, **11**, 149–156.

McDaniel, P. S. (1994). The social behavior and ecology of the black-handed spider monkey (*Ateles geoffroyi*). Unpublished Ph.D. thesis, Saint Louis University, St. Louis, MO.

Meaney, M. J., Stewart, J. and Beatty, W. W. (1985). Sex differences in social play: the socialization of sex roles. In *Advances in the Study of Behavior*, ed. J. S. Rosenblatt, C. Beer, M. C. Busnel and P. J. B. Slater, Orlando, FL: Academic Press, pp. 2–58.

Milton, K. (1981). Estimates of reproductive parameters for free-ranging *Ateles geoffroyi*. *Primates*, **22**, 574–579.

Milton, K. (1993). Diet and social organization of a free-ranging spider monkey population: the development of species-typical behavior in the absence of adults. In *Juvenile Primates: Life History, Development, and Behavior*, ed. M. E. Pereira and L. A. Fairbanks, New York: Oxford University Press, pp. 173–181.

Milton, K. and Hopkins, M. E. (2005). Growth of a reintroduced spider monkey (*Ateles geoffroyi*) population on Barro Colorado Island, Panama. In *New Perspectives in the Study of Mesoamerican Primates*, ed. A. Estrada, P. A. Garber, M. Pavelka and L. Luecke, New York: Springer Verlag, pp. 417–435.

Nishida, T. and Kawanaka, K. (1985). Within-group cannibalism by adult male chimpanzees. *Primates*, **26**, 247–284.

Nunes, A. and Chapman, C. A. (1995). A re-evaluation of factors influencing the sex ratio of spider monkey populations with new data from Maraca Island, Brazil. *Folia Primatol.*, **68**, 31–33.

Pagel, M. and Harvey, P. H. (1993). Evolution of the juvenile period in mammals. In *Juvenile Primates: Life History, Development and Behavior*, ed. M. E. Pereira and L. A. Fairbanks, New York: Oxford University Press, pp. 28–37.

Pereira, M. (1984). Age changes and sex differences in the social behavior of juvenile yellow baboons (*Papio cynocephalus*). Unpublished Ph.D. thesis, University of Chicago, Chicago, IL.

Pereira, M. E. (1988a). Effects of age and sex on intra-group spacing behaviour in juvenile savannah baboons, *Papio cynocephalus cynocephalus*. *Anim. Behav.*, **36**, 184–204.

Pereira, M. E. (1988b). Agonistic interactions of juvenile savanna baboons. I. Fundamental features. *Ethology*, **79**, 195–217.

Pereira, M. E. (1989). Agonistic interactions of juvenile savanna baboons. II. Agonistic support and rank acquisition. *Ethology*, **80**, 152–171.

Pereira, M. E. (1995). Development and social dominance among group-living primates. *Am. J. Primatol.*, **37**, 143–175.

Pereira, M. E. and Altmann, J. (1985). Development of social behavior in free-living nonhuman primates. In *Nonhuman Primate Models for Human Growth and Development*, ed. E. S. Watts, New York: Alan R. Liss, pp. 217–309.

Pereira, M. E. and Fairbanks, L. A., eds. (1993a). *Juvenile Primates: Life History, Development, and Behavior*. New York: Oxford University Press.

Pereira, M. E. and Fairbanks, L. A. (1993b). What are juvenile primates all about? In *Juvenile Primates: Life History, Development and Behavior*, ed. M. E. Pereira and L. A. Fairbanks, New York: Oxford University Press, pp. 3–12.

Pereira, M. E. and Leigh, S. R. (2003). Modes of primate development. In *Primate Life Histories and Socioecology*, ed. P. M. Kappeler and M. E. Pereira, Chicago: University of Chicago Press, pp. 149–176.

Poirier, F. E. and Smith, E. O. (1974). Socializing functions of primate play. *Am. Zool.*, **14**, 275–287.

Pusey, A. E. (1983). Mother-offspring relationships in chimpanzees after weaning. *Anim. Behav.*, **31**, 363–377.

Pusey, A. E. (1990). Behavioral changes at adolescence in chimpanzees. *Behaviour*, **115**, 203–245.

Ramos-Fernández, G. (2001). Patterns of association, feeding competition and vocal communication in spider monkeys, *Ateles geoffroyi*. Unpublished Ph.D. thesis, University of Pennsylvania, Philadelphia.

Ramos-Fernández, G. and Ayala-Orozco, B. (2003). Population size and habitat use of spider monkeys in Punta Laguna, Mexico. In *Primates in Fragments: Ecology and Conservation*, ed. L. K. Marsh, New York: Kluwer, pp. 191–210.

Ramos-Fernández, G., Ayala-Orozco, B., Bonilla-Moheno, M. and Garcia-Frapolli, E. (2005). Conservación comunitaria en Punta Laguna: fortalecimiento de instituciones locales para el desarollo sostenible. Congreso "Casos exitosos de desarollo sostenible del Tropico", Vera Cruz, Mexico.

Ramos-Fernández, G., Vick, L. G., Aureli, F., Schaffner, C. M. and Taub, D. M. (2003). Behavioral ecology and conservation status of spider monkeys in the Otoch Ma'ax Yetel Kooh Protected Area. *Neotrop. Primates*, **11**, 155–158.

Rhine, R. J. (1975). The order of movement of yellow baboons (*Papio cynocephalus*). *Folia Primatol.*, **23**, 73–104.

Robinson, J. G. (1981). Spatial structure in foraging groups of wedge-capped capuchin monkeys *Cebus nigrivittatus. Anim. Behav.*, **29**, 1036–1056.

Ross, C. and Jones, K. E. (1999). Socioecology and the evolution of primate reproductive rates. In *Comparative Primate Socioecology*, ed. P. C. Lee, Cambridge: Cambridge University Press, pp. 73–110.

Sacher, G. A. (1975). Maturation and longevity in relation to cranial capacity in hominid evolution. In *Primate Functional Morphology and Evolution*, ed. R. H. Tuttle, The Hague: Mouton, pp. 417–441.

Sacher, G. and Staffeldt, E. (1974). Relation of gestation time to brain weight for placental mammals: implication for the theory of vertebrate growth. *Am. Nat.*, **108**, 593–615.

Seyfarth, R. M. and Cheney, D. L. (1980). The ontogeny of vervet monkey alarm-calling behaviour: a preliminary report. *Z. Tierpsychol.*, **54**, 37–56.

Seyfarth, R. M. and Cheney, D. L. (1986). Vocal development in vervet monkeys. *Anim. Behav.*, **24**, 1450–1468.

Shimooka, Y. (2003). Seasonal variation in association patterns of wild spider monkeys (*Ateles belzebuth belzebuth*) at La Macarena, Colombia. *Primates*, **44**, 83–90.

Silk, J. B. (1983). Local resource competition and facultative adjustment of sex ratio in relation to competitve abilities. *Am. Nat.*, **121**, 56–66.

Silk, J. B., Samuels, A. and Rodman, P. S. (1981). The influence of kinship, rank, and sex on affiliation and aggression between adult female and immature bonnet macaques (*Macaca radiata*). *Behaviour*, **78**, 111–137.

Smith, E. O. (1978). *Social Play in Primates*. New York: Academic Press.

Stanford, C. (1995). The influence of chimpanzee predation on group size and anti-predator behaviour in red colobus monkeys. *Anim. Behav.*, **49**, 577–587.

Stanford, C. B., Wallis, J. Matama, H. and Goodall, J. (1994). Patterns of predation by chimpanzees on red colobus monkeys in Gombe National Park, Tanzania, 1982–1991. *Am. J. Phys. Anthropol.*, **94**, 213–228.

Symington, M. M. (1987). Sex ratio and maternal rank in wild spider monkeys: when daughters disperse. *Behav. Ecol. Sociobiol.*, **20**, 421–425.

Symington, M. M. (1988). Food competition and foraging party size in the black spider monkey (*Ateles paniscus chamek*). *Behaviour*, **105**, 117–134.

Symington, M. M. (1990). Fission-fusion social organization in *Ateles* and *Pan*. *Int. J. Primatol.*, **11**, 47–61.

Taub, D. M. (1984). Male caretaking behaviour among wild Barbary macaques (*Macaca sylvanus*). In *Primate Paternalism*, ed. D. M. Taub, New York: Van Nostrand Reinhold, pp. 20–55.

Terborgh, J. W. (1983). *Five New World Primates: A Study in Comparative Ecology*. Princeton, NJ: Princeton University Press.

Terborgh, J. W. and Janson, C. H. (1986). The socioecology of primate groups. *Annu. Rev. Ecol. Syst.*, **17**, 111–135.

Valero, A., Schaffner, C. M., Vick, L. G., Aureli, F. and Ramos-Fernández, G. (2006). Intragroup lethal aggression in wild spider monkeys. *Am. J. Primatol.*, **68**, 732–737.

van Roosmalen, M. G. M. (1985). Habitat preferences, diet, feeding strategy and social organization of the black spider monkey (*Ateles paniscus paniscus* Linnaeus 1758) in Surinam. *Acta Amazonica*, **15**, 3–238.

van Roosmalen, M. G. M. and Klein, L. L. (1988). The spider monkeys, genus *Ateles*. In *Ecology and Behaviour of Neotropical Primates*, Vol. 2, ed. R. A. Mittermeier, A. B. Rylands, A. F. Coimbra-Filho and G. A. B. Fonseca, Washington, DC: World Wildlife Fund, pp. 455–537.

van Schaik, C. P. (1983). Why are diurnal primates living in groups? *Behaviour*, **87**, 120–144.

van Schaik, C. P. and Janson, C. H. (2000). *Infanticide by Males and Its Implications*. Cambridge: Cambridge University Press.

van Schaik, C. P. and van Noordwijk, M. A. (1985). Evolutionary effect of the absence of felids on the social organization of the macaques on the island of Simeulue (*Macaca fascicularis fusca*, Miller 1903). *Folia Primatol.*, **44**, 138–147.

van Schaik, C. P. and van Noordwijk, M. A. (1989). The special role of male *Cebus* monkeys in predation avoidance and its effect on group composition. *Behav. Ecol. Sociobiol.*, **24**, 265–276.

Vick, L. G. and Taub, D. M. (1995). Ecology and behavior of spider monkeys (*Ateles geoffroyi*) at Punta Laguna, Mexico. *Am. J. Primatol.*, **36**, 160.

Vick, L. G., Ramos-Fernández, G. and Taub, D. M. (2001). El infanticidio en los monos aranas (*Ateles geoffroyi yucatanensis*) en Punta Laguna, Mexico. 1er Congreso mexicano de Primatología, Mérida, México, September, 2001.

Walters, J. R. (1981). Inferring kinship from behaviour: maternity determinations in yellow baboons. *Anim. Behav.*, **29**, 126–136.

Walters, J. R. (1987). Transition to adulthood. In *Primate Societies*, ed. B. B. Smuts, D. L. Cheney, R. M. Seyfarth, R. W. Wrangham and T. T. Struhsaker, Chicago: Chicago University Press, pp. 358–369.

12 *Demography and group composition of* Ateles

YUKIKO SHIMOOKA, CHRISTINA J. CAMPBELL,
ANTHONY DI FIORE, ANNIKA M. FELTON, KOSEI IZAWA,
ANDRES LINK, AKISATO NISHIMURA, GABRIEL
RAMOS-FERNÁNDEZ AND ROBERT B. WALLACE

Introduction

Spider monkeys are distributed widely throughout Central and South America and studies have been conducted at a variety of sites across the geographic range of the genus (see Table 1.1 in Campbell, this volume). However, detailed information about group composition and demography of spider monkeys remains largely unavailable. Because their fission–fusion social organization allows researchers to observe only a part of a group at any time, short-term surveys can rarely document overall group size and composition. Only a cumulative data set of party composition based on individual identification and longitudinal research can help determine the full composition of a group. Furthermore, the rarity of births and deaths make other demographic variables such as interbirth intervals only available through long-term investigation. In the 1980s, relevant demographic information from wild populations was available only for seven groups from five sites for three *Ateles* species. In this chapter, we present an updated summary of existing data on four *Ateles* species from 18 groups and 13 sites. We analyze both previously published and new data from these sites and compare them in order to re-examine the demographic characteristics of spider monkey groups.

Methods

Demographic data from 18 groups and 13 sites (Table 12.1) were gathered from the literature and augmented with data from a questionnaire sent to spider monkey researchers in 2005. The questionnaire focused on: (1) study object and details (species, study site, study period, total months of observations used for these data), (2) group composition (the number of individuals of each sex and

Spider Monkeys: Behavior, Ecology and Evolution of the Genus Ateles, ed. Christina J. Campbell. Published by Cambridge University Press. © Cambridge University Press 2008.

Table 12.1 *Study groups, sites and data sources*

Species	Study site	Group name	Latitude	Longitude	Study period	Total months of observations used for	Source	Group no.
A. belzebuth belzebuth	La Macarena (Tinigua), Colombia	MB-1	2°40N	74°10W	1987–1989, 1996–2002	52 months	Ahumada (1989, 1990), Izawa (unpub. data), Link (unpub. data)	1
	La Macarena (Tinigua), Colombia	MB-2	2° 40N	74° 10W	1996–2002	20 months	Shimooka (2000, 2003, 2005, unpub. data)	2
	La Macarena (Tinigua), Colombia	MB-3	2° 40N	74° 10W	1996–2002	17 months	Matsushita and Nishimura (2000), Nishimura (unpub. data)	3
	La Macarena (east), Colombia	S group	2° 30N	73° 30W	1967–1968	13 months	Klein (1972)	4
	La Macarena (east), Colombia	0 group	2° 30N	73° 30W	1967–1968	13 months	Klein (1972)	5
	La Maraca Island, Brazil	–	3° 24N	61° 40W	1987–1990	11 months	Nunes and Chapman (1997)	6
	Yasuní (TBS), Ecuador	MQ-1	0° 37N	76° 09W	2005	5 months	Di Fiore, Link and Shimooka (unpub. data)	7
	Yasuní (Catolica), Ecuador	Catolica	0° 42N	76° 28W	2002–2005	24 months	Di Fiore & Link (unpub. data)	8
	Lago Caiman, Parque Nacional Noel Kempff Mercado, Bolivia	Chutolandia	13° 36S	60° 55W	1995–1997	11 months	Wallace (unpub. data)	9
A.belzebuth chamek	La Chonta, Bolivia	Main study group	15° 45S	62° 60W	2003–2004	10 months	Felton (unpub. data)	10
	Cocha Cashu, Manu, Peru	Lake community	11° 51S	71° 19W	1982–1986	21 months	Symington (1988)	11
	Cocha Cashu, Manu, Peru	East community	11° 51S	71° 19W	1982–1986	21 months	Symington (1988)	12

Species	Location	Group	Latitude	Longitude	Years	Duration	Reference	No.
A. paniscus paniscus	Raleighvallen-Voltzberg, Surinam		4° 32N	56° 32W	1977–1978	12 months	van Roosmalen (1985)	13
A. geoffroyi	Barro Colorado Island, Panama[a]	Island group	9° 09N	79° 51W	1968–1980, 1997–2004		Milton (1981, 2005), Campbell (unpub. data)	14
	Santa Rosa, Costa Rica		10° 50N	85° 37W	1983–1989	36 months	Chapman et al. (1989), Chapman (1990)	15
	Otoch Ma'ax Yetel Kooh, (Punta Laguna) Yucatan, Mexico	Eastern group	20° 38N	87° 38W	1997–2002	72 months	Ramos-Fernández et al. (2003)	16
	Otoch Ma'ax Yetel Kooh, (Punta Laguna) Yucatan, Mexico	Western group	20° 38N	87° 38W	1997–2002	72 months	Ramos-Fernández et al. (2003)	17
A. hybridus	Serranía de Las Quinchas, Colombia	AH-1	6° 03N	74° 16W	2005–2006	15 months	Link (unpub. data)	18

[a] The BCI population was entirely extirpated by 1912 and was reintroduced intermittently from 1959 until 1966 (Milton, 2005). Here we analyzed the results from observations recorded for 18 months continuously between 1997 and 1998 in addition to intermittent census data after that. Thirty years have passed since the reintroduction, thus we treated this population the same as other groups.

age category at the end of the period of investigation, and the range of these over the course of the study), (3) records of immigration and disappearance (death or migration) of each sex and cause of death when available, and (4) birth records (interbirth interval when the previous infant survived for more than two years, interbirth interval when the previous infant died before reaching two years of age, infant mortality in the first year for each sex, number of twin birth records, seasonality of infant birth, and number of infants of each sex born during the study period [male/female/unsexed]). Most studies were conducted over consecutive short-term periods, or intermittently but on a long-term basis, thus it was often difficult for researchers to provide information concerning many of these parameters.

Age class was categorized into adult (A), subadult (SA), juvenile, and infant. Estimated ages for each class were as follows: adult ≥ 8 years old, subadult 5–8 years, juvenile 2–5 years, and infant <2 years. Juveniles were distinguished from subadults as those always traveling with their mother. The distinction between adult and subadult is likely to vary slightly among researchers. Thus, in this article, we combined the number of A and SA and used this figure for analysis. We calculated the sex ratio (male/female) of each species by referring to the number of A and SA, excluding juveniles and infants that were still dependent on their mothers. When possible, the sex ratio of infants was also calculated. Additional unpublished data from three groups of La Macarena (Tinigua) concerning the sex ratio within each age category were also used. To examine the bias to either sex among neonates, we calculated the sex ratio of all the neonates born between 1997 and 2002, and that of current A and SA and immature individuals.

Results and discussion

Group size and composition

In the 1980s, when information about spider monkey group size and composition was more limited, group size was documented as 18–42 individuals, typically consisting of 3–5 adult males and 10–18 adult females with an adult sex ratio (male:female) of between 0.22 and 0.42 (Klein, 1972; Milton, 1981; van Roosmalen, 1985; Symington, 1988; Chapman *et al.*, 1989). Our newly compiled and enlarged data set suggests greater variation both in group size and composition than has been previously reported.

Among 18 groups of four *Ateles* species group size varies greatly, from 15 to 56 individuals (Table 12.2). The two largest groups, comprising more than 50 individuals, are reported for *A. belzebuth chamek* from Bolivia. These groups

Table 12.2 *Group composition and the range over years*

Species	Group no.	Adult males	Adult females	Subadult males	Subadult females	Juvenile males	Juvenile females	Infant males	Infant females	Infant unsexed	Total
A. belzebuth belzebuth	1	4 (4)	9 (5–11)	2 (0–2)	4 (4–6)	1 (1)	2 (1–2)	3 (0–3)	4 (1–4)	0 (0–1)	29 (15–29)
	2	5 (5–6)	10 (10–11)	2 (0–2)	0 (0–2)	1 (0–1)	3 (3)	2 (2–4)	7 (5–7)	0 (0–1)	30 (28–30)
	3	3 (3–4)	5 (5–6)	1 (0–1)	2 (0–2)	1 (1–3)	0 (0–2)	3 (1–4)	0 (0–2)	0 (0)	15 (15–19)
	4	5	12	–	–	2	3	–	–	5	27
	5	3 (3)	11 (10–11)	–	–	1 (1)	3 (2–3)	1	–	3 (3)	21 (19–21)
	6	6	8	1	1	2	1	1	2	0	22
	7	5 (5)	9 (9)	1 (1)	5 (5)	2 (2)	0 (0)	2 (1–2)	1 (1)	(0)	25 (24–25)
	8	4 (4–5)	9 (8–9)	0 (0)	2 (2–3)	0 (0)	2 (2–3)	0 (0)	4 (3–4)	0 (0)	21 (19–24)
A. belzebuth chamek	9	15 (14–15)	15 (14–15)	5 (5)	5 (1–5)	4 (4)	6 (6–8)	1 (1)	4 (3–5)	0 (0–2)	55 (53–56)
	10	11 (11–12)	15 (13–15)	7 (7)	5 (5–7)	3 (3–4)	4 (3–4)	1 (1)	7 (3–8)	2 (0–2)	55 (48–55)
	11	5	15	4 subadults total		immature 13 (excluding <1 year)		2			37
	12	5	16	4 subadults total		immature 15 (excluding <1 year)					40
A. paniscus paniscus	13	3	8	–	1	4	0	2	–	–	18
A. geoffroyi	14	4 (4–5)	10 (7–10)	6 (1–6)	3 (3)	1 (1)	4 (3–4)	4 (0–4)	4 (2–4)	–	36 (22–36)
	15	4	18	(1–2)	(3–4)	(5–8)	–	(5–8)	–	–	42
	16	4 (1–4)	7 (5–7)	–	–	0 (0–2)	4 (2–4)	3 (1–3)	2 (2–3)	0 (0)	20 (16–20)
	17	10 (6–10)	13 (13–15)	–	–	2 (1–3)	5 (5–7)	3 (3–4)	7 (3–7)	0 (0)	40 (34–41)
A. hybridus	18	3	11	1	4	2	1	1	0	3	26

Group numbers are taken from Table 12.1. When either the current composition or the range is unavailable, the value is left blank.

contain an extraordinary large number of A and SA males (groups 9 and 10, Table 12.2). Two other groups of this species have been studied in Peru and are also relatively large, but do not contain as many males (groups 11 and 12). All four groups of *Ateles belzebuth chamek* contain 15–16 adult females, a factor that also contributes to the large group size. The smallest group in our sample (*A. belzebuth belzebuth* in Colombia), consisting of 15 individuals, includes only 5 adult females and 3 adult males (group 3). This group is much smaller than the neighboring groups at the same site, but it appears as if this species is more likely to form smaller groups compared with other species (Table 12.3). Lastly, *A. geoffroyi* forms medium size groups (Table 12.3), or sometimes, larger groups with more females like that seen in Santa Rosa NP (group 15). Previously, Coelho *et al.* (1976) observed a very large group of up to 75 individuals at Tikal in Guatemala, although that study was not based on individual identification. In addition, Fedigan and Baxter (1984) observed 10 fully adult males of *Ateles geoffroyi* traveling together at this site. These observations suggest that *A. geoffroyi* may form very large groups under certain environmental conditions, although our knowledge of the causes of intra- and interspecies variation is still limited.

The percentage of As and SAs among the four species ranges from 55% in *A. geoffroyi* to 80% in *A. belzebuth belzebuth* (Table 12.3). There is no clear difference among the four species, but it appears that *A. geoffroyi* groups have slightly more juveniles and infants compared with other species of spider monkeys. This could be due to more disturbed predator communities in Central America, where most of the study sites are in fragmented forest (Wright *et al.*, 1994; Stevenson *et al.*, 2000).

Sex ratios among the A and SA class are highly variable, ranging from 0.22 to 1.00 (Table 12.3). All the groups for which data are available are female-biased, except for the Chutolandia group of *A. belzebuth chamek* (group 9), in which the sex ratio is 1.00 (Figure 12.1). Within populations of each species, there is also large variation in sex ratio, so it is difficult to determine if any species level difference exists. The absolute number of adult males and females is also variable both within and between species (males: 1–15, females: 5–15). The existence of a single-male *A. geoffroyi* group in the eastern group of the Otoch Ma'ax Yetel Kooh population in Mexico is somewhat surprising: for a period of 3.5 years (January 1997–June 2000) there was only one male present in this group (group 16). During that time, another juvenile male matured into a subadult and later into an adult. In 2006–2007, however, this group had 4 adult and subadult natal males without any male immigration from other groups (G. Ramos-Fernández, personal communication). As we discuss later, spider monkeys form a patrilineal social structure in general, thus the number of males in a group is likely to be variable depending on the group history.

Table 12.3 *Mean and range of group size, age composition and sex ratio (A and SA, male/female) for each species*

Species	n	Group size			Percentage of A and SA			A and SA sex ratio		
		Mean	SD	Range	Mean	SD	Range	Mean	SD	Range
A. belzebuth belzebuth	8	23.4	5.3	15–30	69.40	7.30	62.96–80.00	0.54	0.20	0.30–0.83
A. belzebuth chamek	4	46.8	9.6	37–56	67.30	4.53	62.50–72.73	0.77	0.22	0.56–1.00
A. paniscus	1	18		18	66.67		66.67	0.38		0.38
A. geoffroyi	4	34.5	10.0	20–42	60.17	4.64	55.00–64.29	0.58	0.26	0.22–0.77
A. hybridus	1	26		26	73.08		73.08	0.27		0.27

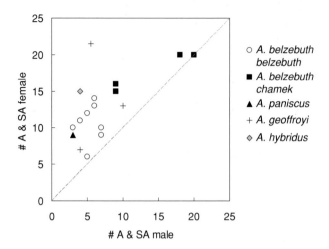

Figure 12.1 Number of adult (A) and subadult (SA) males and females in each group. The number of SAs of each sex was not available for two *A. belzebuth chamek* groups (#11 and 12), thus these two groups only have numbers for adults. The current number of individuals was not available for an *A. geoffroyi* group (#14), thus the average of the range of A and SA was included.

The mean group size of *A. belzebuth chamek* is almost double that of *A. belzebuth belzebuth* (Table 12.3), although apparent interspecific differences in our data may be a result of a data set that is too limited to reveal true interspecific variation. Alternatively, it is possible that there are real group size differences among the various *Ateles* species that result from differences in the carrying capacity of differing habitats. Group size is an important characteristic because it is influenced by environmental factors, group history and demographic patterns. Furthermore group size and composition are important factors that can characterize the social relationships among individuals and between groups. A more detailed examination of the variation among habitats each species occupies may allow us to detect factors that determine group size.

Sexual bias in birth rate, mortality rate and cause of death

As spider monkey group composition is generally female-biased (Figure 12.1, Table 12.1), it is interesting to investigate whether this bias exists from birth. Over the course of a 21-month study, Symington (1987a) found a strongly female-biased sex ratio (M:F = 12:32) at birth in her study population at Cocha Cashu, Peru. On the other hand, Milton and Hopkins (2005) note that on Barro Colorado Island (BCI), natal sex ratio over the years is approximately 1:1. In

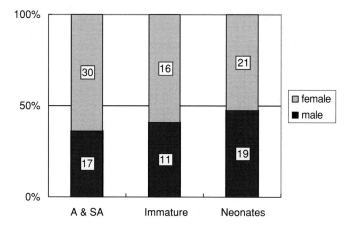

Figure 12.2 Sex ratio declines in *A. belzebuth belzebuth* of La Macarena (Tinigua) populations. Data from three neighboring groups were gathered. A & SA refers to the current number of adults and subadults, and immature refers to the current number of juveniles and infants. Neonate shows the number of all the neonates born between 1997 and 2002.

the same population, a very strong female bias (M:F = 1:5) has been detected in one year, followed by a very strong male bias (M:F = 6:1) in the next cohort (C. J. Campbell, unpublished data). In the case of the La Macarena (Tinigua) population in Colombia, both female and male bias can be detected in neonates over a period of 5 years among three groups: M:F = 5:10 (and 4 unsexed) in MB-1 group, 5:9 in MB-2, 9:2 in MB-3. This yields an overall M:F natal sex ratio for this population of 1:1.1 (excluding unsexed) in total. In the case of Otoch Ma'ax Yetel Kooh population in Mexico (Ramos-Fernández *et al.*, 2003), two study groups had an infant sex ratio (M:F) that ranged from 1:3 to 3:2 and from 3:7 to 4:3, respectively, over the 72 months study period. Again, this indicates M:F natal sex ratio that does not show any consistent bias towards females.

It is therefore possible that the female-biased sex ratio at birth reported by Symington (1987a) may have been a result of small sample size bias in a limited study period. We propose that long-term studies carried out for 20 or more years will reveal that such a bias evens out with a sufficiently long timescale. Such short-term biases should not be ignored, however, as they can have important impacts on group structure. For example, male-biased immature sex ratio will tend to reduce population growth in the short-term future, even if female fecundity remains constant (Strier *et al.*, 2006).

In La Macarena (Tinigua), the male/female sex ratio declines through the course of development (Figure 12.2), thus it is likely that a large bias in sex ratio is brought on after birth. Infant mortality rate in the first year from birth is 20% (9/45) for all the infants, including 5 unsexed ones, 15.8% (3/19) for males and 4.8% (1/21) for females (K. Izawa, A. Nishimura and Y. Shimooka *et al.*, unpublished data). Similarly, infant mortality at Otoch Ma'ax Yetel Kooh has been documented at 20.0% (3/15) in males and 15.0% (3/20) in females (G. Ramos-Fernández, personal communication). Chapman *et al.* (1989) also noted that at Santa Rosa immature males are five times more likely to disappear than females, even though their mothers are always present after the infant's disappearance. Thus it is possible that infant mortality is higher in males than females. Rarely observed potential causes of death are discussed below.

Accidental injuries

Symington (1987b) reports a fall from a tree that caused the death of a subadult male and two females. One female was judged quite old from the degree of wear on her teeth and the other had a leg bone that had been previously broken and then healed. Felton (A. Felton, unpublished data) also found a juvenile male that had died after falling to the ground during a severe storm. Accidental falls can happen both in normal life and in the course of aggression or chasing games, but not all the falls are fatal for the monkeys. Actually, Karesh *et al.* (1998) report that three out of seven immobilized male spider monkeys had healed features on their skeletons, and one female had an existing dislocation of the elbow, but none of them showed abnormality in locomotion. Even so, falls can be a significant cause of mortality, and males that engage in more social interactions than females (Fedigan and Baxter, 1984) or travel further than females (Shimooka, 2005) or perform a behavior similar to "rain dances" (Wallace, 1998) are more likely to fall.

Predation

Julliot (1994) reported that a crested eagle (*Morphinus guianensis*) captured a 6–8 month old spider monkey (sex unknown) at Nouragues Station in French Guiana. At La Macarena (Tinigua), a jaguar (*Panthera onca*) was observed to chase an adult male on the ground, attack the neck with its paw and kill him (Matsuda and Izawa, 2007). Matsuda and Izawa (2007) also observed a puma (*Felis concolor*) attempt to attack an adult female at a salt lick, but she escaped by climbing up a vine. Emmons (1987) and Chinchilla (1997) found the remains of spider monkey victims in the scat of a jaguar and a puma. Campbell (personal communication) also saw strong spider monkey reactions to snakes (*Boa constrictor*), white forest hawks (*Leucopternis albicollis*) and ocelots (*Leopardus pardalis*), and Wallace (1998) noted that harpy eagles (*Harpia harpyja*) and ornate hawk-eagles (*Spizaetus ornatus*) are potential predators. It is generally thought that spider monkeys experience low predation pressure due to their large body size and arboreal nature, but it is likely that the chance of predation is more frequent. Spider monkeys are highly arboreal, but recent observations have revealed that they use the ground more often than we previously expected (Campbell *et al.*, 2005). For example, they use the ground when they eat soil or drink water and to escape attacks from conspecifics or play a chase game. During such behavior, predation risk from terrestrial carnivores is likely to be greatly increased. Whether or not there is an age or sex difference in terrestrial behavior is still unclear, but males that are frequently involved in chases (see below), and who patrol on the ground (Aureli *et al.*, 2006) might be exposed to a higher risk of predation than females.

Lethal aggression by conspecifics

Fedigan and Baxter (1984) demonstrated that juvenile males were twice as likely to receive aggression from their own species than juvenile females at Tikal. Likewise Chapman *et al.* (1989) reported six serious injuries to immature males displayed on their shoulders, legs, hips and tails in Santa Rosa. Vick (this volume) reports a case where a male infant died as a consequence of an attack to his mother by an adult male. There is an observed case of infant killing in La Macarena (Tinigua): a newborn male infant (<2 weeks from birth) experienced lethal aggression and received several large deep lacerations around the tail bottom, thigh and back, and died the next day (Y. Shimooka, unpublished data). The aggression to this infant happened just after the fusion of two parties: one party including several females approached the other party including an adult male and a 5-year-old subadult male. The aggression itself was not clearly observed due to the distance of the incident, but branches were shaken and "growls" that are often emitted by males while grappling were heard. No screams were heard, and a newborn infant displaying fresh cuts fell to the ground shortly after. The mother then came down to the ground without any injuries and searched for her infant for a while, but finally she abandoned the infant on the ground and left the area with other individuals.

Recent observations have reported that not only immature males, but also subadult males are the targets of lethal aggression. For example, Valero *et al.* (2006) report a case at the Otoch Ma'ax Yetel Kooh Reserve in Mexico, where a young adult male (6–7 years old) sustained severe injuries as the result of three months of repeated intragroup aggression from at least three of the oldest adult males in the community. He eventually died after a lethal attack from at least one of the adult males. Campbell (2006) also reported three cases of coalitionary aggression by adult male spider monkeys against subadult males within the community on BCI, Panama. Two of the three cases were followed by the disappearance and presumed death of the victim. In one case, four of the five adult males chased a subadult male into a lake, and when the victim attempted to escape the water on two occasions the adult males pushed him back into the water. These observations provide evidence for intragroup male coalitionary killing. On the other hand, there are no reports of females who have died as a consequence of lethal aggression at present.

Fedigan and Baxter (1984) reported that males chase more than do females, most frequently directing their chases toward adult females, and that females are more often chased than males are, most frequently by adult males (see also Campbell, 2003). In fact, females sometimes show scars on their face, but severe injuries are quite rare. It is likely that the type of aggression differs between sexes; the aggression among males is of higher intensity, and the aggression by males towards females is of lower intensity (A. Link, A. Di Fiore and S. Speher, unpublished data).

These reports suggest that males are more likely to be the target of severe aggression from other male members of the group, which might explain female-biased sex ratios in adult spider monkeys. Males often have injuries or cuts on their face and body, which can be received through aggression among group members or between groups.

Considering that strong male–male relationships are typical in this male-philopatric society (Aureli and Schaffner, this volume), it is surprising that male coalitions are aggressive towards males in the same group. Valero *et al.* (2006) proposed that the tension due to the low availability of mating opportunities might be a proximate trigger for male coalitionary killing of subadult males because intragroup killing occurred when the adult sex ratio (male/female) was higher than the norm for spider monkeys. This is further supported by the cases reported by Campbell (2006). However, it is quite difficult to explain the aggression from males toward male infants, especially when these infants could be related. Even though male infants might be future mating competitors, they are also potential future allies. To let the male infants survive is thus potentially very important when considering the future of the group's ability to effectively defend its territory. Symington (1987a) reports that among high-ranking females, the interbirth interval between the birth of male infant reared successfully to weaning and the birth of a subsequent infant is longer than the interbirth interval following a female infant (36.0 versus 29.0 months). It is thus possible that adult males are aggressive towards male infants in order to induce faster recovery of the mothers' reproductive state, although there are currently no data to test this possibility.

Disease and parasites, or injuries by intergroup aggression may be other causes of death, but these have not yet been reported. These are all possible factors that may lead to a female-biased sex ratio after maturity. It is quite difficult to detect the cause of death in spider monkeys' fission–fusion society, thus only further accumulation of case reports can help us reveal the mechanism behind the female-biased group composition.

Reproductive parameters

Spider monkeys usually bear a single infant at a time. A set of twins (one male and a female) both survived for at least 13 months from birth at La Macarena (Tinigua) (Link *et al.*, 2006; see Campbell and Gibson, this volume for an additional account of twinning). Interbirth interval (IBI) for females that successfully weaned one infant and then gave birth to another appears to vary among species (Table 12.4). In particular, *A. b. belzebuth* appears to have a somewhat longer IBI than other species. These data only include cases where the mother gave birth to two consecutive infants successfully, with no possibility of death of a newborn infant between them. IBI when the previous offspring died before weaning is 17.8 ± 5.6 months ($n = 6$, range: 9–26 months) in La Macarena, Tinigua (K. Izawa, A. Nishimura and Y. Shimooka, unpublished data). Currently, we have no explanation for this apparent species difference in IBI. Campbell *et al.* (2001) found that spider monkeys do experience spontaneous abortion, which could affect estimates of IBI. Hormonal assays by noninvasive methods in the wild are needed to more adequately examine if differences exist among species or sites in basic reproductive hormone profiles or rates of spontaneous abortion. Symington (1987a) suggested that IBI is also affected by a mother's rank and the sex of her previous offspring. Ecological conditions such as the carrying capacity of each habitat may affect the nutritional condition of mothers, and the time mothers need to recover between subsequent births may differ among habitats. The degree of seasonality in reproduction may also affect this value. However, at most of the study sites, the birth season is quite long or nearly indistinguishable (up to 10 months of the year, Table 12.4) and it is thus most likely that reproductive seasonality has at best a relatively small effect on IBI. It is also possible that the time or number of copulations that females need after resumption of ovarian cycling until conception differs among species or habitats. More data on IBI and copulation from longitudinal observation from various sites may shed some light on this topic.

It is difficult to determine how many offspring a female spider monkey may give birth to in her lifetime, or even the average life-span. A captive female (*A. paniscus*) named Buenos came to the Japan Monkey Center in 1961 and died in 2005. She was already an adult when she came to this zoo, so she was estimated to be 52 years old when she died (JMC, 2005). In the wild, spider monkeys will probably not survive as long, but this does indicate that their life-span can be quite long. Given their female-biased immigration patterns (mentioned later), it is more difficult to determine both the life-span and the age at first birth of females in the wild, but we can at least determine the life-span of males through longitudinal observation. The life-span of a female and the estimated number of offspring are important factors, particularly when we refer to

Table 12.4 Reproductive parameters

| Species | Group no. | Infant mortality in first year died/total born | | | | Infant birth sex ratio male/female/ unknown | Seasonality in infant birth (range of records) | Interbirth interval |
		Male	Female	Sex unknown				
A. belzebuth belzebuth	1	1/5	1/10	4/4		5/10/4	Jun–Jan	43.7 ± 5.1
	2	2/5	0/9	1/1		5/9/1	Jun–Feb	(n = 7, range = 38–54)
	3	0/9	0/2	0/0		9/2/0	Feb–Mar, Jul–Dec	
	4&5	–	–	–		–	Sep–Nov, Jan, Apr	
	6	1/5	0/1	–		1/1/0	–	
	7	–	–	–		1/0/0	–	
	8	0/0	1/5	0/0		0/5/0	–	
A. belzebuth chamek	9	–	–	–		–	Nov, Mar–Apr	
	10	0/0	0/5	0/2		0/5/2	Jan–Aug	
	11&12	1/5	3/9	1/1		12/32/0	Nov–Aug	34.5 ± 5.8 (n = 17, range = 25–42)
A. geoffroyi	14	0	0	2/13		19/17/3	May–Feb	31.9 ± 3.0 (n = 7, range = 28–36)
	15	–	–	–		4/4	–	
	16&17	3/15	3/20	–		15/20/0	–	32.0 ± 6.0 (n = 36, range = 24–46)

Table 12.5 *Immigration and disappearance of adult and subadult individuals from the study groups*

Species	Group no.	Immigration Male	Immigration Female	Disappearance Male Death	Disappearance Male Unknown	Disappearance Female Death	Disappearance Female Migration	Disappearance Female Unknown
A. belzebuth belzebuth	1	–	5	–	–	–	1	9
	2	–	5	1[a]	1	–	–	5
	3	–	2	1	–	–	–	2
	5	–	1	–	–	–	–	–
	6	–	–	2	2	–	–	–
	8	–	2	1	–	1	1	0
A. belzebuth chamek	9	–	1	1	–	–	–	1
	10	–	4–5	–	1–2	–	2	–
	11&12	–	–	2[b]	–	3[c]	2	3
A. geoffroyi	14	–	0	3	–	1	–	–
	16&17	–	4	2	–	–	–	17

Cause of death is known: [a]predated by a jaguar, [b]one individual died of old age and one died from a fall, [c]two died of old age, one died just after a stillbirth.

male–male relationships and the number of brothers a male has in a group. For such questions, DNA analysis from noninvasive samples will be also helpful.

Immigration and disappearance

Symington (1988) discusses female dispersal and male philopatry as characteristic of spider monkey communities. In a review of our data on females, there are many reports of both immigration and disappearance (Table 12.5), as well as some observed cases of adult females that reproduce in their natal group (G. Ramos-Fernández, personal communication). Females have been observed to immigrate as nulliparous subadult females (Y. Shimooka, unpublished data), as adult females (Ramos-Fernández *et al.*, 2003), and as mothers with juveniles (Klein, 1972). However, the average age of females at emigration or immigration is not yet clear. In the observation of the MB-2 group at La Macarena (Y. Shimooka, unpublished data), a 4-year-old natal female remained with her mother, while a 5-year-old natal female occasionally moved alone. Since young female immigrants are usually slightly larger than 5-year-old natal females, they are estimated to be around 6 years old. These young immigrants usually keep their distance from other females and prefer the boundary area, because they are often the targets of aggression by conspecifics. In three groups at La Macarena (Tinigua) at least 6 of 12 immigrant females stayed in a group only for a short while and then disappeared (K. Izawa, A. Nishimura and Y. Shimooka, unpublished data). These females were known to be not from neighboring groups, but rather were from other groups farther away. They were observed only a few times and are thought to have left for other groups. When young immigrants remain in a group, they usually do not give birth to an infant for a long period of time: 18 and 33 months for each of the two immigrants of *A. belzebuth belzebuth* (Y. Shimooka, unpublished data), and 14 and 22 months for each of the two immigrants of *A. belzebuth chamek* (Symington, 1987b).

In La Macarena, three immigrations and three disappearances of females with juveniles were reported; all of these six juveniles were females (K. Izawa and Y. Shimooka, unpublished data). In the case of the disappearances, we could not distinguish emigration from death. In three additional cases, females disappeared but their juveniles remained in the group. In two of these cases, the remaining juvenile (one female and one male) died within six months of the mother's disappearance. However, in another case, both a 2.5-year-old male juvenile (Nv) and his 5.5-year-old elder sister (Nn: age was estimated from her body size) survived for at least two years after their mother's disappearance. At first the siblings were inseparable, but after 6 months Nn, the female, began to spend time alone and Nv, the younger male, began to spend time with a female immigrant and with the adult males. In these cases, it is suspected that the mothers died, rather than emigrated as their juveniles were so small (Y. Shimooka, unpublished data).

These observations together suggest that females may transfer from one group to another several times in their lives. In La Macarena (Tinigua), four neighboring groups located along a river have been monitored for six years; no individual has moved between these groups in that time, except for a single case previously observed by Ahumada (1990). This indicates that female *A. belzebuth belzebuth* may travel significant distances in search of an appropriate group for immigration.

To our knowledge there have been no reports of male immigration in any age class. In La Macarena (Tinigua), there were three cases of disappearance and a single case of death for males (A. Nishimura and Y. Shimooka, unpublished data). The three males who disappeared were full adult males and looked older than the other males. It is unlikely that these males emigrated, but probably died within the group, although their bodies were never found.

Together, these observations support the notion that female dispersal is characteristic of *Ateles*. Future studies in an isolated population, perhaps on an island where we can follow females all through their life wherever they go, will provide more information about female life history.

Conclusions

This chapter represents the first attempt to compile and compare demographic data from multiple *Ateles* species and study sites. The data set analyzed here reveals greater variation in spider monkey group demography than was previously known. We found that some interspecies differences are likely to exist, particularly in group size, composition and IBI, although the sample size for each species is still relatively small. If real interspecific differences exist, it could be due to many factors such as phylogeny, habitat disturbance and habitat carrying capacity, but there is currently no way to determine which of these factors, if any, is more important. In order to clarify this point, it is essential to undertake more extensive comparisons within the same species of *Ateles* found in different sites with different habitat characteristics. As the environments that spider monkeys inhabit are quite variable, comparisons of demographic factors across *Ateles* species will be more meaningful if we know more about the likely environmental influences. In such cases, the investigation of parameters that are more easily affected by environment, such as the number of females or the interbirth interval, will be particularly useful.

As noted above, in spider monkeys, where female dispersal and male philopatry are the norm, the number of males in a group depends completely on the number of male infants and their survival rate. In this study, we gathered observations of lethal events, and found a consistent pattern that males receive more lethal aggression from conspecifics than females. Such aggression likely yields a high male mortality, and thus may help explain the female-biased A and SA sex ratio seen across almost all populations of *Ateles*. As each researcher will likely observe only a few cases in which the cause of death can be determined, exchange of information on anecdotal cases between researchers is very important.

Observed cases of reproduction, immigration and emigration are also rare for this genus, so information is still limited. Information from captive studies can provide information on the reproductive biology and allow us to see any phylogenetic variation. However, the most important contribution would be to conduct longitudinal observations in the wild, documenting the entire lives of spider monkeys from birth to death.

In order to understand spider monkey behavior and social relationships, it is important to consider group history. Recently, DNA studies have become available but analysis is very recent, and thus does not yet provide information pertaining to long-term group structure and demography. Unfortunately, there are some sites where fieldwork is difficult to continue due to political instability or environmental disruption. We hope that long-term studies will continue at most of the sites so that we may compile and update this demographic information again in 10 years' time. We look forward to future comparisons from existing and new study sites through the cooperation of researchers.

Acknowledgements

For this study, Y. Shimooka was supported by a grant from the Monbusho COE Program to O. Takenaka (10CE2005), a grant from the Hope project (JSPS) of KUPRI and the grant for the Biodiversity Research of the 21st Century COE (A14). We thank Mr. Yoshimura T. and Mr. Kato A. for the information on captive spider monkeys at the Japan Monkey Center.

References

Ahumada, J. A. (1989). Behavior and social structure of free ranging spider monkeys (*Ateles belzebuth*) in La Macarena. *Field Studies of New World Monkeys, La Macarena Colombia*, **2**, 7–31.

Ahumada, J. A. (1990). Changes in size and composition in a group of spider monkeys at La Macarena. *Field Studies of New World Monkeys, La Macarena, Colombia*, **4**, 57–60.

Aureli, F., Schaffner, C. M., Verpooten, J., Slater, K. Y. and Ramos-Fernández, G. (2006). Raiding parties of male spider monkeys: insights into human warfare? *Am. J. Phys. Anthropol.*, **131**, 486–497.

Campbell, C. J. (2003). Female directed aggression in free-ranging *Ateles geoffroyi*. *Int. J. Primatol.*, **24**, 223–238.

Campbell, C. J. (2006). Lethal intragroup aggression by adult male spider monkeys (*Ateles geoffroyi*). *Am. J. Primatol.*, **68**, 1197–1201.

Campbell, C. J., Aureli, F., Chapman, C. A., *et al.* (2005). Terrestrial behavior of *Ateles* spp. *Int. J. Primatol.*, **26**, 1039–1051.

Campbell, C. J., Shideler, S. E., Todd, H. E. and Lasley, B. L. (2001). Fecal analysis of ovarian cycles in female black-handed spider monkeys (*Ateles geoffroyi*). *Am. J. Primatol.*, **54**, 79–89.

Chapman, C. A. (1990). Association patterns of spider monkeys: the influence of ecology and sex on social organization. *Behav. Ecol. Sociobiol.*, **26**, 409–414.

Chapman, C. A., Fedigan, L. M., Fedigan, L. and Chapman, L. J. (1989). Post-weaning resource competition and sex ratios in spider monkeys. *Oikos*, **54**, 315–319.

Chinchilla, F. A. (1997). La dieta del jaguar (*Panthera onca*), el puma (*Felis concolor*), y el manigordo (*Felis pardalis*) (Carnivora: Felidae) en el Parque Nacional Corcovado, Costa Rica. *Rev. Biol. Trop.*, **45**, 1223–1229.

Coelho, A., Bramblett, C., Quick, L. and Bramblett, S. (1976). Resource availability and population density in primates: a socio-bioenergetic analysis of the energy budgets of Guatemalan howler and spider monkeys. *Primates*, **17**, 63–80.

Emmons, L. H. (1987). Comparative feeding ecology of felids in a neotropical rainforest. *Behav. Ecol. Sociobiol.*, **20**, 271–283.

Fedigan, L. M. and Baxter, M. J. (1984). Sex differences and social organization in free-ranging spider monkeys (*Ateles geoffroyi*). *Primates*, **25**, 279–294.

JMC (2005). *Annual Report, 2004*. Inuyama: Japan Monkey Centre.

Julliot, C. (1994). Predation of a young spider monkey (*Ateles paniscus*) by a crested eagle (*Morphnus guianensis*). *Folia Primatol.*, **63**, 75–77.

Karesh, W. B., Wallace, R. B., Painter, R. L. E., *et al.* (1998). Immobilization and health assessment of free-ranging black spider monkeys (*Ateles paniscus chamek*). *Am. J. Primatol.*, **44**, 107–123.

Klein, L. L. (1972). The ecology and social organization of the spider monkey, *Ateles belzebuth*. Unpublished Ph.D. thesis, University of California, Berkeley.

Link, A., Palma, A. C., Velez, A. and de Luna, A. G. (2006). Costs of twins in free-ranging white-bellied spider monkeys (*Ateles belzebuth belzebuth*) at Tinigua National Park, Colombia. *Primates*, **47**, 131–139.

Matsuda, I. and Izawa, K. (2007). Predation of wild spider monkeys at La Macarena, Colombia. *Primates* (Online First 10.1007/s10329-007-0042-5).

Matsushita, K. and Nishimura, A. (2000). Ecological and sociological studies in wild spider monkeys MB-3 group at La Macarena, Colombia. In *Adaptive Significance of Fission-Fusion Society in* Ateles, ed. K. Izawa, Sendai: Miyagi University of Education, pp. 57–76 (in Japanese).

Milton, K. (1981). Estimates of reproductive parameters for free-ranging *Ateles geoffroyi*. *Primates*, **22**, 574–579.

Milton, K. and Hopkins, M. E. (2005). Growth of a reintroduced spider monkey (*Ateles geoffroyi*) population on Barro Colorado Island, Panama. In *New Perspectives in the Study of Mesoamerican Primates*, ed. A. Estrada, P. A. Garber, M. Pavelka and L. Luecke, New York: Springer Verlag, pp. 417–435.

Nunes, A. and Chapman, C. A. (1997). A re-evaluation of factors influencing the sex ratio of spider monkey populations with new data from Maraca Island, Brazil. *Folia Primatol.*, **68**, 31–33.

Ramos-Fernández, G., Vick, L., Aureli, F., Schaffner, C. M. and Taub, D. M. (2003). Behavioral ecology and conservation status of spider monkeys in the Otoch Ma'ax Yetel Kooh protected area. *Neotrop. Primates*, **11**, 155–158.

Shimooka, Y. (2000). Ecological and sociological studies in wild spider monkeys MB-2 group at La Macarena, Colombia. In *Adaptive Significance of Fission-Fusion Society in* Ateles, ed. K. Izawa, Sendai: Miyagi University of Education, pp. 37–56 (in Japanese).

Shimooka, Y. (2003). Seasonal variation in association patterns of wild spider monkeys (*Ateles belzebuth belzebuth*) at La Macarena, Colombia. *Primates*, **44**, 83–90.

Shimooka, Y. (2005). Sexual differences in ranging of *Ateles belzebuth belzebuth* at La Macarena, Colombia. *Int. J. Primatol.*, **26**, 385–406.

Stevenson, P. R., Quinones, M. J. and Castellanos, M. C. (2000). *Guia de Frutos de los Bosques del Rio Duda, La Macarena, Colombia*. Santafe de Bogota: IUCN.

Strier, K. B., Boubli, J. R., Possamai, C. B. and Mendes, S. L. (2006). Population demography of northern muriquis (*Brachyteles hypoxanthus*) at the Estação

Biológica de Caratinga, Minas Gerasi, Brazil. *Am. J. Phys. Anthropol.*, **130**, 227–237.

Symington, M. M. (1987a). Sex ratio and maternal rank in wild spider monkeys: when daughters disperse. *Behav. Ecol. Sociobiol.*, **20**, 421–425.

Symington, M. M. (1987b). Ecological and social correlates of party size in the black spider monkey, *Ateles paniscus chamek*. Unpublished Ph.D. thesis, Princeton University, NJ.

Symington, M. M. (1988). Demography, ranging patterns, and activity budgets of black spider monkeys (*Ateles paniscus chamek*) in the Manu National Park, Peru. *Am. J. Primatol.*, **15**, 45–67.

Valero, A., Schaffner, C. M., Vick, L., Aureli, F. and Ramos-Fernandez, G. (2006). Intragroup lethal aggression in wild spider monkeys. *Am. J. Primatol.*, **68**, 732–737.

van Roosmalen, M. G. M. (1985). Habitat preferences, diet, feeding strategy and social organization of the black spider monkey (*Ateles paniscus paniscus* Linnaeus 1758) in Surinam. *Acta Amazonica*, **15**, 1–238.

Wallace, R. B. (1998). The behavioral ecology of black spider monkeys in north-eastern Bolivia. Unpublished Ph.D. thesis, University of Liverpool, UK.

Wright, S. J., Gompper, M. E. and DeLeon, B. (1994). Are large predators keystone species in Neotropical forests? The evidence from Barro Colorado Island. *Oikos*, **71**, 279–294.

Part IV
Interactions with humans

13 Spider monkey conservation in the twenty-first century: recognizing risks and opportunities

GABRIEL RAMOS-FERNÁNDEZ AND ROBERT B. WALLACE

Introduction

Spider monkeys (*Ateles* spp.) occur from southeastern Mexico to the southern Amazonia rain forests of central Bolivia and western Brazil (Kellogg and Goldman, 1944; Hall, 1981; Collins and Dubach, 2000a; Wilson and Reeder, 2005). As with many species from tropical forests, their range has been decreasing as these ecosystems are transformed. This is very clear if we compare the distributions published by Kellogg and Goldman (1944) and by Collins and Dubach (2000a). A decrease in the distribution range of all species of *Ateles* during this period suggests that the numbers of all the taxa are declining. However, there are important differences among the taxa in their current distribution range, as well as in the magnitude and causes of the decline in population size.

When reviewing the conservation status of spider monkeys across their current range, the taxonomy of the group must be considered. In this volume, the Collins and Dubach (2000b) taxonomy has been adopted, which recognizes three distinct species of spider monkeys: *Ateles paniscus*, *A. belzebuth*, and *A. geoffroyi* (with a possible fourth species *A. hybridus* – see also Collins, this volume). Other authors (Groves, 1989; Iracilda da Cunha Sampaio *et al.*, 1993; Rylands *et al.*, 2001) have recognized as many as six distinct species with *A. chamek*, *A. hybridus* and *A. marginatus* all upgraded from *belzebuth* subspecies status. As a conservative approach and for the purposes of conservation planning, we prefer to analyze at the subspecies level thereby recognizing all 16 distinct taxa (Table 13.1).

In this chapter we aim to investigate multiple issues concerning the conservation of spider monkeys. We begin by summarizing the formal conservation status of each recognized *Ateles* taxa and then collate and assess the existing reports of population density and geographic distribution for each taxon. Additionally, we review the major threats to spider monkey populations, detail biological reasons why spider monkeys are vulnerable to negative demographic events

Spider Monkeys: Behavior, Ecology and Evolution of the Genus Ateles, ed. Christina J. Campbell.
Published by Cambridge University Press. © Cambridge University Press 2008.

Table 13.1 *Current conservation status and countries of occurrence of all recognized subspecies of spider monkey (*Ateles*)*

Taxa	IUCN category	CITES category	Countries of occurrence
A. *belzebuth belzebuth*	VU (A2acd)	II	Brazil, Colombia, Ecuador, Peru, Venezuela
A. *belzebuth chamek*	LC	II	Bolivia, Brazil, Peru
A. *belzebuth marginatus*	EN (A4c)	II	Brazil
A. *geoffroyi azuerensis*	CR (B1+2abcde, C2a)	II	Panama
A. *geoffroyi frontatus*	LC	I	Costa Rica, Nicaragua
A. *geoffroyi fusciceps*	CR(B1+2abcde, C2a)	II	Ecuador
A. *geoffroyi geoffroyi*	LC	II	Costa Rica, Nicaragua
A. *geoffroyi grisescens*	EN (B1+2abcde, C2a)	II	Panama (possibly Colombia)
A. *geoffroyi ornatus*	EN (A4c)	II	Costa Rica
A. *geoffroyi panamensis*	EN (B1+2abcde, C2a)	I	Costa Rica, Panama
A. *geoffroyi rufiventris*	VU (A1c, B1+2c)	II	Colombia, Panama
A. *geoffroyi vellerosus*	CR (A4c)	II	El Salvador, Guatemala, Honduras, Mexico
A. *geoffroyi yucatanensis*	VU (A4c)	II	Belize, Guatemala, Mexico
A. *hybridus brunneus*	CR (A3cd)	–	Colombia
A. *hybridus hybridus*	CR (A3cd)	–	Colombia, Venezuela
A. *paniscus paniscus*	LC	II	Brazil, French Guiana, Guyana, Suriname, Venezuela

LC: Least concern; VU: Vulnerable; EN: Endangered; CR: Critically endangered. For definitions of categories and criteria used to assign them (in parentheses), see IUCN (2006). For definitions of CITES categories, see UNEP-WCMC (2007).

and habitat disturbances, and develop a list of clear conservation and research priorities for the immediate future.

Current conservation status

The current CITES status and IUCN category of each recognized taxon is summarized in Table 13.1. All spider monkey species are situated on CITES Appendix II (i.e. not currently threatened with extinction but affected by unregulated trade; not considered is *A. hybridus*; UNEP-WCMC, 2007). Two subspecies of *A. geoffroyi* are on CITES Appendix I (i.e. currently threatened with extinction and affected by current or future trade; *A. g. panamensis* and

A. g. frontatus). At the same time, the IUCN Red List of Threatened Species (IUCN, 2006) considers three of the six species of *Ateles* proposed by Rylands *et al.* (2001) as of Least Concern (*A. chamek*, *A. paniscus*, *A. geoffroyi*), while *A. belzebuth* is considered Vulnerable, *A. marginatus* as Endangered, and *A. hybridus* is considered Critically Endangered. At the subspecies level, the IUCN (2006) considers four subspecies as Least Concern, three as Vulnerable, four as Endangered and five as Critically Endangered. Indeed, the brown spider monkey (*A. hybridus brunneus*) appears on the most recent World's 25 Most Endangered Primate List (Mittermeier *et al.*, 2006).

According to the IUCN categories assigned to subspecies in Table 13.1, the population trends for 8 out of 16 subspecies of *Ateles* have been shown to be in decline (criteria A in the definition of threat category). Five subspecies of *A. geoffroyi* also show a limited distribution (criteria B in the definition of threat category).

Potential population size

We now turn to assessing two variables that directly relate to the potential population size of each recognized taxa: population density and geographic distribution. Table 13.2 summarizes all published studies on population densities at point locations for *Ateles*. Although there is considerable variation (2–89 individuals/km^2), the similarities of most density estimates are evident. Spider monkeys are often one of the most abundant primate species at a given site and are a relatively abundant vertebrate species, with most reported densities falling between 10 and 60 animals per km^2. Most of the point locations where densities have been evaluated are within protected areas.

None of the studies included in Table 13.2 includes an assessment of the long-term viability of the studied populations. Recently, conservation biologists have used different variants of population viability analysis (PVA; Schaffer, 1981) to estimate the minimum population size that will guarantee the survival of a population in a specified time period with a given probability. Although there has been no published PVA on any *Ateles* taxa, Strier (1993) published the results of a PVA of a population of the highly endangered *Brachyteles arachnoides*, a closely related species to *Ateles* spp. The main results of this PVA can help to understand which factors could be critical in determining the probability of extinction of *Ateles* populations. The main variables included in this PVA were directly estimated in a small, isolated population of muriquis in a fragmented habitat in Atlantic coastal Brazil. Female age at first reproduction, natal sex ratios, available habitat, inbreeding effects and the role of catastrophes on reproduction and survival rates were combined using the VORTEX algorithm

Table 13.2 *Population density estimates for recognized subspecies of spider monkey* (Ateles)

Taxa	Site	Forest type	Regime	Density estimate (ind/km^2)	Method	Primate density rank	Source
A. belzebuth belzebuth	Maraca, Brazil	Island	Protected area	7	Line transects	17	Nunes, 1995
A. belzebuth belzebuth	La Macarena, Colombia	Lowland	Protected area	12–15	Line transects	14	Klein and Klein, 1976
A. belzebuth belzebuth	Yasuní, Ecuador	Lowland	Protected area	11.5	Line transects	15	Dew, 2001
A. belzebuth chamek	Manu, Peru	Lowland	Protected area	31	Line transects	6	White, 1986
A. belzebuth chamek	Beni, Bolivia	Floodplain	Protected area	24.6	Line transects	11	Garcia and Tarifa, 1988
A. belzebuth chamek	Manu, Peru	Floodplain	Protected area	28	Line transects	7	Symington, 1987
A. belzebuth chamek	Lago Caiman, Bolivia	Lowland	Protected area	32.5	Line transects	5	Wallace et al., 1998
A. belzebuth chamek	Los Fierros, Bolivia	Lowland	Protected area	84.5	Line transects	2	Braza and Garcia, 1987
A. geoffroyi vellerosus	Barro Colorado, Panama	Island	Protected area	2.1–2.3	Complete group count	20	Campbell, 2000
A. geoffroyi vellerosus	Santa Rosa, Costa Rica	Dry successional deciduous forest, 20–400 years old	Protected area	4.5	Line transects	19	Chapman, 1989
A. geoffroyi vellerosus	Yaxchilán, Mexico	Lowland	Protected archaeological site	17	Opportunistic sampling	13	Estrada et al., 2004
A. geoffroyi vellerosus	Tikal, Guatemala	Lowland	Protected area	27.8	Line transects	8	Cant, 1978
A. geoffroyi vellerosus	Tikal, Guatemala	Lowland	Protected area	56.4	Line transects	4	Coelho et al., 1976
A. geoffroyi vellerosus	Corcovado, Costa Rica	Lowland	Protected area	68.45	Line transects and direct counts	3	Weghorst, 2007

Species	Location	Forest type	Protection status	Density	Method	N	Reference
A. geoffroyi vellerosus	Santa Rosa, Costa Rica	Gradient of regeneration, 0–180 years old	Protected area	2–67	Line transects	–	Sorensen and Fedigan, 2000
A. geoffroyi yucatanensis	Yucatán, Mexico	Lowland 30–50 yr. Successional	Protected area	6.3	Line transects	18	Ramos Fernández and Ayala-Orozco, 2003
A. geoffroyi yucatanensis	Yucatán, Mexico	Lowland old growth	Protected area	89.5	Line transects	1	Ramos Fernández and Ayala-Orozco, 2003
A. geoffroyi yucatanensis	Nuchukux, Mexico	Lowland	Not protected	27.1	Line transects	9	Gonzalez-Kirchner, 1999
A. paniscus paniscus	Raleighvallen-Voltzberg, Suriname	Lowland	Protected area	8.2	Line transects	16	van Roosmalen, 1985
A. geoffroyi yucatanensis	Calakmul, Mexico	Lowland	Protected archaeological site	17.2	Opportunistic sampling	12	Estrada et al., 2004
A. hybridus brunneus	Cerro Bran, Colombia	Humid tropical forest	Not protected	19.6–33.3	Point counts and line transects	10	Green, 1978

(Lacy, 1992). The study found that the probability of the population being extinct over the next hundred years was influenced to the greatest extent by the area of available habitat and by environmental catastrophes. Also important were male-biased sex ratios and female age at first reproduction. In the absence of specific PVA applied to *Ateles* populations, these results can serve as guidelines for estimating which populations may be viable in the long term.

In addition to the guidelines provided by PVA to conserve endangered populations, it is increasingly important for researchers and conservationists to recognize that wildlife populations that are not deemed "viable" by PVA are not necessarily doomed, and many of the world's most charismatic large vertebrate species do not have global populations that reach a "viable" status. Nevertheless, the relatively high population densities of *Ateles* suggest that compared with many other conservation target species their chances for reaching PVA "viable" status in some of the larger protected areas of the region are favorable.

The second variable that directly relates to the population size of a given taxon is its true geographical range. An approximate size for the population of each taxon could be obtained by extrapolating the density estimations made in point locations to the total area of the same vegetation type where the taxon is assumed to occur. However, these extrapolations could seriously overestimate population size without a good knowledge of the degree of natural and human-induced heterogeneity in the landscape, as well as the spider monkeys' responses to this heterogeneity (Turner *et al.*, 2001; see below, Habitat degradation). With this caveat in mind, Figure 13.1 presents the distribution of all *Ateles* taxa, as reported by Collins and Dubach (2000a) with modifications for *A. belzebuth chamek*. It is important to note that these distributions were estimated using a simultaneous analysis of genetic variability among different populations and geographic barriers to dispersal (Collins and Dubach, 2000b), thus reflecting how genetically distinct subspecies are distributed geographically. The last column in Table 13.1 summarizes the countries of occurrence of each *Ateles* subspecies.

Major threats

As tropical forests are transformed and human populations grow throughout the range of all *Ateles* species, their population size and geographic distribution are reduced. Both reductions are usually driven by a combination of massive habitat loss driven by land clearance and agricultural expansion (Mittermeier and Cheney, 1987) and habitat degradation mainly in the form of selective logging and other forms of vegetation disturbance (Bernstein *et al.*, 1976; Freese *et al.*,

1982; Johns, 1991a; Michalski and Peres, 2005). In addition to these factors, overhunting also reduces population size (Freese *et al.*, 1982; Mittermeier, 1987, 1991; Peres, 1990, 1991, 1997; Puertas and Bodmer, 1993). The magnitude of these external threats is magnified by the low intrinsic rate of increase of *Ateles* populations, which makes them vulnerable to demographic perturbations (Milton, 1981; van Roosmalen, 1985; Di Fiore and Campbell 2007; see also Shimooka *et al.*, this volume) and by their feeding habits, which make them particularly vulnerable to habitat disturbance (van Roosmalen, 1985; Johns and Skorupa, 1987; Di Fiore and Campbell, 2007).

Habitat loss

Forests cover almost half of the land surface of Latin America (accounting for 22% of the world's forest area) but the region also shows an increasing deforestation rate (0.51% annual from 2000 to 2005, compared with 0.46% from 1990 to 1999; FAO, 2007). This implies that the region loses about 5 million hectares of forest every year (FAO, 2007).

Among the countries where spider monkeys can be found (Table 13.1), Brazil has the largest forest area, with 478 million ha, more than half of the region's surface of forests (FAO, 2007). In particular, the Brazilian Amazonia (originally 400 million forest ha) is the largest and least disturbed area of tropical forest in Latin America, containing the majority or the totality of the range of *A. belzebuth marginatus*, *A. belzebuth belzebuth*, *A. belzebuth chamek* and *A. paniscus* (Figure 13.1; Collins and Dubach, 2000a). Since 1978, at an annual deforestation rate comparable to the rest of Latin America's countries, the Brazilian Amazonia has lost between 1 and 3 million ha per year (INPE, 2005; FAO, 2006). Particularly the eastern and southern portions of the Brazilian Amazonia show the highest rates of deforestation, thus threatening populations of *A. belzebuth marginatus* and *A.belzebuth chamek* the most (see Figure 13.1). Soares-Filho *et al.* (2006) projected different scenarios of forest cover in the Brazilian Amazon in the year 2050 and predicted that under the "business-as-usual" scenario, primate species that have the majority of their current distribution in Brazilian Amazonia would lose between 60% and 100% of their range.

Another important country for the conservation of *Ateles* taxa is Colombia, where three highly endangered subspecies of *Ateles* can be found: *A. hybridus bruneus*, which is endemic to Colombia, *A. hybridus hybridus*, with half of its range in Colombia and half in Venezuela and *A. geoffroyi robustus*, with about 90% of its range within Colombian territory (Defler *et al.*, 2003). Even though Colombia as a whole shows lower deforestation rates than other Latin American countries (0.1% annual from 1990 to 2005; FAO, 2007), it is the

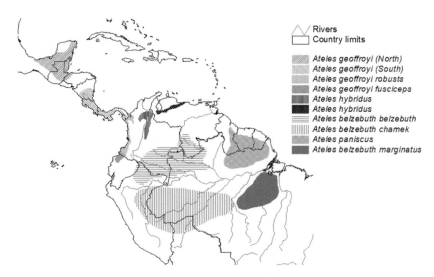

Figure 13.1 Geographical distribution of *Ateles* subspecies, redrawn from Collins and Dubach (2000a) and in reference to important rivers and country limits. *A. geoffroyi (North)* includes *A. g. vellerosus* and *A. g. yucatanensis*, while *A. geoffroyi (South)* includes *A. g. frontatus*, *A. g. geoffroyi*, *A. g. panamensis*, *A. g. azuerensis*, *A. g. grisescens*, and *A. g. ornatus*. Additional point locations for *A. chamek* are indicated for Bolivia (Wallace *et al.*, 1998, 2000, unpublished data; Gómez *et al.*, in press) and Brazil (Iwanaga and Ferrari, 2002).

limited distribution of the *Ateles* subspecies found in this country that make them more vulnerable even to comparatively smaller deforestation rates. In particular, the lowland forest area that lies at the northern end of the Andes in northwestern Colombia is home to *A. hybridus bruneus*. While there are no available deforestation rates for this region, the continuing forest conversion to cattle ranches and agricultural lands within a relatively small area pose a great threat to this subspecies (Defler *et al.*, 2006). The only available population census of *A. hybridus bruneus* in this region was performed in the 1970s (Table 13.2; Green, 1978). The range of *A. geoffroyi robustus* lies in the western slope of the Andean mountain range, but again, there are no available data on how much forest has been lost during the past decade in order to evaluate the degree of threat posed by habitat loss on populations of this subspecies. Defler *et al.* (2003) suggest that hunting pressure on this subspecies is more important than deforestation.

Finally, southeastern Mexico and a large portion of Central America harbor most of the range of all subspecies of *Ateles geoffroyi* (Table 13.1). This region is important because despite a longer history of major forest loss it still maintains large areas of relatively undisturbed forest, but also because it continues to show high deforestation rates: Central America and Mexico average an 0.975% annual deforestation rate, compared with 0.5% annual rate for South America (these figures include temperate and tropical forests; FAO, 2007). Particularly, the tropical forests of southeastern Mexico, which contain most of the range of *A.geoffroyi yucatanensis* and a large portion of the range of *A.geoffroyi vellerosus*, showed an overall annual deforestation rate of 0.9% between 1976 and 2000 (Mas *et al.*, 2004). However, there are important subregional differences both in deforestation rates as well as in the extent of remaining spider monkey habitat within Mexico: while the tropical forests in lowlands near the Gulf of Mexico and the majority of the Isthmus regions have been transformed to pasture and agricultural lands, those in the southeastern Yucatán peninsula and the eastern border with Guatemala still exist in large, continuous extensions (Mas *et al.*, 2004). In fact, the continuous forest extension that includes the southeastern Yucatán peninsula, the northern Guatemala and Belize lowlands (a region commonly known as the Maya forest) is the largest one in Latin America after the Amazon (Vance and Geoghegan, 2002). Nevertheless, the region has been cited as a "hotspot" of tropical deforestation (Achard *et al.*, 1998).

Habitat degradation

Several field studies have demonstrated the potential vulnerability of spider monkeys to habitat disturbance, for example their failure to persist in smaller patches of forest (Lovejoy *et al.*, 1986; Michalski and Peres, 2005), and observed reduced densities in selectively logged or disturbed areas (Bernstein *et al.*, 1976; Freese *et al.*, 1982; Johns, 1991b). With regard to logging, Johns and Skorupa (1987) carried out an analysis to determine which ecological parameters are most useful in predicting the severity of a particular primate species' reaction to logging activities. Their results identified body size and diet type as the best predictors (Johns and Skorupa, 1987). In general, larger frugivorous primate species such as *Ateles* are more susceptible to logging activities (Johns and Skorupa, 1987). Additionally, the preference of spider monkeys for the upper levels of the forest canopy (van Roosmalen and Klein, 1988; Campbell *et al.*, 2005) may increase their susceptibility to logging due to disruptions in many of their aerial pathways.

With regard to habitat degradation in the form of vegetation disturbance and forest fragmentation, it is important to note that a tropical forest that superficially appears as a continuous extension is actually highly heterogeneous, both for natural and anthropogenic causes (Turner *et al.*, 2001). A partially disturbed forest cannot be assumed to contain the same spider monkey density throughout, and neither can an undisturbed forest within which environmental conditions are naturally variable (e.g. water availability, areas naturally disturbed by fire or hurricanes, presence of particular tree species important for the survival of spider monkeys). Indeed, populations with apparently high densities, such as those studied in the Yucatan peninsula in Mexico by Gonzalez-Kirchner (1999) could actually correspond to isolated, less viable populations. This is due to the "crowding effect" of habitat fragmentation (Meffe and Carroll, 1994; Lovejoy *et al.*, 1986) by which a population that originally occupied a large area will temporally exist in high densities in isolated forest remnants. Because some fragments are too small to sustain a viable population in the long term, a temporally high density will be followed by sharp decreases in density. This phenomenon has been observed in several primate species (reviewed in Marsh, 2003).

Michalski and Peres (2005) used a linear regression model to estimate the probability of occurrence of spider monkeys (*A. belzebuth marginatus*) and four other primate species as a function of fragment size in a highly fragmented landscape in southern Brazilian Amazonia. The relevance of this study is that it evaluated the presence and absence of primates in a large number of fragments that varied greatly in size (0.5 to 13 551 ha). The fragments where spider monkeys were absent varied in size from 0.5 to 628 ha, and those in which they were present were 28.6–13 551 ha in size. The probability of occurrence of spider monkeys was 50% or higher in fragments larger than 100 ha, a figure which represents the lower extreme of spider monkey community ranges documented to date (Wallace, this volume). In addition to fragment size, a measure of patch connectivity was a significant predictor of spider monkeys' presence at a given fragment.

When investigating the effect of habitat disturbance upon spider monkey populations it is important to consider the source of disturbance. While forest conversion to cattle ranches and permanent agricultural lands has a strong effect on the presence of spider monkeys (see section on Habitat loss), there are other types of anthropogenic factors that modify the landscape without seriously affecting the species' demographic parameters. This is because spider monkeys can use old stages of successional vegetation, both as feeding areas as well as traveling corridors between patches of old-growth forest (Sorensen and Fedigan, 2000; Ramos-Fernández and Ayala-Orozco, 2003). For example, in the Yucatan peninsula, a large proportion of the habitat of *A. geoffroyi yucatanensis* is also used for slash-and-burn agriculture, in which plots are used for 2–3 years and then abandoned for at least 30 years (García-Frapolli *et al.*, in press). Performed at low densities, in plots no larger than 4 ha, this form of agriculture has shown to be sustainable in terms of the land cover (García-Frapolli *et al.*, in press). In these areas, spider monkeys can use 16–50 year old forest to travel between older fragments of forest but also to feed on tree species that are more abundant in these successional stages (Ramos-Fernández and Ayala-Orozco, 2003). The viability of a population living in such a landscape has been demonstrated by monitoring demographic parameters in two groups studied by Ramos-Fernández *et al.* (2003) over a 6-year period. Similarly Sorensen and Fedigan (2000) found that spider monkeys in Santa Rosa, Costa Rica, returned to forested areas 60–80 years after their conversion to pasture.

An additional factor that will be increasingly important as a source of habitat modification is climate change (reviewed in Hardy, 2003). Patterns of precipitation and temperature variation are important determinants of the fruiting phenology of tropical tree species (Chapman *et al.*, 2005). While this alone would suggest that climate change will impinge upon the ability of spider monkeys to feed successfully in a given area, it is difficult to make an accurate prediction on where, and how, populations will be most affected. Related to this, Korstjens *et al.* (2006) developed a model that correctly predicted the current geographic distribution of spider monkeys based on average annual rainfall and temperature and their annual variation. The strength of this study is that it identified a mechanism by which climatic variables, and thus the availability of food and shelter sites throughout the year, may affect the capacity of spider monkeys to live in a given area. By demonstrating the relationship between time budgets and group size, the authors establish a "maximum ecologically tolerable group size" for spider monkeys in different climatic conditions and, thus, the areas where spider monkeys should occur (Korstjens *et al.*, 2006). Such a model could be useful for predicting the effects of changes in temperature and precipitation patterns upon the future distribution of *Ateles*.

Hunting

Due to their large size and relative palatability *Ateles* is one of the most popular primate hunting targets for many Neotropical rural communities (Peres, 1990; Souza-Mazurek *et al.*, 2000; de Thoisy *et al.*, 2005), and are hence also vulnerable to hunting pressure (Freese *et al.*, 1982; Mittermeier, 1987, 1991; Peres, 1990; de Thoisy *et al.*, 2005; see also Cormier and Urbani, this volume). The threat represented by hunting cannot be underestimated when considering that overhunting is among the principal reasons for catastrophic faunal extinctions after humans arrive to previously uninhabited regions (reviewed in Burney and Flannery, 2005).

In some areas, spider monkeys are subject to very high hunting pressure. For example, Souza-Mazurek *et al.* (2000) found that *A. paniscus* was the most commonly hunted mammal by five Waimiri Atroari villages in central Amazonia, with 421 individuals hunted in one year, accounting for 8% of the total harvest weight. Considering the total population in these five villages, this implies a harvest rate of 1.7 monkeys per consumer-year, a rate that lies at the upper end of the range reported by Redford and Robinson (1987) for *Ateles* spp. in eight different sites throughout Amazonia (0.035 to 1.766 monkeys per consumer-year). It is important to note that 80% of the individuals hunted in the study by Souza-Mazurek *et al.* (2000) were females, who, according to the villagers themselves, have more fat and are more palatable than males. Even though spider monkeys have a female-biased sex ratio (Symington, 1987), the demographic consequences of this high bias toward adult females can be very relevant if one considers the low reproductive rate of spider monkeys (see below). In fact, spider monkeys can be said to be among the first primate species to disappear in a heavily hunted site (Peres, 1990; Souza-Mazurek *et al.*, 2000).

Hunting has been shown to affect *Ateles* population densities. Peres (2000) surveyed hunting levels and population densities of several primate species in 45 sites in Brazilian Amazonia and found that densities of *Ateles* spp. were 83% lower in heavily hunted sites than in sites with lower hunting pressure. Other authors have obtained similar results in various regions, for example *A. belzebuth* in Peru (Bodmer *et al.*, 1997; Nuñez-Iturri and Howe, 2007), *A. geoffroyi* in Costa Rica (Carrillo *et al.*, 2000), *A. paniscus* in French Guiana (de Thoisy *et al.*, 2005), and *A. belzebuth* in Ecuador (Franzen, 2006).

Hunting can also interact synergistically with other anthropogenic disturbances. For example, in fragmented landscapes, where forest fragments are more accessible to hunters, spider monkey populations are much more susceptible to extinction than would be predicted by fragmentation or hunting alone (Peres, 2001). While population density estimates in nonhunted areas suggest that the minimum area that would sustain a viable population of 500 spider monkeys is 113 km^2, this figure increases to 258 km^2 when hunting occurs, if harvest is to be sustainable (Peres, 2001). The large majority of forest fragments throughout southern and eastern Brazilian Amazonia are no larger than 10 ha (Peres, 2001).

At the same time, an overhunted, depressed population of spider monkeys also means that seed dispersal is depressed too (Chapman and Onderdonk, 1998; Link and Di Fiore, 2006; Nuñez-Iturri and Howe, 2007; see also Dew, this volume) and this could have multiple negative consequences for the tree plant community, even in relatively well preserved forests. Muller-Landau (2007) modeled the effect of a reduction of seed dispersal by hunting upon the long-term composition of the plant community. Reduced seed dispersal intensifies kin competition, increases mortality of seeds and seedlings and decreases the proportion of seeds that pass through the digestive system of dispersers. Whether these short-term effects have consequences on long-term reproductive rates and relative abundance of tree species depends to a great extent on the intensity of hunting. If the seed disperser population is simply kept to a constant abundance by hunters (implying decreases of 20% in abundance compared with nonhunting conditions), then the availability of food plants decreases and so the probability of dispersal per seed increases. But in cases where hunting implies large decreases in the seed disperser population (i.e. 80% or more compared with nonhunting abundance), the consequences of greatly diminished seed dispersal include a reduction in reproductive rates that could lead to the local extinction of the tree species (Muller-Landau, 2007). The absence of spider monkeys as seed dispersers also has been shown to affect the genetic structure of their food trees (Pacheco and Simonetti, 2000). *Inga ingoides* seedlings growing under parent plants had more alleles in common in a site where *Ateles paniscus* had been eliminated by hunting than at a nearby site where spider monkeys were still present. This alteration of genetic structure could potentially lead to reduction in tree fitness (Pacheco and Simonetti, 2000).

Disease

Infectious diseases can alter the demographic parameters of a population, to the extent that they can accelerate or bring about its extinction (reviewed in May, 1988; Scott, 1988; Deem *et al.*, 2001). In primate populations, diseases are predicted to decrease the survival of adult females and infants and the number of offspring per capita, all of which are crucial determinants of whether a population will decrease, particularly in situations where other factors, like hunting or habitat loss, are also acting upon the population size (Dobson and Lyles, 1989). As mentioned above, habitat fragmentation can also lead to a temporal increase in population density (Lovejoy *et al.*, 1986; Marsh, 2003), which in turn can lead to an increase in transmission rates and thus augment the longer-term effect of disease on the demographic parameters (Anderson and May, 1979; Scott, 1988).

Wild spider monkeys have been found infected with a range of different parasites (Nunn and Altizer, 2005), many of which are responsible for well-known diseases in humans and in other nonhuman primates. For example, the yellow fever virus, known to devastate populations of closely related howler monkeys (*Alouatta* spp.) throughout Central and South America (reviewed in Crockett and Eisenberg, 1987; Di Bitetti *et al.*, 1994) has been found in *Ateles geoffroyi* in Eastern Panama (Galindo and Srihongse, 1967) and in *Ateles belzebuth chamek* in Bolivia (Karesh *et al.*, 1998). However, it has been more difficult to demonstrate that these parasitic infections actually cause disease, and that these in turn have a negative effect on the demographic parameters of wild animal populations (Smith *et al.*, 2006). As a conservative measure, researchers on wild primate populations should consider the possibility that they may be introducing pathogens into their study population (Wallis and Lee, 1999).

Vulnerability issues

The reproductive biology of *Ateles* makes them particularly vulnerable to negative demographic events. Female spider monkeys do not reach sexual maturity until they are at least 5 years old (Klein, 1971; Eisenberg, 1976; Milton, 1981; van Roosmalen, 1985; Shimooka *et al.*, this volume), are reproductively active in the wild up to at least 22 years of age (Milton, 1981), and have lengthy gestation periods of between 226 and 232 days (Klein, 1971; Eisenberg, 1976; Milton, 1981). Perhaps the most significant reproductive factor to consider, however, is that of the exceptionally prolonged interbirth intervals occurring in *Ateles*, with a range of 2–5 years and 3–4 years being typical in free-ranging populations (Klein, 1971; van Roosmalen, 1985; Symington, 1987; Fedigan and Rose, 1995; Ramos-Fernández *et al.*, 2003; see also Shimooka *et al.*, this volume). These prolonged interbirth intervals are associated with long periods of infant dependency, up to 3 years in free-ranging populations (van Roosmalen, 1985; Di Fiore and Campbell, 2007).

Together, the above reproductive characteristics contribute to a low intrinsic rate of increase in free-ranging *Ateles* populations, which is likely to impose severe limitations on their ability to recover from demographic disturbances. For example, in Robinson and Redford's (1986) review of intrinsic rates of natural increase, *A. belzebuth*, *A. geoffroyi* and *A. paniscus* were lowest in a list of 39 Neotropical mammals which included 13 species of primates. Bodmer *et al.* (1997) found that mammal species with long life-spans, low rates of increase and long generation times made them more vulnerable to extinction by hunting pressure than other species.

The reliance of spider monkeys on ripe fruit has been cited as a key determinant of their vulnerability to habitat loss and degradation (van Roosmalen, 1985; Kinzey, 1997), because it is assumed that only a highly diverse tree community would support a viable population of spider monkeys. However, Symington (1988) suggests that it is the presence of particular keystone species, rather than the total availability of fruiting trees, that explains the differences between the densities of spider monkeys among different populations. Also, frugivory has been noted as one of the factors permitting different primate species in areas with shifting cultivation to survive better than folivorous primates (Cowlishaw and Dunbar, 2000: 239; although the opposite relationship was found in areas subject to selective logging). The results by Michalski and Peres (2005) on the distribution of spider monkeys among fragments of varying size suggest that

Ateles spp. do require relatively undisturbed, large areas of forest, although studies in partially degraded forests, such as those by Ramos-Fernández and Ayala-Orozco (2003) and Sorensen and Fedigan (2000), show that spider monkeys can also cope with some degree of habitat loss and disturbance, provided that they are not hunted.

Conservation and research priorities

Given that a large majority of the habitat for almost all of the subspecies of *Ateles* still exists as large tracts of continuous forest (with the exception of *A. hybridus bruneus*, which has a very limited geographical distribution lying within a highly disturbed region of Colombia: Defler *et al.*, 2006), ensuring the permanence of these areas of continuous forest should be the highest priority in the conservation of spider monkeys. Halting or decreasing deforestation rates in Latin America's tropical forests is a complex task that requires the creation and adequate management and administration of protected areas by national governments, but also the involvement of the international community. To illustrate this, it is useful to compare the different scenarios used by Soares-Filho *et al.* (2006) to predict how much forest in the Brazilian Amazonia will be preserved by the year 2050. One of these scenarios, "business-as-usual," assumes that recent deforestation trends will continue, that all highways currently scheduled to pass through remote regions of Amazonia will be paved and that new private or public protected areas will not be created. This scenario is contrasted with others in which new protected areas are created, private owners are assumed to comply with regulations requiring the protection of some of their land, there is a plan of land use for agroecosystems, and new incentives are provided from international markets for cattle ranchers and soy farmers. These "governance" scenarios (Soares-Filho *et al.*, 2006) increase the amount of original forest cover predicted to remain in 2050 from 53% in the "business-as-usual" scenario to 73% of the original forest cover. Recommendations that will permit the shift from the "business-as-usual" to the "governance" scenarios include the expansion of market pressures for sound land management and the trade of carbon credits derived from the avoidance of carbon emissions by deforestation. Given that the survival of important populations of *Ateles* crucially depends on the forests of the Brazilian Amazonia (see above), these recommendations should be considered as priorities for the conservation of spider monkeys as well.

The creation of protected areas and the enforcement of regulations are also conservation priorities for decreasing deforestation rates (Soares-Filho *et al.*, 2006; Peres, 2005). In order to prioritize forested areas according to their relevance as habitat for more threatened spider monkey taxa, researchers could make use of the wide array of bioclimatic models (reviewed in Pearson and Dawson, 2003), which take into account data on occurrences in point locations (Table 13.2) and from there extrapolate to a larger area of suitable habitat given the physical and biological conditions at the known point locations. Another alternative is the use of time-budget models, such as that developed by Korstjens *et al.* (2006), in which the presence and absence of the species is predicted on the basis of climatic variables such as temperature and precipitation and intervening variables such as the time budget and group size (often available in studies such as those listed in Table 13.2). Both modeling approaches would provide conservation planners with information on where the important populations of more threatened *Ateles* spp. are found and how they could be protected (see for example Margules and Pressey, 2000), although several of the more threatened taxa already have relatively restricted distributions (Table 13.1).

In addition to deforestation, spider monkeys can disappear from a tropical forest, protected or not, due to hunting pressures and habitat degradation. Protected areas in the tropics have proven to be effective in slowing down or halting deforestation, but not necessarily logging, hunting and other extractive activities (Bruner *et al.*, 2001). For example, in an analysis of 8 parks under "Strict Protection" (categories I and II of the IUCN) in Ecuador and Peru, Naughton-Treves *et al.* (2006) found that hunting was still performed in an average the 14% of the total area under protection, compared with an average of 28% of the area of 6 parks under "Multiple Use" (categories III and IV of the IUCN) in the same two countries. Strict regulations and strong enforcement of regulations are two of the highest priorities in the management of protected areas (Bruner *et al.*, 2001). However, because local community survivorship often depends on the resources from protected areas, one crucial component should also be the compensation for the prohibition of extractive activities (Bruner *et al.*, 2001). It should be noted that realistically, from a sustainable finance perspective, compensation may have to take more indirect forms such as support for alternative natural resource management and/or carbon offset initiatives.

Important spider monkey habitat also exists in areas where local populations use the forest in a more or less sustainable way (e.g. Peres, 1994; Bray *et al.*, 2003). Therefore, in addition to creating and efficiently managing nationally protected areas, community-based, local conservation initiatives should also be promoted (reviewed in Gibson *et al.*, 2000). While the success of this approach to conserving biodiversity is highly dependent on local conditions (Molnar *et al.*, 2004), it is clear that the creation of protected areas is fraught with several social and political problems that can be solved if promoting the stewardship of resources by local communities (Brechin *et al.*, 2002). In those cases in which the extractive activities they perform are not sustainable (as in the case of spider monkey hunting in several areas, see Peres 1990, 2001) supporting the development of alternative economic activities then becomes a crucial strategy toward the conservation of spider monkeys and their habitat. One of the possible solutions lies in striking a balance between the intensity with which local communities use several resources simultaneously. An example comes from Western Panama, where a shifting cultivation landscape has proven to provide sustainable hunting yields to indigenous people (Smith, 2005). In this study, all individuals of *A. geoffroyi* were captured in areas of old-growth vegetation, but they were a minority of the total game harvest, which was mainly composed of animals that use young secondary vegetation successfully, such as red brocket deer (*Mazama americana*), agouti (*Dasyprocta punctata*) and collared peccary (*Tayassu tajacu*). If the presence of game species that use secondary vegetation provides humans with bushmeat, this reduces the hunting pressure on spider monkeys and provides an incentive to protecting areas of this kind of vegetation in a shifting cultivation landscape. Such multiple use strategy by indigenous peoples throughout Latin America has been advocated as an alternative way to protect biodiversity in general, and important forest areas in particular (Molnar *et al.*, 2004; García-Frapolli *et al.*, 2007).

Another issue that applies both to protected areas and to conserved areas of forest under other regulations (communally or privately owned, for example) is whether they actually serve to protect what they are intended to protect. This is a problem that requires a wide knowledge of the species to be protected, its habitat requirements and conditions that determine a viable population (Fahrig, 2001). Therefore, one of the most important research priorities should be on the factors that determine the viability of *Ateles* populations in different habitats. As mentioned above (section on Potential population size), one of the factors that Strier (1993) found to be crucial in determining the viability of a decreasing population of *Brachyteles arachnoides* was the area of available habitat. However, as explained above, sometimes it is difficult to determine whether secondary vegetation constitutes available habitat for *Ateles* or not, as it depends crucially on the successional stage and the tree composition. It would be important to understand the relationship between the source and intensity of habitat disturbance and the probability of extinction of a population of spider monkeys. Moreover, in spatially heterogeneous habitats, factors such as habitat patch size, shape and connectivity can all be crucial determinants of a species' ability to use a given landscape in a continuous or more limited fashion (Turner *et al.*, 2001, p. 217; Tischendorf and Fahrig, 2000) and thus to survive in viable populations. Therefore, whether a viable population of spider monkeys will survive in a degraded habitat is closely related to the issue of habitat connectivity: in fragmented landscapes, what are the minimum conditions of connectivity between fragments, at various spatial scales, that would permit genetic exchange between subpopulations in different fragments? In the case of spider monkeys, it is necessary to know how a single spider monkey group can use different habitat fragments, how individuals migrate between groups in different habitat fragments and how populations might be linked by infrequent migration events. These are unresolved questions that may need in-depth studies of diet and ranging patterns, together with genetic analyses to determine the exchange between subpopulations (see Di Fiore *et al.*, this volume; Wallace, this volume; Collins, this volume).

Finally, there is a pressing need to establish the frequency with which reported infections actually become important diseases in wild populations of *Ateles*, when these diseases have actual demographic consequences. It is necessary that, when handling wild spider monkeys, researchers include health studies and develop ties with veterinarians such that if die-offs or disease events occur at their study sites, colleagues can be quickly called in to investigate.

Conclusions

In general, if existing protected areas in the region can be managed efficiently and financed sustainably in the future, then given their relatively high natural population densities most spider monkey species would have a number of guaranteed and significant populations under protection, particularly compared with other larger vertebrates and indeed other primate species in the region. Of course the reality is that even this relatively straightforward challenge will require a considerable effort by the conservation community over the coming decades.

At the genus level, the future of wild *Ateles* populations is tightly linked to that of continuous extensions of tropical forests, particularly in Brazilian Amazonia. However, the subspecies that face the bleakest scenarios in the near future, due to habitat loss and hunting, are distributed outside of Brazil: *A. hybridus brunneus, A. h. hybridus, A. geoffroyi azuerensis, A. g. fusciceps, A. g. grisescens* and *A. g. vellerosus* (Table 13.1). All IUCN and CITES categories for these subspecies are established with information from different times and degree of certainty (IUCN, 2006; UNEP-WCMC, 2007). Therefore, they should be continuously updated, given the rapid increases in habitat loss and degradation, as well as the demonstrated negative effect of hunting on many wild populations.

The conservation of *Ateles* is embedded within larger issues that involve ecosystem functioning and landscape-level processes, which can help conservationists focus on the right level when planning their actions. Their important role as seed dispersers suggests that it is not enough to justify conservation actions based on the threats faced by a particular population or taxon. It is necessary to refer to the potential long-term consequences of their local extinction upon tree species diversity and ecosystem structure. Similarly, their requirement of habitat fragments above a certain size that lie within a vegetation matrix of a particular quality suggests that it is not enough to ensure the permanence of forest fragments if the connections to other fragments are not ensured. In fact, spider monkeys could serve as indicators of fragment connectivity, given their flexibility in using secondary vegetation, but also their requirement of relatively well preserved old-growth forest fragments.

Finally, the conservation of *Ateles* is crucially dependent on the degree to which local communities in forested areas can fulfill their demands for agricultural lands and animal protein in a sustainable way. National governments and international organizations should do everything they can to create protected areas and enforce regulations within them, but ultimately the survival of these areas, with all their biodiversity, will depend on the quality of the much larger forested areas in which they are embedded.

References

Achard, F., Eva, H., Glinni, A., *et al*. (1998). *Identification of Deforestation Hot Spot Areas in the Humid Tropics*. Trees Publ. Series B. Research Report No. 4. Brussels: Space Application Institute, Global Vegetation Monitoring Unit, Joint Research Centre, European Commission.

Anderson, R. M. and May, R. M. (1979). Population biology of infectious diseases. *Nature*, **280**, 361–367.

Bernstein, I. S., Balcaen, P., Dresdale, L., *et al*. (1976). Differential effects of forest degradation on primate populations. *Primates*, **17**, 401–411.

Bodmer, R. E., Eisenberg, J. F. and Redford, K. H. (1997). Hunting and the likelihood of extinction of Amazonian mammals. *Conserv. Biol.*, **11**, 460–466.

Braza, F. and Garcia, J. E. (1987). Rapport préliminaire sur les singes de la région montagneuse de Huanchaca, Bolivie. *Folia Primatol.*, **49**, 182–186.

Bray, D. B., Merino-Pérez, L., Negreros-Castillo, P., *et al*. (2003). Mexico's community-managed forests as a global model for sustainable landscapes. *Conserv. Biol.*, **17**, 672–677.

Brechin, S. R., Wilshusen, P. R., Fortwangler, C. L. and West, P. C. (2002). Beyond the square wheel: toward a more comprehensive understanding of biodiversity conservation as social and political process. *Soc. Nat. Res.*, **15**, 41–64.

Bruner, A. G., Cullison, R. E., Rice, R. E. and da Fonseca, G. A. B. (2001). Effectiveness of parks in protecting tropical biodiversity. *Science*, **291**, 125–128.

Burney, D. A. and Flannery, T. F. (2005). Fifty millennia of catastrophic extinctions after human contact. *Trends Ecol. Evol.*, **20**, 395–401.

Campbell, C. J. (2000). The reproductive biology of black-handed spider monkeys (*Ateles geoffroyi*): integrating behavior and endocrinology. Unpublished Ph.D. thesis, University of California, Berkeley.

Campbell, C. J., Aureli, F., Chapman, C. A., *et al.* (2005). Terrestrial behavior of *Ateles* spp. *Int. J. Primatol.*, **26**, 1039–1051.

Cant, J. G. H. (1978). Population survey of the spider monkey *Ateles geoffroyi* at Tikal, Guatemala. *Primates*, **19**, 525–535.

Carrillo, E., Wong, G. and Cuarón, A. D. (2000). Monitoring mammal populations in Costa Rican protected areas under different hunting restrictions. *Conserv. Biol.*, **14**, 1580–1591.

Chapman, C. A. (1989). Primate populations in northwestern Costa Rica: potential for recovery. *Primate Conserv.*, **10**, 37–44.

Chapman, C. A. and Onderdonk, D. A. (1998). Forests without primates: primate/plant codependency. *Am. J. Primatol.*, **45**, 127–141.

Chapman, C. A., Chapman, L. J., Struhsaker, T. T., Zanne, A. E., Clark, C. J. and Poulsen, J. R. (2005). A long-term evaluation of fruiting phenology: importance of climate change. *J. Trop. Ecol.*, **21**, 1–14.

Coelho, A. M., Jr, Coelho, L. S., Bramblett, C. A., Bramblett, S. S. and Quick, L. B. (1976). Ecology, population characteristics and sympatric association in primates: a sociobioenergetic analysis of howler and spider monkeys in Tikal, Guatemala. *Yrbk. Phys. Anthropol.*, **20**, 96–135.

Crockett, C. M. and Eisenberg, J. F. (1987). Howlers: variations in group size and demography. In *Primate Societies*, ed. B. B. Smuts, D. L. Cheney, R. M. Seyfarth, R. W. Wrangham and T. T. Struhsaker, Chicago: University of Chicago Press, pp. 54–68.

Collins, A. C. and Dubach, J. M. (2000a). Biogeographic and ecological forces responsible for speciation in *Ateles*. *Int. J. Primatol.*, **21**, 421–444.

Collins, A. C. and Dubach, J. M. (2000b). Phylogenetic relationships of spider monkeys (*Ateles*) based on mitochondrial DNA variation. *Int. J. Primatol.*, **21**, 381–420.

Cowlishaw, G. and Dunbar, R. (2000). *Primate Conservation Biology*. Chicago: University of Chicago Press.

Deem, S. L., Karesh, W. B. and Weisman, W. (2001). Putting theory into practice: wildlife health in conservation. *Conserv. Biol.*, **15**, 1224–1233.

Defler, T. R., Morales, A. L. and Rodríguez, J. V. (2006). Brown spider monkey, *Ateles hybridus brunneus* (Gray, 1872). In *Primates in Peril: The World's 25 Most Endangered Primates 2004–2006*, ed. R. A. Mittermeier, C. Valladares-Pádua, A. B. Rylands, *et al.* Report to IUCN/SSC Primate Specialist Group (PSG), International Primatological Society (IPS) and Conservation International (CI), Washington, DC, p. 23.

Defler, T. R., Rodríguez, J. M. and Hernández-Camacho, J. I. (2003). Conservation priorities for Colombian primates. *Primate Conserv.*, **19**, 10–18.

Dew, J. L. (2001). Synecology and seed dispersal in woolly monkeys (*Lagothrix lagotricha poeppigii*) and spider monkeys (*Ateles belzebuth belzebuth*) in Parque Nacional Yasuni, Ecuador. Unpublished Ph.D. thesis, University of California, Davis.

Di Bitetti, M. S., Placci, G., Brown, A. D. and Rode, D. I. (1994). Conservation and population status of the brown howling monkey (*Alouatta fusca clamitans*) in Argentina. *Neotrop. Primates*, **2**, 1–4.

Di Fiore, A. and Campbell, C. J. (2007). The atelines: variation in ecology, behavior and social organization. In *Primates in Perspective*, ed. C. J. Campbell, A. Fuentes, K. C. MacKinnon, M. Panger and S. K. Beader, New York: Oxford University Press, pp. 155–185.

Dobson, A. P. and Lyles, A. M. (1989). The population dynamics and conservation of primate populations. *Conserv. Biol.*, **3**, 362–380.

Eisenberg, J. F. (1976). Communication mechanisms and social integration in the black spider monkey, *Ateles fusciceps robustus*, and related species. *Smithson. Contrib. Zool.*, **213**.

Estrada A., Luecke, L., van Belle, S., Barreta, E. and Meda, M. R. (2004). Survey of black howler (*Alouatta pigra*) and spider (*Ateles geoffroyi*) monkeys in the Mayan sites of Calakmul and Yaxchilan, Mexico and Tikal, Guatemala. *Primates*, **45**, 33–39.

Fahrig, L. (2001). How much habitat is enough? *Biol. Conserv.*, **100**, 65–74.

FAO (Food and Agriculture Organization of the United Nations) (2006). *Tendencias y perspectivas del sector forestal en América Latina y el Caribe*. Rome: FAO (in Spanish).

FAO (Food and Agriculture Organization of the United Nations) (2007). *State of the World's Forests*. Rome: FAO.

Fedigan, L. M. and Rose, L. M. (1995). Interbirth interval variation in three sympatric species of Neotropical monkey. *Am. J. Primatol.*, **37**, 9–24.

Franzen, M. (2006). Evaluating the sustainability of hunting: a comparison of harvest profiles across three Huaorani communities. *Environ. Conserv.*, **33**, 36–45.

Freese, C. H., Heltne, P. G., Castro, R. N. and Whitesides, G. (1982). Patterns and determinants of monkey densities in Peru and Bolivia, with notes on distributions. *Int. J. Primatol.*, **3**, 53–90.

Galindo, P. and Srihongse, S. (1967). Evidence of recent jungle yellow-fever activity in eastern Panama. *Bull. World Health Organ.*, **36**, 151–161.

Garcia, J. E. and Tarifa, T. (1988). Primate survey of the Estacion Biologica Beni, Bolivia. *Primate Conserv.*, **9**, 97–100.

García-Frapolli, E., Ayala-Orozco, B., Bonilla-Moheno, M., Espadas-Manrique, C. and Ramos-Fernández, G. (2007). Biodiversity conservation, traditional agriculture and ecotourism: land cover/land use change projections for a natural protected area in the northeastern Yucatan Peninsula, Mexico. *Landscape Urban Plan*, **83**, 137–153.

Gibson, C. C., McKean, M. A. and Ostrom, E. (2000). *People and Forests: Communities, Institutions, and Governance*. Cambridge, MA: MIT Press.

Green, K. M. (1978). Primate censusing in northern Colombia: a comparison of two techniques. *Primates*, **19**, 537–550.

Gómez, H., Ayala, G. and Wallace, R. B. (in press). Biomasa de primates y ungulados en bosques amazonicos preandinos en el Parque Nacional y Area Natural de Manejo Integrado Madidi (La Paz, Bolivia). *Mastozoología Neotropical*.

Gonzalez-Kirchner, J. (1999). Habitat use, population density and subgrouping pattern of the Yucatan spider monkey (*Ateles geoffroyi yucatanensis*) in Quintana Roo, Mexico. *Folia Primatol.*, **70**, 55–60.

Groves, C. P. (1989). *A Theory of Human and Primate Evolution*. Oxford: Clarendon Press.

Hall, E. R. (1981). *The Mammals of North America*, Vol. 1. New York: John Wiley & Sons.

Hardy, J. T. (2003). *Climate Change: Causes, Effects and Solutions*. New York: John Wiley & Sons.

INPE (Instituto Nacional de Pesquisas Espaciais) (2005). Monitoramento da floresta amazónica brasileira por satélite / Monitoring of the Brazilian Amazon forest by satellite: 1988–2005. Sao Paulo: INPE, Sao José dos Campos (in Portuguese).

Iracilda da Cunha Sampaio, M., Cruz Schneider, M. P. and Schneider, H. (1993). Contribution of genetic distance studies to the taxonomy of *Ateles*, particularly *Ateles paniscus paniscus* and *Ateles paniscus chamek*. *Int. J. Primatol.*, **14**, 895–903.

IUCN (2006). *2006 IUCN Red List of Threatened Species*. Online, www.iucnredlist. org: accessed June 17, 2007.

Iwanaga, S. and Ferrari, S. F. (2002). Geographic distribution and abundance of woolly (*Lagothrix cana*) and spider (*Ateles chamek*) monkeys in southwestern Brazilian Amazonia. *Am. J. Primatol.*, **56**, 57–64.

Johns, A. D. (1991a). Vertebrate responses to selective logging: implications for the design of logging systems. *Phil. Trans. Roy. Soc. Lond.*, **335**, 437–442.

Johns, A. D. (1991b). Forest disturbance and Amazonian primates. In *Primate Responses to Environmental Change*, ed. H. O. Box, London: Chapman and Hall.

Johns, A. D. and Skorupa, J. P. (1987). Responses of rain-forest primates to habitat disturbance: a review. *Int. J. Primatol.*, **8**, 157–191.

Karesh, W. B., Wallace, R. B., Painter, R. L. E., *et al.* (1998). Immobilization and health assessment of free-ranging black spider monkeys (*Ateles paniscus chamek*). *Am. J. Primatol.*, **44**, 107–123.

Kellogg, R. and Goldman, E. A. (1944). Review on the spider monkeys. *Proc. US Natl. Mus. Nat. Hist.*, **96**, 1–45.

Kinzey, W. G. (1997). Synopsis of New World primates (16 genera). In *New World Primates: Ecology, Evolution, and Behavior*, ed. W. G. Kinzey, New York: Aldine de Gruyter, pp. 169–324.

Klein, L. L. (1971). Observations on copulation and seasonal reproduction of two species of spider monkeys, *Ateles belzebuth and A. geoffroyi. Folia Primatol.*, **15**, 233–248.

Klein, L. L. and Klein, D. (1976). Neotropical primates: aspects of habitat usage, population density, and regional distribution in La Macarena, Colombia. In *Neotropical Primates: Field Studies and Conservation* ed. R. W. Thorington, Jr. and P. G. Heltne, Washington, DC: National Academy of Sciences, pp. 70–78.

Korstjens, A. H., Verhoeckx, I. L. and Dunbar, R. I. M. (2006). Time as a constraint on group size in spider monkeys. *Behav. Ecol. Sociobiol.*, **60**, 683–694.

Lacy, R. C. (1992). VORTEX: a computer simulation model for population viability analysis. *Wildlife Res.*, **20**, 45–65.

Link, A. and Di Fiore, A. (2006). Seed dispersal by spider monkeys and its importance in the maintenance of neotropical rain-forest diversity. *J. Trop. Ecol.*, **22**, 235–246.

Lovejoy, T. E., Bierregaard, R. O., Jr., Rylands, A. B., *et al.* (1986), Edge and other effects of isolation on Amazon forest fragments. In *Conservation Biology: The Science of Scarcity and Diversity*, ed. M. E. Soulé, Sunderland, MA: Sinauer Associates, pp. 257–285.

Margules, C. R. and Pressey, R. L. (2000). Systematic conservation planning. *Nature*, **405**, 243–253.

Marsh, L. K. (2003). *Primates in Fragments: Ecology and Conservation*. New York: Kluwer Academic / Plenum Press.

Mas, J. F., Velazquez, A., Diaz-Gallegos, J. R., *et al.* (2004). Assessing land use/cover changes: a nationwide multidate spatial database for Mexico. *Int. J. App. Earth Obs. Geoinform.*, **5**, 249–261.

May, R. M. (1988). Conservation and disease. *Conserv. Biol.*, **2**, 28–30.

Meffe, G. K. and Carroll, C. R. (1994). *Principles of Conservation Biology*. Sunderland, MA: Sinauer Associates.

Michalski, F. and Peres, C. A. (2005). Anthropogenic determinants of primate and carnivore local extinctions in a fragmented forest landscape of southern Amazonia. *Biol. Conserv.*, **124**, 383–396.

Milton, K. (1981). Estimates of reproductive parameters for free-ranging *Ateles geoffroyi*. *Primates*, **22**, 574–579.

Mittermeier, R. A. (1987). Effects of hunting on rain forest primates. In *Primate Conservation in the Tropical Rain Forest*, ed. C. W. Marsh and R. A. Mittermeier, New York: Alan R. Liss Inc., pp. 109–146.

Mittermeier, R. A. (1991). Hunting and its effect on wild primate populations in Suriname. In *Neotropical Wildlife Use and Conservation*, ed. J. G. Robinson and K. H. Redford, Chicago: University of Chicago Press, pp. 93–107.

Mittermeier, R. A., and Cheney, D. L. (1987). Conservation of primates and their habitats. In *Primate Societies*, ed. B. B. Smuts, D. L. Cheney, R. M. Seyfarth, R. W. Wrangham and T. T. Struhsaker, Chicago: University of Chicago Press, pp. 477–490.

Mittermeier, R. A., Valladares-Padua, C., Rylands, A. B., *et al.* (2006). *Primates in Peril: The World's 25 Most Endangered Primates 2004–2006*. Unpublished report, Arlington, VA: IUCN/SSC Primate Specialist Group–International Primatological Society–Conservation International.

Molnar, A., Scherr, S. J. and Khare, A. (2004). *Who Conserves the World's Forests? A New Assessment of Conservation and Investment Trends*. Washington DC: Forest Trends – Ecoagriculture Partners.

Muller-Landau, H. C. (2007). Predicting the long-term effects of hunting on plant species composition and diversity in tropical forests. *Biotropica*, **39**, 372–384.

Naughton-Treves, L., Alvarez-Berríos, N., Brandon, K., *et al.* (2006). Expanding protected areas and incorporating human resource use: a study of 15 forest parks in Ecuador and Peru. *Sustainability: Science, Practice, & Policy*, **2**, 32–44.

Nunes, A. (1995). Foraging and ranging patterns in white-bellied spider monkeys. *Folia Primatol.*, **65**, 85–99.

Nunn, C. L. and Altizer, S. (2005). The Global Mammal Parasite Database: an online resource for infectious disease records in wild primates. *Evol. Anthropol.*, **14**, 1–2.

Nuñez-Iturri, G. and Howe, H. F. (2007). Bushmeat and the fate of trees with seeds dispersed by large primates in a lowland rain forest in Western Amazonia. *Biotropica*, **39**, 348–354.

Pacheco, L. F. and Simonetti, J. A. (2000). Genetic structure of a mimosoid tree deprived of its seed disperser, the spider monkey. *Conserv. Biol.*, **14**, 1766–1775.

Pearson, R. G. and Dawson, T. P. (2003) Predicting the impacts of climate change on the distribution of species: are bioclimate envelope models useful? *Glob. Ecol. Biogeog.*, **12**, 361–371.

Peres, C. A. (1990). Effects of hunting on Western Amazonian primate communities. *Biol. Conserv.*, **54**, 47–59.

Peres, C. A. (1991). Humboldt's woolly monkeys decimated by hunting in Amazonia. *Oryx*, **25**, 89–95.

Peres, C. A. (1994). Indigenous reserves and nature conservation in Amazonian forests. *Conserv. Biol.*, **8**, 586–588.

Peres, C. A. (1997). Effects of habitat quality and hunting pressure on arboreal folivore densities in Neotropical forests: a case study of howler monkeys (*Alouatta* spp.). *Folia Primatol.*, **68**, 199–222.

Peres, C. A. (2000). Evaluating the impact and sustainability of subsistence hunting at multiple Amazonian forest sites. In *Hunting for Sustainability in Tropical Forests*, ed. J. G. Robinson and E. L. Bennett, New York: Columbia University Press, pp. 31–56.

Peres, C. A. (2001). Synergistic effects of subsistence hunting and habitat fragmentation on Amazonian forest vertebrates. *Conserv. Biol.*, **15**, 1490–1505.

Peres, C. A. (2005). Why we need megareserves in Amazonia. *Conserv. Biol.*, **19**, 728–733.

Puertas, P. and Bodmer, R. E. (1993). Conservation of a high diversity primate assemblage. *Biodiv. Conserv.*, **2**, 586–593.

Ramos-Fernández, G. and Ayala-Orozco, B. (2003). Population size and habitat use of spider monkeys at Punta Laguna, Mexico. In *Primates in Fragments: Ecology and Conservation*, ed. L. K. Marsh, New York: Kluwer Academic / Plenum, pp. 191–209.

Ramos-Fernández, G., Vick, L. G., Aureli, F., Schaffner, C. and Taub, D. M. (2003). Behavioral ecology and conservation status of spider monkeys in the Otoch Ma'ax Yetel Kooh protected area. *Neotrop. Primates*, **11**, 157–160.

Redford, K. H. and Robinson, J. G. (1987). The game of choice: patterns of indian and colonist hunting in the Neotropics. *Am. Anthropol.*, **89**, 650–667.

Robinson, J. G. and Redford, K. H. (1986). Intrinsic rate of natural increase in Neotropical forest mammals: relationship to phylogeny and diet. *Oecologia*, **68**, 516–520.

Rylands, A. B., Mittermeier, R. A. and Konstant, W. R. (2001). Species and subspecies of primates described since 1990. *Neotrop. Primates*, **9**, 75–78.

Scott, M. E. (1988). The impact of infection and disease on animal populations: implications for conservation biology. *Conserv. Biol.*, **2**, 40–56.

Shaffer, M. L. (1981). Minimum population sizes for species conservation. *BioScience*, **31**, 131–134.

Smith, D. A. (2005). Garden game: shifting cultivation, indigenous hunting and wildlife ecology in Western Panama. *Hum. Ecol.*, **33**, 505–537.

Smith, K. F., Sax, D. F. and Lafferty, K. D. (2006). Evidence for the role of infectious disease in species extinction and endangerment. *Conserv. Biol.*, **20**, 1349–1357.

Soares-Filho, B. S., Nepstad, D. C., Curran, L. M., *et al.* (2006). Modeling conservation in the Amazon basin. *Nature*, **440**, 520–523.

Sorensen, T. C. and Fedigan, L. M. (2000). Distribution of three monkey species along a gradient of regenerating tropical dry forest. *Biol. Conserv.*, **92**, 227–240.

Souza-Mazurek, R. R., Pedrinho, T., Feliciano, X., *et al.* (2000). Subsistence hunting among the Waimiri Atroari Indians in central Amazonia, Brazil. *Biodiv. Conserv.*, **9**, 579–596.

Strier, K. B. (1993). Viability analyses of an isolated population of muriqui monkeys (*Brachyteles arachnoides*): implications for primate conservation and demography. *Primate Conserv.*, **14–15**, 43–52.

Symington, M. (1987). Sex ratio and maternal rank in wild spider monkeys: when daughters disperse. *Behav. Ecol. Sociobiol.*, **20**, 421–425.

Symington, M. (1988). Environmental determinants of population densities in *Ateles*. *Primate Conserv.*, **9**, 74–79.

Tischendorf, L. and Fahrig, L. (2000). On the usage and measurement of landscape connectivity. *Oikos*, **90**, 7–19.

de Thoisy, B., Renoux, F. and Julliot, C. (2005). Hunting in northern French Guiana and its impact on primate communities. *Oryx*, **39**, 149–157.

Turner, M. G., Gardner, R. H. and O'Neill, R. V. (2001). *Landscape Ecology in Theory and Practice*. New York: Springer-Verlag.

UNEP-WCMC (2007). *United Nations Environment Program – World Conservation and Monitoring Centre Species Database: CITES-Listed Species*. Online: accessed June 17, 2007.

Vance, C. and Geoghegan, J. (2002). Temporal and spatial modelling of tropical deforestation: a survival analysis linking satellite and household survey data. *Agricult. Econ.*, **27**, 317–332.

van Roosmalen, M. G. M. (1985). Habitat preferences, diet, feeding strategy, and social organization of the black spider monkey (*Ateles p. paniscus* Linnaeus 1758) in Surinam. *Acta Amazonica*, **15**, 1–238.

Wallace, R. B., Painter, R. L. E., Rumiz, D. I. and Taber, A. B. (2000). Primate diversity, distribution and relative abundances in the Reserva Vida Silvestre Rios Blanco y Negro, Department Santa Cruz, Bolivia. *Neotrop. Primates*, **8**, 24–28.

Wallace, R. B., Painter, R. L. E. and Taber, A. B. (1998). Primate diversity, habitat preferences and population density estimates in Noel Kempff Mercado National Park, Santa Cruz, Bolivia. *Am. J. Primatol.*, **46**, 197–211.

Wallis, J. and Lee, D. R. (1999). Primate conservation: the prevention of disease transmission. *Int. J. Primatol.*, **20**, 803–826.

Weghorst, J. A. (2007). High population density of black-handed spider monkeys (*Ateles geoffroyi*) in Costa Rican lowland wet forest. *Primates*, **48**, 108–116.

White, F. (1986). Census and preliminary observations on the ecology of the black-faced black spider monkey (*Ateles paniscus chamek*) in Manu National Park, Peru. *Am. J. Primatol.*, **11**, 125–132.

Wilson, D. E. and Reeder, D. M. (2005). *Mammal Species of the World: A Taxonomic and Geographic Reference*, Vol. 1. Baltimore, MD: Johns Hopkins University Press.

14 The ethnoprimatology of spider monkeys (Ateles spp.): from past to present

LORETTA A. CORMIER AND BERNARDO URBANI

Introduction

Human and nonhuman primates share a relatively recent history of inter-action in the New World in comparison with the Old World. The earliest known platyrrhine fossils only date back to the late Oligocene in Bolivia (e.g. Rosenberger *et al.*, 1991; Takai *et al.*, 2000). The earliest definitive evidence of human beings in South America does not occur until approximately 30 million years later around 12 500 years ago in Chile (Dillehay, 1989, 1997). Current evidence cannot reliably place humans in Amazonia earlier than 11 000 years ago (Roosevelt *et al.*, 1991, 1996). Nevertheless, the roughly 10 000 years of human–nonhuman primate sympatry provides a long history of interaction among numerous Neotropical primate species and diverse human cultures.

The term "ethnoprimatology," coined by Sponsel (1997), is a newly emerging subdiscipline of anthropology that bridges primatology and cultural anthropol-ogy. Primatologists tend to focus their research on understanding the behav-ior and ecology of a particular primate species or subspecies. Perhaps the most studied aspect of human–nonhuman interactions by primatologists has involved human development and deforestation of primate habitats over the last 500 years. With so many of the world's primate species endangered or threat-ened, such an approach is logical, meaningful, and most certainly critical for understanding the consequences of human behavior to the quite literal survival of many nonhuman primate species. It is likely that human influence of primate habitats extends even further back in time (e.g. pre-Columbian Mesoamerica; see Kirch, 2005). For example, Hershkovitz (1984) argued that the distribution of some species of spider monkeys, howler monkeys and capuchins may have been influenced by pre-Columbian trade.

Examination of the widely varying cultural attitudes, beliefs and behaviors toward monkeys is generally beyond the purview of primatological studies.

Spider Monkeys: Behavior, Ecology and Evolution of the Genus Ateles, ed. Christina J. Campbell. Published by Cambridge University Press. © Cambridge University Press 2008.

Cultural anthropologists tend to have a similar narrow focus on the lifeways, beliefs and behaviors of particular cultures or ethnic groups. The limitation in many ethnographic works of cultural anthropologists for primatology is that "monkeys" may only enter the literature as a generalized dietary category or type of pet. Ethnoprimatology is an attempt to bridge the expertise of the two disciplines in order to better understand the relationship between different human cultures and primate species.

The biggest challenge in addressing the ethnoprimatology of spider monkeys (genus *Ateles*) is that specific information about the relationship between specific cultures and specific primate species is often not available in the literature for varying reasons. Here we focus on three types of evidence: spider monkeys in the archaeological record, spider monkeys as human food, and spider monkeys in the mythology and cosmologies of human groups. The limited evidence of spider monkeys in the archaeological record may be, in part, a reflection of the general difficulty in recovering archaeological remains of primates in the tropical forest environments. Fleagle and Kay (1997, p. 3) described the status of the platyrrhine fossil record by saying, "until recently, a large shoe box could contain the primate fossils from all of South America and the Caribbean from the last 30 million years." Although the general fossil record has strengthened considerably, it is noteworthy that, still, very little evidence exists of *Ateles* – and primates in general – in the archaeological record over the last 10 000 years.

In the ethnographic record, perhaps the most commonly studied aspect of human–nonhuman primate interactions involves human hunting practices and the relative role of nonhuman primate game in the diet. Urbani's (2005) review of the literature on primates in the diets of New World peoples yields 56 case studies. While this provides a solid basis for drawing dietary comparisons, it also makes clear that primate predation has been understudied in the New World, and the primate hunting practices of many cultures remain unknown. Even less attention has been given to indigenous belief systems involving primates. Symbolic structures, rituals, myths and cosmologies frequently include monkeys in the New World, which is apparent in surveys of South American mythology (e.g. Bierhorst, 2002; Cormier, 2006; Wilbert, 1978, 1980; Wilbert and Simoneau, 1982, 1983, 1984, 1988, 1989a, 1989b, 1989c, 1990a, 1990b, 1991, 1992). However, far fewer descriptions exist that differentiate *Ateles* from the generic monkey in myth and cosmology, and fewer still differentiate the species of spider monkey (although it can be inferred from the location of the ethnic group). Nonetheless, the evidence that is available suggests that this is a viable area for future research.

Spider monkeys in the Pre-Columbian period: the archaeological record

Atelines, in general, are poorly known from the archaeological record. Apart from the reason given above, it remains unclear why these primates are infrequently found in this context as for different traditional societies of the Neotropics, the atelines are among the preferred food items and at the same time are highly selected as hunting game (see Urbani, 2002, and 2005 for a current ethnographical review). A plausible explanation may lie with bone dislocation and the fact that in some cases only selected body parts are transported from the forest to the settlements. In any case, even with more precise information, as is discussed below, any interpretation on the use of atelines, and particularly *Ateles*, by pre-Columbian societies is still speculative. For example, Urbani and Gil (2001) reported the presence of howler monkeys (*Alouatta seniculus*) bones from *c*. 3000 years ago in a cave site of northeastern Venezuela, a region inhabited from pre-Columbian times by human populations of a probable Carib language affiliation (Tarble, 1985). The faunal remains of this site were associated with human tools, and the bones were found dislocated, a fact that might suggest a potential alimentary context. Nevertheless, no evidence of fire was found either in this archaeological site or on the bones. Therefore, other explanations might be that they were cooked and/or brought as raw meat to the cave settlement or that the monkeys were used for another purpose besides food, perhaps as pets. Furthermore, Urbani and Gil (2001) found that, for this region, Spaniard chroniclers indicated that *piaches* (shamans) of these Carib societies were recognized by having "*un idolillo sentado en forma de mono, que dicen que es su dios*" (like a little seated monkey idol, that was said to be their god; Ruíz Blanco, 1965/1672: 41). Therefore, it is possible that primates may also have had a cosmological meaning for the pre-Columbian and postcontact societies in this part of Venezuela. This evidence seems to suggest that most of the interpretation of pre-Columbian atelines' uses and social constructions are approximate inferences.

Despite the fact that little evidence exists of *Ateles* in the archaeological record, some bone remains of this primate genus, along with other Neotropical primate species, have been recovered. In Teotihuacán, central Mexico, Valadez-Azúa and Childs-Rattray (1993) found a single primate tooth of *Alouatta palliata* dated between AD 400 and AD 650. The material was recovered in the *Barrio de los Comerciantes* (Neighborhood of the Merchants), a part of this pre-Hispanic city inhabited mainly by people of the Gulf of Mexico, today the state of Veracruz. This could be suggestive that in earlier times primates were traded from the Atlantic coast to central Mexico, a region with no feral primate populations. More recently, also in Teotihuacán, bones of *Ateles geoffroyi* that might have come from the same Mexican lowland region were found (D. Platas, personal communication). The manner in which *Ateles* and *Alouatta* were utilized by the people of this city remains unknown; however, as indicated by Friar Bernardino de Sahagún in 1555, the Amerindian merchants of this part of central Mexico used to have monkeys' hands as amulets to increase their luck in selling all their products (Sahagún, 1986/1555; Urbani, 1999).

In another part of Mesoamerica, capuchin and howler monkey teeth as well as a spider monkey proximal right ulna with cut marks have recently been found in the principal area of Sitio Drago in Isla Colón, Bocas del Toro region, Panama (T. Wake, personal communication). The *Ateles* evidence seems to indicate that this primate species was eaten. Considering that currently only night monkeys (*Aotus zonalis*), howler monkeys (*Alouatta palliata*) and white-faced capuchin monkeys (*Cebus capucinus*) inhabit this island (Urbani, 2003), it might be possible to suggest that *Ateles* was traded or brought to this place from the mainland during the pre-Columbian period; or that in the past this genus populated this island and now is locally extinct.

Probably one of the most conspicuous pre-Columbian representations of a primate in the New World is the well-known monkey figure of the Nazca lines complex in southern Peru (Figure 14.1). This geoglyph is a clear example of cultural/ideological interexchange between human societies of the arid Pacific coast and the lowland Amazonia of Peru (Lathrap, 1973). Considering the body configuration, particularly the long tail and arms, it is plausible that it is a representation of a spider monkey. Archaeological material culture that resembles spider monkeys is also scarce, and in fact most of the pottery that depicts monkeys is so stylistic that it is not possible to properly categorize them into a particular genus. However, in Mesoamerica, specifically in Mexico, there is pottery that might suggest the representation of *Ateles* (Preuss, 1901). For example, *Hun Chuen*, the Mayan god for the spider monkeys (see below), is constantly represented in the Mayan pottery, particularly in funerary vases (Coe, 1978; Braakhuis, 1987; Bruner and Cucina, 2005). Furthermore, among the pre-Columbian cultures that populated the central portion of Mexico, the Aztecs were singular in the representation of figurines that seem to resemble spider monkeys. For instance, a stone primate with a bird mask dated to approximately AD 1500 and found in a pyramid of Mexico City has characteristics of the spider monkey (extended tail and arms as well as a pronounced ventral area) and was identified to be a pregnant monkey named *Ehecatl-Quetzacoalt*, the Aztec god of wind (Bray, 2002; Figure 14.2). In addition, figurines of primates were recently found in the site of El Carrizal, near the locality of Ciudad Ixtepec, Oaxaca state (Mexico), a region that is historically and exclusively inhabited by spider monkeys, *Ateles geoffroyi* (Ortiz-Martínez and Rico-Gray, 2005; M. Winter,

Figure 14.1 The Nazca lines monkey. Maximum length: approximately 100 m. (Redrawn from an aerial photograph courtesy of S. Urbani.)

personal communication). These figurines are possible depictions of spider monkeys that were probably made by the people of the Mixe-Zoque ethnic group during the Kuak Phase dated 200 BC (M. Winter, personal communication).

Finally, the distribution of spider monkeys, *tucha* (in Yucatecan Maya), could in part be ecologically associated with Mayan archaeological sites (T. Urquiza Haas, personal communication). Densities of *ramón* trees (*Brosimum alicastrum*) are higher surrounding archaeological ruins (Coelho *et al.*, 1976; White and Hood, 2004; Estrada *et al.*, 2004; T. Urquiza-Haas, personal communication). This tree species was apparently used by the pre-Hispanic Mayans and its presence at archaeological sites appears to reflect precontact human horticultural practices (Puleston, 1982; Peters, 1983). *Brosimum alicastrum* seems to be one of the preferred feeding trees for spider monkeys in the Mayan archaeological sites (Cant, 1990; Ponce-Santillo *et al.*, 2004), and the densities of this primate genus are higher in such pre-Hispanic sites (Estrada *et al.*, 2004). Thus it is possible that the current distribution and densities of *Ateles geoffroyi* around forested archaeological ruins might be related to the pre-Hispanic human modification of the floristic composition of these sites, and particularly of *B. alicastrum*.

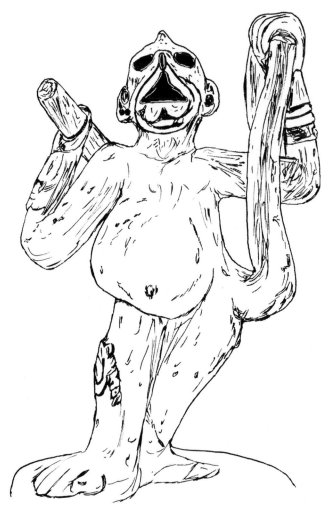

Figure 14.2 An Aztec spider monkey representation. Measurements:
60 × 37 × 33 cm. (Redrawn from Bray, 2002: 181.)

Spider monkeys as food: an ethnographical exploration

Evidence of spider monkey hunting for subsistence purposes was reported in
early Spaniard chronicles. In Venezuela, and particularly in the Guayana region
and the Orinoco Basin, two accounts suggest that *Ateles* was probably among
the preferred monkeys hunted by indigenous human populations. In his classic

work *Historia natural, civil y geográfica de las naciones situadas en las riveras del río Orinoco* (1749), the Jesuit priest Joseph Gumilla said,

> *Los Indios Tunevos gustan mucho de los Monos negros . . . y al ver gente, baxan con furia hasta las últimas ramas de los árboles, sacudiéndolas . . . con eso los Cazadores que los matan á su gusto . . . todas aquellas Naciones [de indígenas de esta región] comen de ellos; ni hay en qué escrupilizar; [ya que] los Monos grandes, solo se mantienen de frutas silvestres, muy sanas y sabrosas; de las quales se mantienen tambien los Indios durante su montería; y en los viages que los Padres hacen por aquellas y otras selvas, observan los frutales en que están comiendo los Monos y Micos, y á todo seguro comen y se mantienen de aquellas frutas* [The Tunevo Amerindians highly prefer the black monkeys (*Ateles belzebuth*) . . . that when encountered people angrily come down to the lowest branches of the trees in order to shake the branches . . . then, the hunter easily kills them . . . all the Native nations eat them without hesitation; since those large monkeys only survive with healthy and tasty wild fruits, then the Amerindians also use these same fruits when travelling; [moreover] during the trips that the friars conduct in these forests, they also observe which fruiting trees the monkeys feed on in order to eat the same fruits.]
>
> (Gumilla, 1984/1749: 261–262)

In addition, in 1785, Friar Ramón Bueno wrote in his historical treaty,

> *La marimonda es un mono negro, con el rabo muy largo [y] se mantiene en los palos más altos. Los indios las comen. . . . Matadas con escopeta pocas caen; las más quedan muertas y pegadas de la punta del rabo en alguna rama, que los indios suben luego a desprenderlas, por serles muy gustosas asadas* [The marimonda (*Ateles belzebuth*) is a black monkey with a very long tail (that) lives in the higher canopy. The Amerindians ate them . . . When hunted with a firearm, they tend to stay in the branches hanging dead from their tails. Then the Amerindians climbed the trees in order to take them, because they are considered very tasty when cooked.]
>
> (Bueno, 1965/1785: 107)

More recently, after a comprehensive review of current ethnographical accounts of hunting practices of Neotropical primates among 56 indigenous societies (Urbani, 2005), it was found that 22 had *Ateles* as one of their mammal game species (Table 14.1). It was found that *Ateles* is not only hunted in Amazonian sites (12/22; 55%) as mainly expected, but also in non-Amazonian sites (10/22; 45%: Guianas, and Mesoamerican forests). Moreover, in the cases

Table 14.1 *Indigenous societies that hunt spider monkeys (Ateles spp.) for subsistence*

Human society	*Ateles*[a,b]	Indigenous name	Study site	References
Amahuaca	NA/"years"; 3rd	NA	Ucayali, Amazonas, Peru	Carneiro (1970)
Surinamese Carib	2/15 months; 4th	NA	Bigi Poika, Suriname	Mittermeier (1991)
Cashinahua	NA/>13 years; NA	*isu*	Maneya-Xumuya, Cujanja River, Peru	Kensiger et al. (1975)
Colombian Barí	5/4 months; NA	NA	Southern Perijá Range, Colombia	Beckerman (1980)
Venezuelan Barí	54/12 months; NA	*sugshaa*	Southeastern Perijá Range, Venezuela	Lizarralde (2002)
Guaymí	NA/4 months; 1st	NA	Western Costa Rica	González-Kichner and Sainz de la Maza (1998)
Huambisa	NA/10 months; NA	*wáshi*	Cenepa-Santiago Rivers, Amazonia, Peru	Berlin and Berlin (1983)
Huaorani	10/11 months; 13th	NA	Quehueiri-ono, Shiripuno River, Napo, Ecuador	Mena et al. (2000)
Lacandón	NA/15 months; NA	*ma'k*	Norte del Najá, Chiapas, Mexico	Baer and Merrifield (1972)
Makuna	4/36 days; 4th	*gake* (?)	Pira-Paraná-Apaporis Rivers, Vaupés, Colombia	Århem (1976)
Matsigenka	14/1 year; 2nd	*osheto*	Manu Biosphere Reserve, Peru	Shepard (2002)
Miskito	NA/>5 years; NA	NA	Coco–La Mosquitía Rivers, Nicaragua–Honduras	Conzemius (1984)
Nambiquará	NA/148 days; NA	*hotasu*	Guapore-Chapada Parecis, Matto Grosso, Brazil	Setz (1991)
Paaca Nova	NA/NA; NA	NA	Guaporé River, Rondônia, Brazil	von Graeve (1989)
Piro	32/10 months; 2nd	NA	Manu River, Peru	Alvard and Kaplan (1991), Alvard (1993, 1995)
Shipibo-Conibo	NA/>5 years; NA	*iso*	Ucayali River, Amazonas, Peru	Eakin et al. (1980), Campos (1977)
Sumu	NA/>5 years; NA	NA	Coco–La Mosquitía Rivers, Nicaragua–Honduras	Conzemius (1984)
Surinamese Carib	2/15 months; 4th	NA	Bigi Poika, Suriname	Mittermeier (1991)
Tirio	11/15 months; 6th	NA	Sipaliwini-Pouso Tirio area, Suriname	Mittermeier (1991)
Waimiri-Atroari	42/13 months; 1st	*kwata*	Alalau River-BR 174, Roraima-Amazonas, Brazil	Souza-Mazurek et al. (2000)
Brazilian Yanomamö	17/5 months; 2nd	*paxo*	Catrimani River, Roraima-Amazonas, Brazil	Saffirio and Scaglion (1982)
Venezuelan Yanomamö	1/217 days; 10th	NA	Toropo-teri, Padamo River, Amazonas, Venezuela	Hames (1979)
Ye'kwana	13/216 days; 5th	NA	Toki, Padamo River, Amazonas, Venezuela	Hames (1979)

Notes:

[a]Number of *Ateles* hunted/Length of the ethnographic study.

[b]Rank number of this primate species in relation to all game mammal species hunted by each of the listed indigenous societies.

NA = not available.

in which the total number of hunted mammal species were known for an indigenous group, *Ateles* was often selected as one of the first five preferred meat
choices (9/13; 77%).

In western Costa Rica, 74% of Guaymi men interviewed said they have
hunted primates (González-Kirchner and Sainz de la Maza, 1998). Moreover,
despite the fact that monkey consumption is relatively low, no more than one per
week, 51% of the members of this ethnic group have eaten monkeys. Among
the primate species in Guaymi lands (*Saimiri oerstedii*, *Cebus capuchinus*,
Alouatta palliata and *Ateles geoffroyi*), Central American spider monkeys are
considered the most delicious of all primate meats in 96.6% of all interviews.
Guaymi villagers keep spider monkey infants as pets. It is interesting to note
that the Guaymi villagers considered spider monkeys as the second preferred
primate pet after capuchins; but the first type of monkey sold to tourists and
creole traders above any other primate. Finally, according to informants, as
a consequence of hunting, *A. geoffroyi* seemed to be particularly rare in the
Guaymi reserves (González-Kirchner and Sainz de la Maza, 1998).

Also, in Mesoamerica, the Lacandón of Chiapas (Mexico) used to hunt monkeys (*Ateles geoffroyi* and *Alouatta pigra*) with a special arrow before the intro-
duction of firearms (Baer and Merrifield, 1972). Monkeys were boiled with
green plantains and served with *tamales* (boiled corn bread). Their fat was
used to fry beans, after obtaining approximately one cup per monkey (Baer and
Merrifield, 1972). In addition, the Miskito from Nicaragua and Honduras has
A. geoffroyi and *Cebus capucinus* as their preferred monkey meat (Conzemius,
1984). In South America, black spider monkeys (*A. paniscus*) were the second
most frequent primate species (following howler monkeys) found in the kitchen
remains of the Surinamese Tirio (Mittermeier, 1991). The author indicated that
this ethnic group also used *Ateles'* teeth for making necklaces and arm bones
for combs.

In Brazil, Nambiquara villagers living in forest settlements hunt black spider monkeys. In fact, *Ateles paniscus* seems to be hunted specifically during
the rainy season (Setz, 1991). In the same region, the Wari' (Paaca-Nova)
prefer spider monkeys and howlers for their meat (von Graeve, 1989). In the
Roraima region of Brazil, together with peccaries (*Tayassu tajacu* and *T. pecari*)
and tapirs (*Tapirus terrestris*), spider monkeys (*A. paniscus*) account for the
87% of the complete game weight taken during a year by the Waimiri-Atroari
(Souza-Mazurek *et al.*, 2000). Eighty percent of all spider monkeys catches
were females. This is the opposite of what was found for other game mammal genera, where males were preferred. *Ateles paniscus* females were highly
selected because they were considered to have more fat and to be more "tasty"
than male individuals. Indeed, the Waimiri-Atroari consumed annually four

times more spider monkeys than any other Amerindian group (Souza-Mazurek *et al.*, 2000).

The Cashinahua men of eastern Peru "usually hunt along parallel ridges or along the valley floors on either side of a ridge [in order to] encounter . . . a band of spider monkeys, *isu*, [then, the hunters] can call the others through whistled messages so as to maximize the kill" (Kensinger *et al.*, 1975: 27). This hunting description is of particular interest because it seems to suggest that this indigenous group knows the tendency of spider monkeys to navigate using topographic features of the environment as was described by Suárez (2003) in Ecuadorian Amazonia. In the Venezuelan Guayana, the Shirián of the upper Paragua River not only perform primate calls to locate monkeys, but also know the fruiting pattern of feeding trees used by *pasha* (*Ateles belzebuth*) in order to find them (B. Urbani, unpublished data). The Amahuca from the Peruvian Amazonia considered spider monkeys and tapirs to be the tastiest meats (Carneiro, 1970). In addition, the author described that in some instances hunters actually drink the blood of spider monkeys for good luck in hunting this animal (Carneiro, 1970).

In the Peruvian Amazon, the Piro take on average 4.7 hours to obtain a spider monkey (*Ateles belzebuth chamek*, published as *A. paniscus*) and in 100% of the cases in which this monkey taxon was encountered, it was successfully obtained (Alvard, 1993). It was found that for all spider monkey kills, 86% are adults and 14% represent immature individuals (Alvard, 1995). The author also found that there was a 1:3 kill ratio in favor of female spider monkeys for both age-structure classes. To locate and hunt *A. b. chamek*, the Piro not only uses bows and shotguns but also dogs (Alvard and Kaplan, 1991). Also in the Amazon region of Peru, among the Matsigenka, woolly monkeys (*Lagothrix lagothricha*) and black spider monkeys (*A. belzebuth chamek*, published as *A. paniscus*) are the preferred primate preys (Shepard, 2002). Hunters of this indigenous group might travel from 5 to 8 km to find spider monkeys. They imitate their calls to locate the animals and climb trees to recover wounded monkeys. Hunting is biased toward females; however, adult males appeared to be preferred. During the rainy season, spider monkeys are considered to be more tasty and full of fat. When spider monkey females are killed, the offspring are often kept as pets, and in fact escaped *Ateles* pets are not hunted, probably because of the bonds created between the monkeys and the keepers (Shepard, 2002).

Among the Venezuelan Barí of the Perijá Mountain Range, brown spider monkeys (*Ateles hybridus*) are the favorite prey (Figure 14.3). The Barí name, *sugshaa*, for the spider monkeys actually means meat (Lizarralde, 2002). Female spider monkeys are preferentially selected, probably due to the extra weight of their offspring, making it more difficult for them to escape. Orphaned

Figure 14.3 Brown spider monkeys (*Ateles hybridus*) obtained by a Barí hunting party. (Photograph courtesy of M. Lizarralde.)

infants are often kept as pets, being the most common pet species among the Barí. The hunting of *A. hybridus* is directly linked with the knowledge of the feeding trees used by these monkeys. Indeed the main objective of hunting expeditions is to find capuchin and brown spider monkeys. The importance of this species to the Barí can be seen in the descriptions by Lizarralde (2002) of one Barí hunter who apparently killed 54 spider monkeys in one year and of a party of five hunters returning to the settlement with 16 *Ateles hybridus* after a single day's hunt. The Barí make necklaces from the teeth of brown spider monkeys, which are thought to provide humans with monkey attributes such as tree climbing and velocity (Lizarralde, 2002).

Creole communities (*campesinos, riberinhos, caboclos* in Latin America) also hunt spider monkeys. Over a period of half a year in public markets at Iquitos, Peru, 198 *Ateles* sp. were sold (approximate calculation by Urbani, 2005 after data from Castro *et al.*, 1975). On the other hand, some *campesino* communities in the central part of the Perijá Mountain Range, Venezuelan–Colombian border, have taboos for eating brown spider monkeys (*Ateles hybridus*) (B. Urbani, unpublished data). Here, these monkeys are considered to be so similar to humans, like *carajitos* (little children), that they are not eaten, not even hunted. In the Maceo and Remedios counties, Antioquia province, Colombia, *Ateles hybridus* is hunted for food, and also is used for treating malaria by the *campesinos* (A. L. Morales-Jiménez, personal communication). A. L. Morales-Jiménez also communicated that *Ateles fusciceps* (a.k.a. *A. geoffroyi robustus*) is hunted by Amerindian and Afro-Colombian communities in the Pacific coast region of this country. In the region of Santarém, on the right bank of the Tapajós River (state of Pará, Brazil), hunters, from settlements along the road BR-163 (mainly migrants from the Brazil's *nordeste*), prefer *Ateles belzebuth marginatus* individuals for subsistence, and keep and sell their infants (A. L. Ravetta, personal communication). Likewise along the Trombetas River, also in the state of Pará, *A. paniscus* is one of the favorite hunted primate taxa for the inhabitants of the riverine settlement. The *ribeirinhos* also keep spider monkey offspring as pets when the mothers are killed for food (L. de Carvalho-Oliveira, personal communication).

In sum, it is found that spider monkeys are among the preferred primates and mammals in Central and South American indigenous societies. In general, *Ateles*, especially females, are considered more tasty and fatty. The hunters know the ecology and behavior of these monkeys, particularly their use of space, feeding trees and calls, and sometimes have a special interest in killing them during the rainy season. The bias in the hunting pattern of spider monkeys toward female individuals seems to particularly affect their demography, and eventually their social structure (Urbani, 2005). Hunting of females also helps in the capture of infants that are normally kept as pets.

In addition to the preferences that are found for spider monkeys as food, some groups also have avoidances and taboos for this taxon. Such avoidances are conditioned by cultural beliefs and intersect the domains of subsistence activities and symbolic life. The availability of *Ateles* species as a food source for a given society does not completely explain the degree to which they will be utilized. For example, among the Matsigenka, howler monkeys (*Alouatta*) are considered among the most abundant primates in the Manu National Park (see also Terborgh, 1983), but the similarly sized spider monkeys and woolly monkeys are taken at a rate ten times higher than howlers (Shepard, 2002).

Preferences and avoidances can vary widely among groups. Among societies that hunt *Ateles* species, and where food taboos or avoidances have been reported, some are reported as applying to "monkeys" in general, some apply to certain monkey species, and can be characterized as applying to persons in a particular ritual or social status. In the first case, the Parakanã of Pará, Brazil, have *Ateles belzebuth marginatus* in their area, but are reported to have a taboo on all monkey species (Milton, 1991). The Parintintin of the Madeira Basin do not have a taboo on monkeys, but they do avoid eating them (including the local *A. b. chamek*) due to their physical similarity to human beings (Kracke, 1978). The Kalapalo of the Upper Xingu in Mato Grosso express the opposite sentiment. They consider all land mammals (which contrasts with water creatures) to be disgusting to eat, with the exception of monkeys (including the local *A. b. marginatus*), because of their physical similarity to human beings (Basso, 1973).

Some ethnographic accounts include differentiation among primate species, with taboos applied to some species, but not to others. For example, the Cashinahua hunt spider monkeys, but do not consider owl monkeys, howler monkeys or squirrel monkeys to be edible (Kensinger *et al.*, 1975). Among the Matis of Amazonas, Brazil, *Ateles belzebuth chamek* (published as *A. paniscus*) is one of the primate species hunted, but not one of the species to which taboos apply (Erikson, 1997, 2001). According to Erikson, the smaller *Saguinus mystax* and *Saimiri sciureus* are primarily hunted for their teeth and are taboo for those who have recently undergone ritual facial tattooing.

Another type of taboo that occurs in the literature relates to the couvade. The couvade is a widespread folk belief in Amazonia that involves ritual restrictions surrounding a pregnancy or postpartum period, which apply to both the mother and the father of the child (Rivière, 1974). The Yanomamö are described as having a general couvade taboo on eating monkeys, including spider monkeys (McDonald, 1977), as are the Tukano (Reichel-Dolmatoff, 1978). For some other societies, the couvade restrictions are described as applying only to certain species. The Huaorani hunt all the primate species occurring in their area, including *Ateles*, but the couvade restrictions apply only to *Alouatta* and *Lagothrix* species (Rival, 1998). The Shipibo also hunt spider monkeys (Campos, 1977; Eakin *et al.*, 1980), but their couvade applies only to *Cebus albifrons* (Behrens, 1986). Similarly, the Sirionó hunt spider monkeys, but the couvade applies only to owl monkeys and howler monkeys (Holmberg, 1985/1950; McDonald, 1977). The Wapishana of Guyana and Roraima, Brazil, hunt eight species of monkeys (*Alouatta seniculus*, *Ateles paniscus*, *Cebus apella*, *Cebus olivaceus*, *Chiropotes satanas*, *Pithecia pithecia*, *Saguinus midas* and *Saimiri sciureus*), but here, the couvade applies only to spider monkeys (Henfry, 2002).

Spider monkeys in mythology and cosmology

Animal myths and metaphors

Animal myths and metaphors are common in the symbologies and cosmologies of human cultures around the world. The greatest problem in understanding the specific role of *Ateles* among New World peoples is that ethnographers often employ the Western folk term of "monkey." The category can be generally understood to refer to primates with tails such as the platyrrhines. The use of the term may unintentionally lead to the assumption that a similar perception of "monkeyness" exists in the indigenous folk taxonomies, which may very well be unwarranted. For example, among the Guajá, individual primate species in their area are differentiated with specific names, but are grouped into higher level categories based on what they view as shared ancestry among species considered to be siblings to one another (Cormier, 2003a). For example, *Cebus apella* and *Cebus kaapori* are considered to be sibling species, which might suggest an ethnobiological correspondence with Western Linnean taxonomy. However, *Aotus* is considered to be a sibling species to *Potus flavus*, the kinkajou, because both are nocturnal animals with prehensile tails. The Guajá view themselves as siblings to *Alouatta belzebul*, for the loud vocalizations of howler monkeys are considered to be similar to human singing.

In a broad view of perceptions of humanity and nonhuman animal life, Viveiros de Castro (1998, 1999) has made the observation that it is common in the cosmologies of indigenous Amazonian peoples to view animals as former human beings. He has recounted multiple myths that involve the transformation of human beings into animals. Examples of such a transformation of a human into a spider monkey can be found among the Aguaruna (Brown, 1984), the Barí (Lizarralde, 2002), the Desana (Reichel-Dolmatoff, 1978), the Matskigenka (Shepard, 2002), and the Yanomamö (Wilbert and Simoneau, 1990a).

Shepard (2002) has contrasted Matsiguenka thought with Western thought as a kind of "devolution." Rather than the popular Western view of the humanity representing an evolutionary "stage" following an earlier, less differentiated nonhuman primate "stage," contemporary monkeys are transformed beings who were human in a prior form of their existence. Shepard's use of the term "devolution" is not meant literally, but intended to suggest how such a belief might be interpreted in terms of the Western folk perception of human evolution. That is, the Western folk interpretation of evolution that suggests that

"monkeys" evolved into "apes," which evolved into "humans," as if humanity is a predestined evolutionary end-game of advancement and improvement
upon inferior primate forms. In the same vein, what Vivieros de Castro (1998,
1999) is suggesting is that the underlying premise in Amazonian thought is
that the primordial undifferentiated state is "primitive humanity," which gave
rise to differentiated animal species. Thus, human beings are not so much to be
considered a kind of animal as animals are considered to be a kind of person.

Such personification of animals should not be mistaken for a conservation
ethic. Descola (1998) has warned that the tendency to use indigenous societies
to project Western values related to animal rights often bears little relation to the
actual beliefs of the peoples themselves. Pet-keeping falls into this category.
Keeping Neotropical monkey species as pets appears to be a common and
widespread practice that has been documented in numerous human groups
(e.g. the Ache [Hill and Hawkes, 1983], the Barí [Lizarralde, 2002], the Guajá
[Cormier, 2003a], the Huaorani [Rival, 1993], the Kagwahív [Kracke, 1978],
the Matsigenka [Shepard, 2002], the Mekranoti Kayapo [Werner, 1984], the
Upper Putumayo River Indians [Hernández-Camacho and Cooper, 1976], the
Wayãpi [Campbell, 1989], and the Yanomamö [Smole, 1976]). The practice
of pet-keeping among indigenous groups may not correspond to the way dogs
and cats are kept as companion animals in Western groups. Fausto (1999) has
described the prey/pet paradox in Amazonia where the same species that are
hunted as food may also be kept as pets. For the Parakanã and several other
Amazonian groups, Fausto describes shamans referring to their spirit familiars
as "pets," arguing that the pet relation is one of symbolic control. Among the
Jivaró, Taylor (2001) argues that pet-taming is viewed as a kind of mothering
and further, that taming of pets is linked to what they view as the taming of
women through marriage. Erikson (2000) views Amazonian pet-keeping as a
counterbalance to hunting and as a means to assuage one's conscience in killing
animals and to appease the divinities who oversee animals. Cormier (2003a,
2003b) argues that the Guajá emphasis on howler monkeys as food involves
a kind of symbolic cannibalism related to their belief that howlers are human
siblings. They are desirable as both food and pets because of their similarity to
humans.

Although the folk perceptions of the "evolutionary progression" of the
human/animal divide differ in Western and Amazonian thought, one shared
feature is that among animals, nonhuman primates are often positioned as
occupying the closest node of relatedness in the human/animal divide (see
Cormier, 2003a, 2006). While other animals also serve in this role, nonhuman primates are particularly amenable to anthropomorphic characterizations
due to their physical and behavioral similarities to human beings. In Amazonia, monkeys may serve to illustrate the continuities between humans and

animals, or alternatively, they may serve as a means to accentuate differences in the nature/culture divide. Two broad ways in which spider monkey myths, metaphors and symbols can be categorized are as either involving transformation or contagion. The first involves the changing of human beings into spider monkeys. The second involves either the conference of either positive or negative spider monkey attributes to human beings, or less commonly, the conference of human attributes to spider monkeys.

Transformation

In some myths, the transformation of spider monkeys into human beings is performed by a creator divinity and serves as a cautionary tale for inappropriate human behavior. Reichel-Domaltoff (1976) has described one general role of animals in myths as being to serve as metaphors for survival when they are punished for disobeying rules of adaptive significance. For the Huaorani of Ecuador, a similar argument has been made for monkeys (Rival, 1996). Here, she indicates that a number of Huaorani myths involve a theme of social catastrophes caused by monkeys who overstep their boundaries in either trying to behave too much like human beings or in behaving too differently from human beings. Three examples of cautionary tales involving spider monkeys can be found among the Barí, the Matsigenka, and the Sirionó.

In Barí mythology, a time is referred to when no monkeys existed. *Sabasebaa*, the creator divinity, accompanied a Barí searching for food in the forest when they encountered another Barí eating fruit in a tree. They asked them to toss down fruit, but they tossed down only the peels. In anger, the creator divinity transformed them into spider monkeys and instructed the Barí to eat them (Lizarralde, 2002).

A Matsigenka myth regarding spider monkeys also has an element of cautionary tale where humans who are not following appropriate cultural expectations are transformed into monkeys. Here, *Yavireri*, the first shaman, transformed humans into all of the existing forms of animals. *Yaniri*, the howler monkey, and *Osheto*, the spider monkey, were brothers-in-law. *Yaniri* was lazy and borrowed beans from *Osheto* rather than raising his own crops. After *Yaniri* borrowed beans several times from *Osheto* and ate them rather than planting them, *Osheto* became angry and punched *Yaniri* in the throat, creating the enlarged larynx characteristic of howler monkeys (Shepard, 2002). In another tale, two impolite guests at a party were transformed into the woolly and the spider monkey.

A creation myth of monkeys among the Sirionó involves an element of punishment for inappropriate behavior. The mythical jaguar was delousing the son of the creator divinity/moon (*Yási*), and bit him in the head and killed him. The Moon questioned all the animals about who had killed his son and they replied that they did not know. The mythical spider monkey (*Erubát*) and the mythical howler monkey (*Tendí*) subsequently were at a drinking festival, where *Erubát* declared that he wanted to have a red coat like the howler monkey. In anger, the Moon declared that the spider monkey would be black. The Moon then grabbed the howler monkey by the neck and pulled his throat into its contemporary shape, becoming the explanation for why howlers howl (Holmberg, 1985/1950). Among the Jivaroan-speaking Aguaruna of Peru, a primordial spider monkey, *Tsewa*, is responsible for transforming a human being into the contemporary spider monkey (Brown, 1984). The Macro-Ge speaking Bororo of Brazil have a similar figure. *Júkorámo-dogédu* is a mythical monkey who created people and the forest (Wilbert and Simoneau, 1983).

The depiction of contemporary monkeys as human/monkey hybrids also appears in several Amazonian myths. Two very similar myths occur among the Mundurucú and Lokono Arawak (Drummond, 1977) involving the marriage of a man to a female howler monkey. Their offspring give rise to contemporary howlers. Human/monkey marriages that result in hybrid offspring can also be found among the Warao (Wilbert, 1980), although the species is not indicated; *Ateles* does not ovelap the Warao land. One such hybrid was found for spider monkeys among the Warí (Conklin, 2001). The Warí believe that spider monkeys have partial human origins and are descended from the union of a Warí man and a male spider monkey. Interestingly, the Warí believe that all the original spider monkeys were male.

Contagion

Another broad category of belief that can be found among indigenous people in the Neotropics involves a kind of contagion through contact with differing monkey species. Traits that are either desirable or undesirable for human beings are viewed as epitomized in the behavior of differing monkey species. On the one hand, it can be considered a kind of indigenous primatology to the degree that behavioral characteristics among monkey species are contrasted. But it also involves symbolic reflexivity, for it is an anthropomorphic attribution of human behaviors to nonhuman primate species.

Among the Bororo, Crocker (1985) suggests a kind of magical contagion from eating monkeys, which are considered to epitomize speed and grace. On the other hand, among the Matsigenka (consistent with their mythology described above), howler monkeys are considered to be lazy and capuchins are considered to be thieves and it is believed that these traits can be conferred to a person by eating these monkeys (Shepard, 2002). Lizarralde (2002) describes also contrasting types of monkey contagion among the Barí. The Barí keep spider monkeys as pets and believe that wearing spider monkey tooth necklaces can confer the manual dexterity of spider monkeys to the wearer. In contrast, howler monkey teeth are not worn as necklaces and are not kept as pets. They are believed to be of low intelligence and slow speed. This is echoed in another example, which bridges the couvade and contagion. Vilaça (2002) reports a Warí shaman telling parents that their child was turning into a monkey because the parents had not followed the appropriate protocol for eating capuchins. Similarly, the Sirionó belief that if one eats the young of a spider monkey, howler monkey or "yellow" monkey, your lips will turn white (Priest, 1966).

In a few myths, contagion works in the opposite direction where the behavior of monkeys is influenced by their contact with humans. The Yanomamö have a myth involving behavioral changes in spider monkeys due to contact with human menstrual blood. A young woman is forced by her husband to travel during menstruation and she and her sister turn into mountains. The husband is attacked by bands of monkeys (including a band of spider monkeys) who are behaving in this aggressive manner because they have smelled human menstrual blood and it has enraged them. The myth continues with a transformation of a human being into a spider monkey. The spider monkeys drag the man into a tree, stick a blowgun into his back to make a tail, and force him to make spider monkey calls. He then changes into a spider monkey (Wilbert and Simoneau, 1990a).

Aguaruna beliefs include a gender dimension for spider monkeys (Brown, 1984). Spider monkeys are considered to be the quintessential game animal due both to their behavior and physical characteristics. They are admired for what is seen as cleverness and caution when hunted and desired as food due to the thick layer of fat they acquire when wild fruits are seasonally available. Contagion here is due to the susceptibility of spider monkeys (*anem*) to hunting magic songs. These songs are linked to human sexuality and male attraction of females, which has also been described by Taylor (2001) for hunting in general for another Jivaroan-speaking people.

A more animistic contagion exists among some Amazonian groups where monkeys are believed to have supernatural or shamanic powers that they use to intervene in human affairs. Among the Bororo, monkeys are associated with *bope*, a principle of both organic and spiritual transformation (Crocker, 1985). Part of becoming a shaman involves being surprised in the forest and spoken to by a monkey, usually a howler monkey. Among the Warí, some animals (including spider monkeys) are considered to possess spirits, and illness can be a manifestation of an attempt by an animal to incorporate a human being into their species (Vilaça, 2002). The Matsigenka attribute differential supernatural powers to pygmy marmosets and spider monkeys (Shepard, 2002). Pygmy marmosets (*Cebuella pygmaea*) are considered to be magical and potentially dangerous because they may lead a hunter astray in the forest. Spider monkeys are one of the Matsigenka animal spirits that are capable of stealing the souls of children and causing illness. Spider monkeys (*Ateles paniscus*) are also considered to have magical properties among the Wapishana, who use their pelts in magical killings (Henfry, 2002).

Among the Maya, a myth exists which involves human to monkey transformation, cautionary tale, and monkey animistic contagion. In the *Popol Vuh* (Anonymous, 1994/1554–1558), the sacred book of the Maya, it is said that a woman pregnant by *Hun Hunahpu* (the Mayan Great Hunter) gave birth to twins, *Xbalanque* and *Hunahpu*. *Hun Chuen* and *Hun Baatz*, the two previous sons of *Hun Hunahpu*, were so jealous of the newborns that they used to make them work extremely hard. One day, the new twins requested *Hun Chuen* and *Hun Baatz* to look for food on a tree; these older brothers-in-law climbed the tree that suddenly became so huge that they were not able to come down. Then, *Hun Chuen* transformed into a spider monkey and *Hun Baatz* into a howler monkey. Subsequently, *Hun Chuen* and *Hun Baatz* acquired the ability of diviners and artisan gods. Moreover, the *Popol Vuh* (Anonymous, 1994/1554–1558) explained that during the third creation of the Maya, the first humans were created with the ability of speaking; however, they died during a flood. The survivors became spider monkeys (Braakhuis, 1987). In the fourth creation, humankind fully began.

Discussion

Varied human cultures and numerous monkey species have shared space in the Neotropics over at least the last 10 000 years. Although *Ateles* is poorly known from the archaeological record, it is likely from what evidence is available that they have long had a place in indigenous diets and symbolism. As previously discussed, one problem in the ethnoprimatology of spider monkeys is that "monkey" is often used as a generic category with inadequate attention to differences in indigenous perceptions of species. While sufficient evidence exists to draw some conclusions, the role of spider monkeys in human cultures remains largely unexplored.

Spider monkeys are among the preferred food species in areas where they are present. A bias toward hunting females sometimes exists, and that the "fatted" state associated with the eating of seasonal fruits is a frequent explanation for why spider monkeys are hunted as food. Many indigenous groups have extensive knowledge of the behavior and ecology of spider monkeys. One additional example comes from the Murui (Witoto) from the Caquetá River of Colombia who refer to *Ateles belzebuth* as *Mehéku*. This word resembles the locomotion pattern of this primate genus, brachiation, and at the same time results from the word *Mehére* that means "heavy" (Townsend and Ramírez, 1995). Numerous human groups demonstrate an understanding of the biology of this primate genus and interact and recall it in accordance with their own social and ecological contexts.

Neotropical monkeys are also very common figures in mythology, cosmology and symbolism among the varied New World peoples. Beliefs regarding spider monkeys are categorized into those involving "transformation" and those involving "contagion." The first involves myths where spider monkeys are transformed human beings. The second involves either positive or negative attributes that can be conferred to human beings through contact with spider monkeys (and sometimes vice versa). Given how frequently the generic category of "monkey" appears in Central and South American myths, it is clear that spider monkey symbolism also remains an area that needs further exploration.

Less attention is given here to the role of spider monkeys as pets. The keeping of monkeys as pets among indigenous peoples is perhaps so commonly observed that it has not been problematized as an area of research. Spider monkey pet-keeping remains a potential area of research in ethnoprimatology, not only among indigenous peoples, but also for city-dwellers. In urban Latin American societies, spider monkeys have been thought of as animals with lust-oriented behaviors, but at the same time are especially selected as pets. For instance, in the Amazonian city of Iquitos in lowland Peru, many locals believe that spider monkeys behave differently toward women than to men (D. Urdaneta, personal communication). In addition, *Ateles geoffroyi* is particularly common as a primate pet in Mexico City (Duarte-Quiroga and Estrada, 2003). Perhaps the most famous Latin American primate pet was Frida Kalho's Mexican spider monkey named Fulang Chang, who lived and died in the Coyoacán neighborhood of this Latin American capital city.

Acknowledgements

We would like to thank Christina Campbell for her kind invitation to participate in this volume. Special thanks for their communications, in alphabetical order, to Leonardo de Carvalho-Oliveira, Andrea Cucina, Alba Lucía Morales-Jimenez, Teresita Ortíz-Martínez, Diana Platas, André Luis Ravetta, Diego Urdaneta, Tania Urquiza-Haas, Thomas Wake and Marcus Winter;

and to Manuel Lizarralde and Susana Urbani for providing the photographs used in Figure 14.3 and Figure 14.1, respectively. We appreciate the valuable suggestions of Paul A. Garber, Christina Campbell, and the anonymous reviewers. Thanks to Margaret Enrile for her cooperation. B. Urbani was funded by a Fulbright-OAS Fellowship and currently by a UIUC Assistantship.

References

Alvard, M. (1993). Testing the "Ecologically noble savage" hypothesis: interpecific prey choice by Piro hunters of Amazonian Perú. *Hum. Ecol.*, **21**, 355–387.

Alvard, M. (1995). Intraspecific prey choice by Amazonian hunters. *Curr. Anthropol.*, **36**, 789–818.

Alvard, M. and Kaplan, H. (1991). Procurement technology and prey mortality among indigenous Neotropical hunters. In *Human Predators and Prey Mortality*, ed. M. C. Stiner, Boulder: Westview Press, pp. 79–104.

Anonymous (1994/1554–1558). *Popol Vuh*. Mexico City: Fondo de Cultura Económica (Edition under the supervision of A. Recinos).

Århem, K. (1976). Fishing and hunting among the Makuma: economy, ideology and ecological adaptation in the northwest Amazon. *Gotenborgs Etnografiska Museum Årstryck* (Sweden), pp. 27–44.

Baer, P. and Merrifield, W. R. (1972). *Los Lacandones de México. Dos Estudios*. Mexico City: Instituto Nacional Indigenista–Secretaría de Educación Pública.

Basso, E. B. (1973). *The Kalapalo Indians of Central Brazil*. New York: Holt, Rinehart and Winston.

Beckerman, S. (1980). Fishing and hunting by the Barí in Colombia. *Working Papers on South American Indians*, **2**, 68–109.

Behrens, C. A. (1986). Shipibo food categorization and preference: relationships between indigenous and western dietary concepts. *Am. Anthropol.*, **88**, 647–658.

Berlin, B. and Berlin, E. A. (1983). Adaptation and ethnozoological classification: theoretical implications of animal resources and diet of the Aguaruna and Huambisa. In *Adaptive Responses of Native Amazonians*, ed. R. B. Hames and W. T. Vickers, New York: Academic Press, pp. 301–325.

Bierhorst, J. (2002). *The Mythology of South America*. New York: Oxford University Press.

Braakhuis, H. E. M. (1987). Artificers of the days: functions of the howler monkey gods among the Mayas. *Bijdragen tot de Taal-, Land-, en Volkenkunde*, **143**, 25–53.

Bray, W. (2002). *Aztecs*. London: Royal Academy of Arts.

Brown, M. F. (1984). The role of words in Aguaruna hunting magic. *Am. Ethnol.*, **3**, 545–558.

Bruner, E. and Cucina, A. (2005). *Alouatta, Ateles*, and the Mesoamerican cultures. *J. Anthropol. Sci.*, **83**, 111–118.

Bueno, R. (1965/1785). *Tratado Histórico*. Caracas: Academia Nacional de la Historia.

Campbell, A. T. (1989). *To Square with Genesis, Causal Statements and Shamanic Ideas in Wayãpí*. Edinburgh: Edinburgh University Press.

Campos, R. (1977). Producción de pesca y caza en una comunidad Shipibo en el río Pisqui. *Amazonia Peruana*, **1**, 53–74.

Cant, J. G. H. (1990). Feeding ecology of spider monkeys (*Ateles geoffroyi*) at Tikal, Guatemala. *Hum. Evol.*, **5**, 269–281.

Carneiro, R. L. (1970). Hunting and hunting magic among the Amahuaca of the Peruvian Montaña. *Ethnology*, **9**, 331–341.

Castro, N., Revilla, J. and Neville, M. (1975). Carne de monte como una fuente de proteínas en Iquitos, con referencia especial a monos. In *La Primatología en el Perú*. (1990), ed. N. Castro-Rodríguez, Lima: Proyecto Peruano de Primatología, pp. 17–35.

Coe, M. D. (1978). *Lords of the Underworld: Masterpieces of Classic Mayan Ceramics*. Princeton, NJ: Princeton University.

Coelho, A. M., Jr., Bramblett, C. A., Quick, L. B. and Bramblett, S. S. (1976). Resource availability and population density in primates: a sociobioenergetic analysis of the energy budgets of Guatemalan howler and spider monkeys. *Primates*, **17**, 63–80.

Conklin, B. A. (2001). *Consuming Grief: Compassionate Cannibalism in an Amazonian Society*. Austin: University of Texas Press.

Conzemius, E. (1984). *Estudio etnográfico sobre los indios Miskitos y Sumus de Honduras y Nicaragua*. San José (Costa Rica): Libro Libre.

Cormier, L. A. (2003a). *Kinship with Monkeys: The Guajá Foragers of Eastern Amazonia*. New York: Columbia University Press.

Cormier, L. A. (2003b). Animism, cannibalism, and pet-keeping among the Guajá of Eastern Amazonia. *Tipití: J. Soc. Anthropol. Lowland South America*, **1**, 71–88.

Cormier, L. A. (2006). A preliminary review of Neotropical primates in the subsistence and symbolism of indigenous lowland South American peoples. *Eco. Environ. Anthropol.*, **2**, 14–31.

Crocker, J. C. (1985). *Vital Souls, Bororo Cosmology, Natural Symbolism, and Shamanism*. Tucson: University of Tucson Press.

Descola, P. (1998). Estrutura ou sentimento: a relação com o animal na Amazônia. *Mana*, **4**, 23–45.

Dillehay, T. D. (1989). *Monte Verde, A Late Pleistocene Settlement in Chile*. Washington, DC: Smithsonian Institution Press.

Dillehay, T. D. (1997). *Monte Verde: A Late Pleistocene Settlement in Chile*. Vol. 2: *The Archaeological Context and Interpretation*. Washington DC: Smithsonian Institution Press.

Drummond, L. (1977). Structure and process in the interpretation of South American myth: the Arawak dog spirit people. *Am. Anthropol.*, **79**, 842–868.

Duarte-Quiroga, A. and Estrada, A. (2003). Primates as pets in Mexico City: an assessment of the species involved, source of origin, and general aspects of treatment. *Am. J. Primatol.*, **61**, 53–60.

Eakin, E., Lauriault, E. and Boonstra, H. (1980). *Bosquejo etnográfico de los Shipibo-Conibo del Ucayali*. Lima: Ignacio Prado Pastor Ediciones.

Erikson, P. (1997). On Native American conservation and the status of Amazonian pets. *Curr. Anthropol.*, **38**, 445–446.

Erikson, P. (2000). The social significance of pet-keeping among Amazonian Indians. In *Companion Animals and Us*, ed. P. Poberseck and J. Serpell, Cambridge: Cambridge University Press, pp. 7–26.

Erikson, P. (2001). Myth and material culture: Matis blowguns, palm trees, and ancestor spirits. In *Beyond the Visible and the Material: The Amerindianization of Society in the Work of Peter Rivière*, ed. L. M. Rival and N. L. Whitehead, Oxford and New York: Oxford University Press, pp. 101–121.

Estrada, A., Luecke, L., Van Belle, S., Barrueta, E. and Rosales-Meda, M. (2004). Survey of black howler (*Alouatta pigra*) and spider (*Ateles geoffroyi*) monkeys in the Mayan sites of Calakmul and Yaxchilan, Mexico and Tikal, Guatemala. *Primates*, **45**, 33–39.

Fausto, C. (1999). Of enemies and pets: warfare and shamanism in Amazonia. *American Ethnologist*, **26**, 933–956.

Fleagle, J. G. and Kay, R. F. (1997). Platyrrhines, catarrhines, and the fossil record. In *New World Primates: Ecology, Evolution, and Behavior*, ed. W. Kinzey, Chicago: Aldine, pp. 2–23.

González-Kirchner, J. P. and Sainz de la Maza, M. (1998). Primate hunting by Guaymi Amerindians in Costa Rica. *Hum. Evol.*, **13**, 15–19.

Gumilla, J. (1984/1749). *Historia Natural, Civil y Geográfica de las naciones situadas en las riveras del Orinoco*. Santander de Chilicao (Colombia): Edit. Carvajal S. A. (Facsimilar Edition).

Hames, R. B. (1979). A comparison of the efficiencies of the shotgun and the bow in Neotropical forest hunting. *Hum. Ecol.*, **7**, 219–252.

Henfrey, T. B. (2002). Ethnoecology, resource use, conservation, and development in a Wapishana community in South Rupununi, Guyana. Unpublished Ph.D. thesis, University of Kent, Canterbury, UK.

Hernández-Camacho, J. and Cooper, R. W. (1976). The nonhuman primates of Colombia. In *Neotropical Primates, Field Studies and Conservation*, ed. R. W. Thorington, Jr. and P. B. Heltne, Washington DC: National Academy of Sciences, pp. 35–69.

Hershkovitz, P. (1984). Taxonomy of squirrel monkeys genus *Saimiri* (Cebidae, Platyrrhini): a preliminary report with description of a hitherto unnamed form. *Am. J. Primatol.*, **7**, 155–210.

Hill, K. and Hawkes, K. (1983). Neotropical hunting among the Aché, of Eastern Paraguay. In *Adaptive Responses of Native Amazonians*, ed. R. B. Hames and W. T. Vickers, New York: Academic Press, pp. 139–188.

Holmberg, A. R. (1985/1950). *Nomads of the Long Bow, The Siriono of Eastern Bolivia*. Prospect Heights, IL: Waveland Press.

Kensinger, K. M., Rabineau, P., Tanner, H., Ferguson, S. G. and Dawson, A. (1975). *The Cashinahua of Eastern Peru. Studies in Anthropology and Material Culture*, Vol. 1. Providence: The Haffereffer Museum of Anthropology at Brown University Press.

Kirch, P. V. (2005). Archaeology and global change: the Holocene record. *Annu. Rev. Env. Resour.*, **30**, 409–440.

Kracke, W. H. (1978). *Force and Persuasion, Leadership in an Amazonian Society.* Chicago: University of Chicago Press.

Lathrap, D. W. (1973). The antiquity and importance of long distance trade relationships. *World Archaeol.*, **5**, 170–186.

Lizarralde, M. (2002). Ethnoecology of monkeys among the Barí of Venezuela: perception, use and conservation. In *Primates Face to Face: Conservation Implications of Human and Nonhuman Primate Interconnections*, ed. A. Fuentes and L. D. Wolfe, Cambridge: Cambridge University Press, pp. 85–100.

McDonald, D. R. (1977). Food taboos: a primitive environmental protection agency (South America). *Anthropos*, **72**, 735–748.

Milton, K. (1991). Comparative aspects of diet in Amazonian forest-dwellers. *Philos. Trans. Roy. Soc. Lond. B*, **334**, 253–263.

Mena, V. P., Stallings, J. R., Regalado, B. J. and Cueva, L. R. (2000). The sustainability of current hunting practices by the Huaorani. In *Hunting for Sustainability in Tropical Forests*, ed. J. G. Robinson and E. L. Bennett, New York: Columbia University Press, pp. 57–78.

Mittermeier, R. A. (1991). Hunting and its effect on wild primates populations in Suriname. In *Neotropical Wildlife Use and Conservation*, ed. J. G. Robinson and K. H. Redford. Chicago: University of Chicago Press, pp. 93–107.

Ortiz-Martínez, T. and Rico-Gray, V. (2005). Monos araña (*Ateles geoffroyi*) habitando una selva baja caducifolia en el distrito de Tehuantepec, Oaxaca, México. *Programa y libro de resúmenes del II Congreso Mexicano de Primatología*: no page number.

Peters, C. M. (1983). Observations of Maya subsistence and ecology of a tropical tree. *Am. Antiquity*, **48**, 610–615.

Ponce-Santillo, G., Cano, E., Andresen, E. and Cuaron, A. (2004). Seed dispersal by two species of primates and secondary dispersal by dung beetles of the Ramon tree (*Brosimum alicastrum*) in Tikal National Park, Guatemala. *Folia Primatol.*, **75**(S1), 318.

Preuss, K. T. (1901). Der Affe in der mexikanischen Mythologie. *Ethnologisches Notizblatt*, **2**, 66–76.

Priest, P. N. (1966). Provision for the aged among the Siriono of Bolivia. *Am. Anthropol.*, **68**, 1245–1247.

Puleston, D. (1982). The role of ramón in Maya subsistence. In *Maya Subsistence: Studies in Memory of Dennis E. Puleston*, ed. K. Flannery, New York: Academic Press, pp. 349–366.

Reichel-Dolmatoff, G. (1976). Cosmology as ecological analysis: a view from the rain forest. *Man*, **11**, 307–318.

Reichel-Dolmatoff, G. (1978). Desana animal categories, food restrictions, and the concept of colour energies. *J. Latin Am. Lore*, **4**, 243–291.

Rival, L. M. (1993). The growth of family trees: understanding Huaorani perceptions of the forest. *Man*, **28**, 635–652.

Rival, L. M. (1996). Blowpipes and spears: the social significance of Huaorani technological choices. In *Nature and Society: Anthropological Perspectives*, ed. P. Descola and G. Pálsson, New York: Routledge, pp. 145–164.

Rival, L. M. (1998). Domestication as a historical and symbolic process: wild gardens and cultivated forests in the Ecuadorian Amazon. In *Advances in Historical Ecology*, ed. W. Balée, New York: Columbia University Press, pp. 232–250.

Rivière, P. (1974). The couvade: a problem reborn. *Man*, **9**, 423–435.

Roosevelt, A. C., Housley, R. A., Imazio Da Silveira, M., Maranca, S. and Johnson, R. (1991). Eighth millennium pottery from a prehistoric shell midden in the Brazilian Amazon. *Science*, **254**, 1621–1624.

Roosevelt, A. C., Lima da Costa, M., Lopes Machado, C., *et al.* (1996). Paleoindian cave dwellers in the Amazon: the peopling of the Americas. *Science*, **272**, 373–384.

Rosenberger, A. L., Hartwig, W. C. and Wolff, R. G. (1991). *Szalatavus attricuspis*, an early platyrrhine primate. *Folia Primatol.*, **56**, 225–233.

Ruíz Blanco, P. M. (1965/1672). *Conversión de Piritu*. Caracas: Academia Nacional de la Historia.

Saffirio, G. and Scaglion, R. (1982). Hunting efficiency in acculturated Yanomama villages. *J. Anthropol. Res.*, **38**, 315–328.

Sahagún, B. de. (1986/1555). *Historia de las Indias*. Caracas: Biblioteca Ayacucho.

Setz, E. Z. F. (1991). Animals in the Nambiquara diet: methods of collection and processing. *J. Ethnobiol.*, **11**, 1–22.

Shepard, G., Jr. (2002). Primates in Matsigenka subsistence and worldview. In *Primates Face to Face: Conservation Implications of Human and Nonhuman Primate Interconnections*, ed. A. Fuentes and L. D. Wolfe, Cambridge: Cambridge University Press, pp. 101–136.

Smole, W. J. (1976). *The Yanoama Indians: A Cultural Geography*. Austin: University of Texas Press.

Souza-Mazurek, R., Pedrinho, T., Feliciano, X., *et al.* (2000). Subsistence hunting among the Waimiri Atroari Indians in central Amazonia, Brazil. *Biodivers. Conserv.*, **9**, 579–596.

Sponsel, L. E. (1997). The human niche in Amazonia: explorations in ethnoprimatology. *New World Primates: Ecology, Evolution, and Behavior*, ed. W. G. Kinzey, Chicago: Aldine Press, pp. 143–165.

Suárez, S. A. (2003). Spatio-temporal foraging skills of white-bellied spider monkeys (*Ateles belzebuth belzebuth*) in the Yasuní National Park, Ecuador. Unpublished Ph.D. thesis, State University of New York, Stony Brook.

Takai, M., Anaya, F., Shigehara, N. and Setoguchi, T. (2000). New fossil materials of the earliest *Branisella boliviana* and the problem of platyrrhine origins. *Am. J. Phys. Anthropol.*, **111**, 263–281.

Tarble, K. (1985). Un nuevo modelo de expansión Caribe para la época prehispánica. *Antropológica*, **63–64**, 45–81.

Taylor, A. C. (2001). Wives, pets, and affines: marriage among the Jivaro. In *Beyond the Visible and Material: The Amerindianization of Society in the Work of Peter Rivière*, ed. N. L. Whitehead and L. M. Rival, Oxford: Oxford University Press, pp. 45–56.

Terborgh, J. (1983). *Five New World Primates. A Study in Comparative Ecology*. Princeton, NJ: Princeton University Press.

Townsend, W. R. and Ramírez, V. M. (1995). *Cultural teachings as an ecological data base: Murui (Witoto) knowledge about primates*. Center for Latin American Studies, University of Florida. Online (www.latam.ufl.edu/latinoamericanist95/townsend.html).

Urbani, B. (1999). Nuevo mundo, nuevos monos: sobre primates neotrop. en los siglos XV y XVI. *Neotrop. Primates*, **7**, 121–125.

Urbani, B. (2002). Neotropical ethnoprimatology: an annotated bibliography. *Neotrop. Primates*, **10**(1), 24–26.

Urbani, B. (2003). Utilización del estrato vertical por el mono aullador de manto (*Alouatta palliata*, Primates) en Isla Colón, Panamá. *Antropo.*, **4**, 29–33.

Urbani, B. (2005). The targeted monkey: a re-evaluation of predation on New World primates. *J. Anthropol. Sci.*, **83**, 89–109.

Urbani, B. and Gil, L. (2001). Consideraciones sobre restos de primates de un yacimiento arqueológico del Oriente de Venezuela (América del Sur): Cueva del Guácharo, estado Monagas. *Munibe (Antropologia-Arkeologia)*, **53**, 135–142.

Valadez-Azúa, R. and Childs-Rattray, E. (1993). Restos arqueológicos relacionados con monos mexicanos encontrados en "El Barrio de Los Comerciantes" de la antigua ciudad de Teotihuacan. In *Estudios Primatológicos en México*, Vol. 1, ed. A. Estrada, E. Rodríguez-Luna, R. López-Wilchis and R. Coates-Estrada, Xalapa, Mexico: Asociación Mexicana de Primatología and Biblioteca Universidad Veracruzana, pp. 215–229.

Vilaça, A. (2002). Making kin out of others in Amazonia. *J. Roy. Anthropol. Inst.*, **8**, 347–365.

Viveiros de Castro, E. (1998). Cosmological deixis and Amazonian perspectivism. *J. Roy. Anthropol. Inst.*, **4**, 469–488.

Viveiros de Castro, E. (1999). The transformation of objects into subjects in Amerindian ontogenies. Paper presented at the American Anthropological Association invited session: Re-animating Religion: A Debate on the New Animism, Chicago.

von Graeve, B. (1989). *The Pacaa Nova. Clash of Cultures on the Brazilian Border*. Petersburg (Canada): Broadview Press.

Werner, D. (1984). *Amazon Journey: An Anthropologist's Year Among Brazil's Mekranoti Indians*. New York: Simon and Schuster.

White, D. A. and Hood, C. S. (2004). Vegetation patterns and environmental gradients in tropical dry forests of the northern Yucatan Peninsula. *J. Vegetat. Sci.*, **15**, 151–160.

Wilbert, J. (1978). *Folk Literature of the Gê Indians*, Vol. I. Los Angeles: UCLA Latin American Center Publications.

Wilbert, J. (1980). *Folk Literature of the Warao Indians*. Los Angeles: UCLA Latin American Center Publications.

Wilbert, J. and Simoneau, K. (1982). *Folk Literature of the Toba Indians*, Vol. I. Los Angeles: UCLA Latin American Center Publications.

Wilbert, J. and Simoneau, K. (1983). *Folk Literature of the Bororo Indians*, Vol. I. Los Angeles: UCLA Latin American Center Publications.

Wilbert, J. and Simoneau, K. (1984). *Folk Literature of the Gê Indians*, Vol. II. Los Angeles: UCLA Latin American Center Publications.

Wilbert, J. and Simoneau, K. (1988). *Folk Literature of the Mocoví Indians*. Los Angeles: UCLA Latin American Center Publications.

Wilbert, J. and Simoneau, K. (1989a). *Folk Literature of the Ayoreo Indians*. Los Angeles: UCLA Latin American Center Publications.

Wilbert, J. and Simoneau, K. (1989b). *Folk Literature of the Cadeveo Indians*. Los Angeles: UCLA Latin American Center Publications.

Wilbert, J. and Simoneau, K. (1989c). *Folk Literature of the Toba Indians*, Vol. II. Los Angeles: UCLA Latin American Center Publications.

Wilbert, J. and Simoneau, K. (1990a). *Folk Literature of the Yanomamö Indians*. Los Angeles: UCLA Latin American Center Publications.

Wilbert, J. and Simoneau, K. (1990b). *Folk Literature of the Yaruro Indians*. Los Angeles: UCLA Latin American Center Publications.

Wilbert, J. and Simoneau, K. (1991). *Folk Literature of the Cuiva Indians*. Los Angeles: UCLA Latin American Center Publications.

Wilbert, J. and Simoneau, K. (1992). *Folk Literature of the Sikuani Indians*. Los Angeles: UCLA Latin American Center Publications.

Index